Vitamins are essential micronutrients available to animal organisms through the diet. This book takes a fresh approach to vitamin-binding proteins, with emphasis on the nature of the binding of the vitamin ligand to a protein and its sequela. The role of vitamin-binding proteins as initiators of the metabolic response is evaluated. Experts in the field from around the world present a state-of-the-art account of their work on the interaction of vitamins with specific intracellular systems through the appropriate binding proteins and how this interaction results in the biological action of vitamins. This is the first comprehensive book dealing with the subject. The book will be of interest to research workers and postgraduate students in the fields of biochemistry and nutrition.

T0292192

Intercellular and intracellular communication

General editor: Professor B. Cinader

Vitamin receptors
Vitamins as ligands in cell communication

Intercellular and intracellular communication

Vitamin receptors:
Vitamins as ligands in cell communication

EDITED BY

KRISHNAMURTI DAKSHINAMURTI

*Professor of Biochemistry and Molecular Biology, Faculty of Medicine,
University of Manitoba, Winnipeg, Canada*

CAMBRIDGE
UNIVERSITY PRESS

CAMBRIDGE UNIVERSITY PRESS
Cambridge, New York, Melbourne, Madrid, Cape Town, Singapore,
São Paulo, Delhi, Dubai, Tokyo

Cambridge University Press
The Edinburgh Building, Cambridge CB2 8RU, UK

Published in the United States of America by Cambridge University Press, New York

www.cambridge.org
Information on this title: www.cambridge.org/9780521122399

First published 1994
This digitally printed version 2009

A catalogue record for this publication is available from the British Library

Library of Congress Cataloguing in Publication data

Vitamin receptors: vitamins as ligands in cell communication/edited
by Krishnamurti Dakshinamurti.
    p.   cm.
Includes index.
ISBN 0 521 39280 2 (hardback)
1. Vitamins—Receptors. 2. Carrier proteins.   I. Dakshinamurti,
Krishnamurti, 1928–
QP771.V5734   1994

574.19′26—dc20   93–4680   CIP

ISBN 978-0-521-39280-8 Hardback
ISBN 978-0-521-12239-9 Paperback

CONTENTS

CONTRIBUTORS

P. Radhakantha Adiga *Center for Reproductive Biology and Molecular Endocrinology and Department of Biochemistry, Indian Institute of Science, Bangalore 560 012, India*

David H. Alpers *Division of Gastroenterology, Washington University School of Medicine, St. Louis, MO 63110, U.S.A.*

Richard G.W. Anderson *Departments of Pediatrics, Pharmacology and Cell Biology, University of Texas Southwestern Medical Center, Southwestern Medical School, Dallas, TX 75235, U.S.A.*

Jasbir Chauhan *Department of Biochemistry and Molecular Biology, Faculty of Medicine, University of Manitoba, Winnipeg, Canada R3E0W3.*

Krishnamurti Dakshinamurti *Department of Biochemistry and Molecular Biology, Faculty of Medicine, University of Manitoba, Winnipeg, Canada R3E0W3.*

Vincent Giguère *Division of Endocrinology, Hospital for Sick Children, and Department of Molecular and Medical Genetics, Faculty of Medicine, University of Toronto, Toronto, Canada M5G 1X8.*

Barton A. Kamen *Departments of Pediatrics, Pharmacology and Cell Biology, University of Texas Southwestern Medical Center, Southwestern Medical School, Dallas, TX 75235, U.S.A.*

J. Wesley Pike *Sr. Division of Biochemistry, Ligand Pharmaceutical, 9393 Towne City Drive, San Diego, CA 92121–3016, U.S.A.*

Bellur Seetharam *Division of Gastroenterology, Medical College of Wisconsin, Milwaukee, WI, U.S.A.*

Dianne Robert Soprano *Department of Biochemistry and Fels Institute for Cancer Research and Molecular Biology, Temple University School of Medicine, Philadelphia, PA 19140, U.S.A.*

Steven Weitman *Departments of Pediatrics, Pharmacology and Cell Biology, University of Texas Southwestern Medical Center, Southwestern Medical School, Dallas, TX 75235, U.S.A.*

PREFACE

Contemporary research has allowed us to discern a linguistic structure in biological interactions in which molecules in solution and in or on membranes react with one another, sequentially, and thus form sentences of a language. Interlocking ligand–receptor systems constitute an intracellular and intercellular communication system in which molecules, such as cytokines, hormones, metabolic products, vitamins and foreign macromolecules can serve as words and convey signals – messages – through combination with membrane structures, i.e. receptors. These signals can initiate synthesis of other factors which can combine with receptors and form the sentences of the molecular language. Thus, molecular language consists of sequential interactions between receptors and ligands, which are either cell-bound or secreted by one cell and taken up by membrane structures of another cell. A succession of these interactions at the membranes of cells and organelles coordinate cell metabolism within the same and between different cells and organs.

The analogy between language and molecular communication can also be detected in the evolutionary diversification of words, i.e. in the existence of superfamilies of molecules, having a common two-chain structure and involved in recognition of different ligands. Members of these families show homologies in a variety of cell types and among cells of different animals, from invertebrates to vertebrates.

Within cells, some types of organelles acquire macromolecules from the cytoplasm. The site-specificity of these acquisitions depends on receptors, which recognize appropriate proteins by signals that may consist of amino acid sequences or of post-translational modifications.

In communication, occurring between different types of cells, receptors can be activated through soluble factors, and hence at a distance. Receptor–ligand interaction can also occur between membranes of different cell types, i.e. via adhesion molecules that play a role in structural development of organs, exemplified by neural cell adhesion and embryological development under the influence of 'master' cells.

Once a ligand has combined with the membrane receptor of a cell, a change in membrane configuration can ensue and activate a series of metabolic events. Alternatively, ligands and receptors can enter the cell through pathways which start in 'coated pits', and can then be delivered to lysosomes for degradation or be recycled to the surface. This process of endocytosis mediates and regulates the entrance and half-life of various types of molecules, such as growth factors, viruses, toxins and nutrients.

Cell communication is regulated by activation events and by processes which limit the period during which a given stimulus can affect biochemical reactions that are initiated via a particular receptor. This limitation is achieved by processes such as endocytosis, recycling, and affinity changes in receptors, and through disassociation of macromolecular complexes with which the ligand binding site is associated.

Function of each cell type involves internal exchange between cytoplasm and various membrane-bound compartments, i.e. organelles. The function of each organ involves communication between different cell types; that of the organism as a whole involves communication between different organs.

The vast majority of molecules involved in biological sentences, whether receptors or soluble ligands, are autologous. However, the impact of external waves of energy can trigger the sentences of the neuronal system and external molecules can trigger sentences in the immune and nutritional systems.

Past volumes of this series have dealt with various aspects of molecular language, reviewing hormone receptors, cellular interactions in plants, tumor biology, neurology and psychiatry. The fifth volume of our series examined the steps by which molecules can enter the cell interior and be transported with and without receptor to the locations within the cell in which they interact or in which they are being disposed of. It thus dealt with traffic within the cell and with the interrelation between pathways of endocytosis and exocytosis. The present (sixth) volume of this series considers vitamins in terms of the processes by which they enter the system and the receptors by which they affect and activate the system. It deals with nutrients, i.e. with molecules, synthesized in the external world, and hence with an instance of communication initiated by encounters with externally generated molecules. This particular feature is common to the nutritional and the immune systems.

Vitamin receptors deals with chemically diverse molecules, water- and lipid-soluble vitamins, taken up in the gastrointestinal tract through specific saturable membrane components. It surveys binding mechanisms, intracellular transport and, in some instances, transport to the nucleus and the interaction with sections of the genome, resulting in transcription of specific proteins. So far, there is little indication of a common ancestral receptor molecule from which vitamin receptors have descended; they may be derived from different lines of descent. There are indications that some vitamin receptors may have common molecular antecedents with receptors of the endocrine system. The identification of receptor structures remains an important challenge for the future development of analysis of the language of communication in which vitamins are involved.

Bernhard Cinader

INTRODUCTION

This volume attempts to assemble the current information on the ligand (vitamin) – receptor (binding protein) interaction as it applies to each of the vitamins, from the site of enteric absorption to the site of vitamin function. For the absorption and transport of most vitamins, specific proteins, which have a recognition factor incorporated into them, are involved. At the plasma membrane of the cell, a second recognition factor has to function before internalization of the vitamin. Intracellular vitamin-binding proteins and their interaction with cellular systems resulting in the biological action of the vitamins provides the last step in the successive stages of communication. Thus, we have the concept of a ligand, an extraorganismic molecule, reacting with a specific protein entity, the receptor, facilitating absorption and transport of the ligand to its site of action with the resultant sequence of physiological events to which we refer as 'vitamin function'. The term 'receptor', in the context of vitamins, is used to refer to proteins with a recognition function and not to receptors in the traditional pharmacological sense. It includes proteins which bind to the vitamin ligand and function as extracellular or intracellular transporters as well as the smaller number of vitamin-binding nuclear proteins. In the latter case, vitamins might act as small molecular modulators of protein–DNA interaction, a situation analogous to that of some hormones. The designation 'receptors' for these categories of vitamin-binding proteins appears to be appropriate as it connotes a function, more than ligand binding.

Vitamins are required by all protozoans and metazoans that have been studied. They do not conform to any single chemical classification. Some vitamins, such as retinol (vitamin A), cholecalciferol (vitamin D), α-tocopherol (vitamin E) and phylloquinone (vitamin K), are lipid-soluble; others, such as thiamine (vitamin B_1), riboflavin (vitamin B_2), niacin, pyridoxine (vitamin B_6), panthothenic acid, lipoic acid, biotin, folic acid, cyanocobalamin (vitamin B_{12}) and ascorbic acid (vitamin C), are water-soluble. While the unlimited availability of vitamins may have permitted the evolutionary loss of the ability of animals to synthesize vitamins, the ability to absorb vitamins from the environment or from digested food is essential. The modes of uptake, intra- and intercellular transport of these two broad categories of vitamins are different, based on their solubility characteristics. Lipoprotein binders are implicated in the transport of the lipid-soluble vitamins. Among the water soluble-vitamins the bindings of cyanocobalamin and folate, respectively, to specific proteins were recognized early. The intestinal absorption of other water-soluble vitamins was assumed to be through simple diffusion. Increasing evidence has accumulated to indicate that the absorption of many water-soluble vitamins in the gastrointestinal tract is through saturable mechanisms. In addition, the concentrations of many water-soluble vitamins in the brain are much higher than in plasma. This is true of placental transport as well, implicating specific transporters for individual vitamins across the blood–brain and placental barriers, respectively.

Absorption of vitamins by most cells occurs through membrane-associated transport systems which are present in low concentrations. These receptor-like proteins may consist of an extracellular vitamin-binding domain linked to a transmembrane anchor. The fertilized egg and embryo obtain vitamins from the external environment using cell-surface vitamin receptor/transporter proteins. The vitamins are used directly without storage. In contrast, the eggs of birds and reptiles contain both the vitamins and their respective vitamin-binding proteins deposited in the yolk for transport to the oocyte.

The physiological effects of vitamins follow their entry into the cell, metabolic transformation and association with specific apoenzymes, resulting in accentuation of a metabolic pathway. In this respect vitamins are metabolic initiators. Where the vitamin function is other than as a coenzyme or prosthetic group of an enzyme its effect is through regulation of cellular protein synthesis. The mechanisms of action of lipid-soluble vitamins are similar to those of the steroid hormones. Water-soluble vitamins such as biotin and folic acid also seem to have a similar function.

The transcriptional patterns of selected genes are often altered by external signals that the cell receives. Thus changes in nutritional supplies, interaction with hormones, viral infection or exposure to environmental factors such as metals, light, high temperature or radiation can either induce or repress the transcription of specific genes or gene sets. In many cases the regulation of transcription (up or down) is mediated by alterations in the availability of essential transcription factors or by change in the factors' ability to bind to specific regulatory sequence motifs.

The nuclear vitamin D receptor has been characterized as belonging to the steroid family of receptors. Retinol and retinoic acid and their respective binding proteins may serve unique vital functions within different tissues and organs at different times during development. The large number of retinoic acid receptor isoforms expressed in a subtype-specific fashion during development and in differentiated tissues is consistent with the diversity of retinoid action. Folate binding proteins might be involved in transmission of messages for cell division as the accumulation of folate is essential for purine and thymidine synthesis. Biotin has been shown to mimic the actions of insulin in inducing glucokinase mRNA and repressing phosphoenolpyruvate carboxykinase mRNA. The effect of biotin is seen in the starved as well as the diabetic animal in a way that does not involve the synthesis or secretion of insulin. The presence of nuclear binding proteins for biotin and pyridoxal phosphate is indicated, although their functions are not yet known.

We are still at a preliminary stage in our exploration of the groups of vitamin binding proteins. This volume aims to focus attention on the protein-bound vitamin function. The role of some vitamins as the prosthetic group of an enzyme or as a coenzyme participating in enzymatic catalysis is well recognized. This traditional aspect of vitamin function is not considered here. The emphasis is rather on the nature of the binding of the vitamin ligand to a protein, and its sequela.

Krishnamurti Dakshinamurti

1

Serum and Cellular Retinoid-Binding Proteins

Dianne Robert Soprano

Introduction

Retinoids include both the natural forms of vitamin A and a large number of synthetic analogues which may or may not display the biological activity of vitamin A. The major natural forms of vitamin A include retinol, retinal and retinoic acid (see Figure 1). All of these forms of vitamin A have a conjugated double bond system which renders them extremely hydrophobic and thus quite water insoluble.

Vitamin A is an essential nutrient in the diet for normal growth (Wolback & Howe, 1925, 1933), differentiation (Wolback & Howe, 1925, 1933), reproduction (Thompson *et al.*, 1964) and vision (Wald, 1968). Vitamin A from the diet is absorbed and transported to the liver. Once in the liver, vitamin A can either be stored or transported to individual cells of target tissues where vitamin A elicits its biological functions. The target tissues include not just the eye but virtually all tissues of the body. In addition, once within the target cell, vitamin A must be transported to its appropriate site(s) of action. Finally, vitamin A may be transported from the target tissue back to the liver or to another target tissue. In each case, an extremely hydrophobic molecule must be transported through an aqueous environment in order to perform its physiological roles.

Since the isolation of the first retinoid binding protein, retinol-binding protein (RBP) in 1968 (Kanai *et al.*, 1968), much information has been obtained related to the identification, characterization and most recently the molecular biology of a number of proteins which bind retinoids. These binding proteins include serum RBP (Goodman, 1984; Soprano & Blaner, 1993); two cellular retinol-binding proteins, CRBP-I (Chytil & Ong, 1984) and CRBP-II (Ong, 1984); two cellular retinoic acid-binding proteins, CRABP-I (Chytil & Ong, 1984) and CRABP-II (Bailey & Siu, 1988; Giguère *et al.*, 1990*a*); four nuclear retinoic acid receptors, α-RAR (Giguère *et al.*, 1987; Petkovich *et al.*, 1987; Zelent *et al.*, 1989), β-RAR (Zelent *et al.*, 1989; Benbrook *et al.*, 1988; Brand *et al.*, 1988), γ-RAR (Zelent *et al.*, 1989; Krust *et al.*, 1989; Giguère *et al.*, 1990*b*) and RXR (Mangelsdorf *et al.*, 1990); and two unique retinoid binding proteins, cellular retinal-binding protein, CRALBP (Futterman *et al.*, 1977), and interphotoreceptor retinol-binding protein, IRBP (Bridges, 1984), found only in visual tissue. It is becoming more apparent that except for retinyl esters the important biologically active forms of vitamin A may always be bound both intracellularly and extracellularly to an appropriate binding protein in order for vitamin A to be transported through aqueous environments.

Figure 1.1. Chemical formulae for major naturally occurring retinoids. (*a*) All-*trans* retinol; (*b*) all-*trans* retinal; (*c*) all-*trans* retinoic acid.

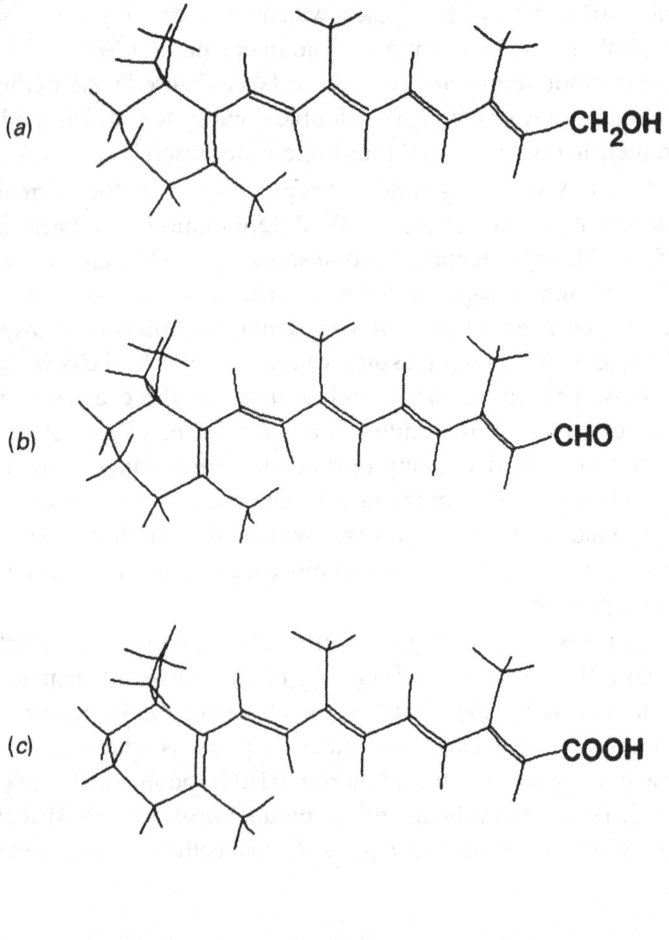

(*a*)   —CH₂OH

(*b*)   —CHO

(*c*)   —COOH

Table 1.1. *Retinoid-binding proteins*

Name	M_r	Endogenous ligand	Site of action	Function
Retinol-binding protein, RBP	21 000	all-*trans* retinol	blood	intercellular transport of retinol
Cellular retinol-binding protein, type I, CRBP-I	15 700	all-*trans* retinol	within cells of vitamin A responsive tissues	intracellular transport of retinol
Cellular retinol-binding protein, type II, CRBP-II	15 600	all-*trans* retinol, all-*trans* retinal	intestinal mucosa cells	absorption and transport of dietary vitamin A within enterocytes
Cellular retinoic acid-binding protein, type I, CRABP-I	15 500	all-*trans* retinoic acid	within cells of vitamin A responsive tissues	intracellular transport of retinoic acid
Cellular retinoic acid-binding protein, type II, CRABP-II	15 500	all-*trans* retinoic acid	developing embryo and adult skin cells	intracellular transport of retinoic acid

The focus of this chapter will be on RBP, CRBP-I, CRBP-II, CRABP-I and CRABP-II and their role in the intercellular and intracellular transport of vitamin A. Table 1 contains a summary of the general properties of each of these retinoid-binding proteins. I will discuss each of these retinoid-binding proteins in the context of their role in the normal metabolism of vitamin A within experimental animals and humans. (The RARs are believed to be involved in the transactivation of retinoic-acid-responsive genes similar to steroid or thyroid hormone receptors and are the subject of Chapter 2 in this volume.) Finally, the discussion of the specialized retinoid-binding proteins found in visual tissue is beyond the scope of this chapter; however, the topic has been reviewed by Chader (1989) and Bridges (1984).

Role of retinoid-binding proteins in the absorption of vitamin A

The natural sources of vitamin A in the diet include retinyl esters, derived from animal sources, and certain plant carotenoid pigments, such as β-carotene. Within the intestinal mucosal cells β-carotene is converted to retinol in a two-step enzymatic process catalyzed by β-carotene, 15,15'-dioxygenase and retinaldehyde reductase. The first step involves the cleavage of β-carotene at the central double bond by a dioxygenase mechanism to yield two molecules of retinal. In the second step retinal is reduced to retinol (for review see Goodman & Blaner, 1984). On the other hand, retinyl esters are hydrolyzed in the intestinal lumen and the resulting retinol is absorbed into the mucosal cells. Therefore, retinol is the major form of dietary vitamin A in the enterocyte. This retinol is ultimately esterified with free fatty acids and incorporated into chylomicrons.

CRBP-II is a major protein in the small intestine. In fact, more than 1% of the soluble protein isolated from the rat jejunal mucosa is CRBP-II. CRBP-II was initially purified in 1984 by Ong from rat neonatal pups. This protein was found to have a relative molecular mass (M_r) of 15 600 and to bind all-*trans* retinol. More recently, two forms of CRBP-II termed CRBP-(II)A and CRBP-(II)B have been reported (Schaefer *et al.*, 1989). The primary amino acid sequence of these two forms of CRBP-II has been found to be identical. Further analysis of these two forms has demonstrated that CRBP-(II)B is acetylated at the amino terminal while CRBP-(II)A is not. Future experiments are required to determine whether these two forms of CRBP-II have different biological functions.

Both a cDNA (Li *et al.*, 1986) and a genomic clone (Demmer *et al.*, 1987) have been isolated for CRBP-II. Analysis of these clones has provided the primary amino acid sequence of CRBP-II and the structure of the CRBP-II gene. This gene contains four exons which span 0.65 kilobases and three introns with an aggregate length of 19.5 kilobases. CRBP-II has been found to be located on mouse chromosome 9 and human chromosome 3 (Demmer *et al.*, 1987). Analysis of the promoter of CRBP-II *in vitro* by transfection of the CRBP-II promoter linked to a reporter gene into CV-1 cells has demonstrated that CRBP-II gene expression is dramatically up-regulated by retinoic acid in the presence of RXR but not RAR (Mangelsdorf *et al.*, 1991). It remains to be determined whether this up-regulation of CRBP-II gene expression by retinoic acid occurs *in vivo* and is physiologically important.

Analysis of the level of CRBP-II (Ong, 1984) and CRBP-II mRNA (Li *et al.*, 1986; Levin *et al.*, 1987) has revealed an interesting developmental pattern of expression and tissue distribution. CRBP-II and CRBP-II mRNA are present in the fetal rat liver by the sixteenth day of gestation and fetal rat intestine by the nineteenth day of gestation. However, CRBP-II and CRBP-II mRNA levels in the liver abruptly decline after parturition and remain undetectable throughout adulthood. In the intestine, however, CRBP-II and CRBP-II mRNA levels remain high, such that the level of CRBP-II in the intestine of the adult animal is 500-fold higher than any other tissue in the body which has been examined. Immunohistochemical localization studies have extended these findings by clearly demonstrating that CRBP-II is localized in the absorptive cells of the rat and human small intestine, from the duodenum to the ileum (Crow & Ong, 1985; Ong & Page, 1987). Furthermore, immunoreactive CRBP-II is not observed in the crypts of Lieburkuhn but rather appears abruptly at the crypt–villus junction. Therefore, the amount of immunoreactive CRBP-II continues to increase as the cells migrate up the villi such that the highest concentration of CRBP-II is found in the cells at the tip of the villi. Quick & Ong (1989) have extended these observations by demonstrating that the concentration of CRBP-II in the small intestine may be related to the nutritional requirement of the animal. They found that the concentration of CRBP-II rose in pregnant dams near the end of pregnancy (about day 17) and continued to rise during the lactation period. CRBP-II concentrations in the small intestine of the dam reached a peak at day 21 postpartum of approximately 3-fold greater than the concentration of CRBP-II in nulliparous, non-pregnant female control rats (Quick & Ong, 1989). This pattern of CRBP-II localization in the intestinal cells suggests that CRBP-II plays an important role in the absorption of retinol and the formation of retinyl esters ultimately to become incorporated into the chylomicron.

Two additional findings have further extended the suggestion that CRBP-II plays an important role in the absorption of dietary retinol. First, Ong and colleagues have demonstrated that CRBP-II binds not only retinol but also retinal (MacDonald & Ong, 1987). Furthermore, retinal bound to CRBP-II has been found to be available for reduction by microsomal preparations from the rat intestinal mucosa to retinol (Kakkad & Ong, 1988). This retinol remains bound to CRBP-II; in contrast to free retinol, the CRBP-II bound retinol is not oxidized to retinal. Second, retinol bound to CRBP-II can be esterified by microsomal preparations from rat small intestine mucosal cells in an acyl-CoA-independent fashion (Ong et al., 1987). Taken together, these results suggest that CRBP-II may play an essential role in the accepting of retinal generated from the cleavage of β-carotene and favoring its reduction to retinol; and in the binding of retinol directly absorbed into the enterocyte. CRBP-II then appears to direct the bound retinol to the appropriate esterifying enzymes for the production of retinyl esters which are ultimately incorporated into chylomicrons. Thus, CRBP-II appears to play an active and specific role in the intestinal metabolism of β-carotene as well as the esterification of retinol for transport to the liver via the chylomicron.

Role of retinoid-binding proteins in the transport of vitamin A between tissues

Retinyl esters incorporated into chylomicrons are ultimately taken up by the hepatocytes of the liver as part of the chylomicron remnants. A detailed discussion of these processes was reviewed by Goodman (1965). Once within the liver, retinol can either be stored as retinyl esters within the stellate cells or be transported from the liver to the target tissues of the body via the plasma.

Vitamin A is transported in plasma as retinol bound to RBP. RBP is a single polypeptide chain which is synthesized as a larger-molecular-mass precursor which is cotranslationally processed (Soprano *et al.*, 1981). This cotranslational processing results in the removal of the signal peptide and the formation of the mature approximately 21 kDa protein which is ultimately secreted. The complete primary amino acid sequence of the mature human RBP (Rask *et al.*, 1979), rat RBP (Sundelin *et al.*, 1985c) and rabbit RBP (Sundelin *et al.*, 1985c) has been reported. Human, rat and rabbit RBP each contain 182 amino acids with the human and rat displaying 86% sequence homology and the human and rabbit displaying 92% sequence homology.

RBP has one binding site for one molecule of retinol (Muto & Goodman, 1972). Normally, RBP is secreted as the RBP–retinol (holo-RBP) complex (Muto & Goodman, 1972). Three-dimensional analysis of RBP has demonstrated it to have an 8-stranded beta-barrel core that completely encapsulates the retinol molecule (Newcomer *et al.*, 1984). *In vitro*, RBP has been demonstrated to bind various retinoids including retinal, retinoic acid, *cis*-isomers of retinol and retinyl acetate. However, *in vivo* only all-*trans* retinol has been found to be bound to RBP. In plasma, holo-RBP strongly interacts with another plasma protein called transthyretin (TTR) (previously termed prealbumin) and normally circulates as a 1:1 molar RBP–TTR complex (Kopelman *et al.*, 1976; Van Jaarsveld *et al.*, 1973). It is believed that the binding of holo-RBP to TTR reduces glomerular filtration and renal catabolism of RBP. The normal level of RBP in human plasma is about 40–60 μg ml^{-1}.

cDNA clones have been isolated for human RBP (Colantuoni *et al.*, 1983), rat RBP (Sundelin *et al.*, 1985c) and rabbit RBP (Lee *et al.*, 1992). Nucleotide sequence analysis of the cDNA clones has confirmed, with slight differences, the previously reported amino acid sequence of mature RBP along with the deduction of the amino acid sequence of the signal

Table 1.2. *Relative levels of RBP mRNA in adult tissues*

Tissues	Relative RBP mRNA level
Liver	100%[a]
Epididymal fat pad	20%[a]
Kidney	5–10%[a]
Lungs	1–3%[a]
Spleen	1–3%[a]
Brain	1–3%[a]
Stomach	1–3%[a]
Heart	1–3%[a]
Skeletal muscle	1–3%[a]
Lacrimal gland	0.1–0.03%[a]
Large intestines	ND[b]
Small intestines	ND
Testes	ND
Pancreas	ND

[a] Relative levels of RBP mRNA are taken from Lee *et al.* (1992), Soprano *et al.* (1986*a*) and Makover *et al.* (1989*a*).
[b] ND, non-detectable by Northern blot assay with the limit of detection at approximately 1% of that of the liver as reported in Soprano *et al.* (1986*a*).

peptide. In addition, genomic clones have also been isolated for human RBP (D'Onofrio *et al.*, 1985) and rat RBP (Laurent *et al.*, 1985). Southern blot analysis and analysis of the genomic clones has demonstrated that the human RBP gene is a single copy gene which spans a region of approximately 10 kilobases. Both rat and human RBP genes consist of six exons and five introns and demonstrate very similar organization. Laurent *et al.* (1985) have observed that each exon of the rat RBP gene corresponds closely to distinct structural elements in the protein structure.

The hepatocytes of the liver are a major site of RBP synthesis and secretion (Soprano *et al.*, 1986*a*). However, using RBP cDNA as a probe, it has been demonstrated that along with the liver there are several extrahepatic sites of RBP mRNA and presumably synthesis of RBP in the adult rat. Table 1.2 contains a summary of the tissues and the relative level of RBP mRNA in each tissue. As can be seen the relative level of RBP mRNA ranges from undetectable (sensitivity of Northern blot assay in early studies was only to a level of RBP mRNA which was approximately 1% of that observed in the liver) to 20% of the level of RBP mRNA in the liver (Soprano *et al.*, 1986*a*; Makover *et al.*, 1989*b*;

Martone *et al.*, 1988). More recently, the lacrimal gland, using the more sensitive assay of RNase protection, has been found to contain RBP mRNA at a level of 0.01 – 0.03% of that observed in the liver (Newcomer *et al.*, 1984). It is likely that some of the tissues previously examined by Northern blot which were found to have an undetectable level of RBP mRNA may actually contain low levels of RBP mRNA. Though the amount of RBP mRNA within many of these tissues is small compared with that in the liver, the composite amount of RBP mRNA and potentially the amount of RBP secreted from extrahepatic tissues into the plasma is quite substantial.

Further analysis of kidney mRNA has demonstrated that kidney poly (A+) RNA can be translated in rabbit reticulocytes *in vitro* to yield *immunoprecipitable* RBP, indicating that kidney-derived RBP mRNA is capable of translation (Soprano *et al.*, 1986a). In addition, *in situ* hybridization has demonstrated that RBP mRNA is localized in the S3 segment of the proximal tubule cells of the kidney and is not distributed homogenously throughout the kidney (Makover *et al.*, 1989a). Therefore the amount of RBP mRNA within an individual cell of the S3 segment of the proximal tubule cells of the kidney is not small with respect to the amount observed in a liver hepatocyte. It is quite likely that RBP mRNA within other extrahepatic tissues may also be localized within individual cell type(s) of the tissue.

Several studies have examined the expression of RBP during embryonic development of rats (Soprano *et al.*, 1986b; Makover *et al.*, 1989b; Sklan & Ross, 1987). RBP mRNA can be first observed as early as seven days of gestation localized in the visceral endodermal cells. From the ninth day of gestation until the twentieth day of gestation, RBP mRNA is observed in the visceral yolk sac endoderm. Quantitation of RBP mRNA in the visceral yolk sac from the fourteenth to the twentieth day of gestation indicates that RBP mRNA levels are relatively constant throughout this period of development, averaging approximately 50% of the level observed in the adult liver. In addition, explant cultures of visceral yolk sac have been demonstrated to synthesize immunoprecipitable RBP *in vitro*. This indicates that in the visceral yolk sac RBP mRNA is translated to immunoreactive RBP. In the developing liver, RBP mRNA was first observed at the tenth day of gestation. RBP mRNA levels in the fetal liver increased steadily throughout gestation, reaching a level of 70% of that observed in the adult liver at birth. Finally, RBP mRNA and synthesis of RBP is induced in F9 teratocarcinoma cells which are differentiated to embryoid bodies (which contain visceral endoderm) (Soprano *et al.*, 1988a). F9 teratocarcinoma cells are believed to be very similar to fetuses of 3–5 days' gestation. Taken together, these results suggest that RBP mRNA expression begins very early in fetal development, initially in the visceral endoderm and later in the fetal liver. It is quite likely that the very early fetal expression of RBP mRNA and synthesis of RBP reflects the essential requirement of the developing fetus for retinoids.

RBP synthesis and secretion by extrahepatic tissues has been demonstrated using *in vitro* systems including translation of kidney RNA (Soprano *et al.*, 1986*a*), synthesis and secretion of RBP by explant culture of visceral yolk sac (Soprano *et al.*, 1986*b*) and synthesis and secretion of RBP by F9 teratocarcinoma cells differentiated to embryoid bodies (Soprano *et al.*, 1988*a,b*). However, it has been difficult to demonstrate *in vivo* synthesis and secretion of RBP by an extrahepatic tissue which expresses RBP mRNA. This is because once RBP is synthesized and secreted by these extrahepatic tissues the RBP enters the blood and is then indistinguishable from that produced by the liver. Study of the lacrimal gland and the lacrimal gland fluid has offered a unique system to address this question. This is because the lacrimal gland fluid is isolated from the blood by an efficient blood–lacrimal barrier and the lacrimal gland contains proteins primarily synthesized in the lacrimal gland. Recent studies have demonstrated that the lacrimal gland contains RBP mRNA and that the lacrimal gland fluid contains RBP, by using a specific monoclonal antibody (Lee *et al.*, 1992). This is the first demonstration that an extrahepatic site of RBP mRNA actually synthesizes and secretes RBP *in vivo*, implying that other organs which contain RBP mRNA can also synthesize and secrete RBP.

The function of RBP synthesized in each of the extrahepatic tissues is not known. However, RBP synthesized in each of these extrahepatic tissues may play an important role(s) in the overall transport and delivery of retinol in the body. Studies of Lewis and colleagues have demonstrated that there is quantitative recycling of retinol within the body of animals (Lewis *et al.*, 1981). Therefore, we have hypothesized that when retinol leaves an extrahepatic tissue, a new molecule of RBP is synthesized locally, retinol is added to this molecule in the microsomes, and the holo-RBP is secreted into the plasma for delivery of retinol back to the liver or to other extrahepatic tissues (Soprano *et al.*, 1986*a*). Four examples are described below. First, we have proposed that RBP synthesized in the visceral yolk sac (which is a site of true placentation in the rat (Steven & Morris, 1975) may be involved in the delivery of retinol from the mother to the developing fetus (Soprano *et al.*, 1986*b*). Second, we have suggested that RBP synthesized in the kidney may be involved in the recycling of retinol which is glomerular-filtered back to the liver and other extrahepatic tissues (Soprano *et al.*, 1986*a*). Third, we have hypothesized that RBP synthesized in adipose tissue may be involved in the transport of retinol taken up from the chylomicron or chylomicron remnant by the adipose tissue back to the liver and extrahepatic target tissues (Makover *et al.*, 1989*a*). Fourth, we have hypothesized that RBP synthesized in the lacrimal gland may be involved in the transport of retinol to the cornea of the eye via the lacrimal gland fluid (Lee *et al.*, 1992). Further research is necessary to explore these and other possible hypotheses.

As previously stated, retinyl esters from the diet are transported to the liver via the chylomicron remnant and taken up by the liver hepatocytes. Within the liver, excess retinol is stored principally in the stellate cells as retinyl esters. In cases of insufficient dietary retinol, the stored retinyl esters are released from the stellate cells of the liver (for review, see Hendriks *et al.*, 1987). Recently, it has been suggested that RBP may play a role in the transfer of retinol from the hepatocytes to the stellate cells of the liver (Blomhoff *et al.*, 1985, 1988; Gjoen *et al.*, 1987). However, stellate cells do not contain detectable amounts of RBP mRNA (Yamada *et al.*, 1987) and contain very small amounts of RBP (Blaner *et al.*, 1985). Therefore, at this time we cannot exclude other possible mechanisms for the transfer of retinol, which include: transfer of retinol mediated by direct contact of the hepatocytes and the stellate cells; and transfer between these cell types which involves CRBP. More definitive studies are required to define the mechanism of retinol transfer between the hepatocytes and the stellate cells of the liver.

The plasma level of RBP is highly regulated within individuals and remains quite constant except in extremes of vitamin A nutriture or in disease states. One factor which controls the rate of secretion of RBP and not the rate of biosynthesis of RBP is retinol. The secretion of RBP from the liver is specifically blocked in rats which have been depleted of their retinol stores, resulting in the accumulation of apo-RBP in the liver and a concomitant decline in the plasma level of RBP (Muto *et al.*, 1972). Upon repletion of retinol-depleted rats with retinol, RBP is rapidly secreted from the liver into the plasma (Smith *et al.*, 1973). On the other hand, the *in vivo* rate of RBP synthesis in the liver (Soprano *et al.*, 1982) and the level of RBP mRNA in the liver (Soprano *et al.*, 1986*a*) are not influenced by the retinol status of rats. Similar findings have also been found in the visceral yolk sac (Soprano *et al.*, 1988*b*) and the lacrimal gland (Lee *et al.*, 1992). Analysis of RBP secretion in primary hepatocytes and in Hep G2 cells demonstrates that a low level of RBP secretion occurs in the absence of retinol and that the rate of RBP secretion is stimulated in a dose-dependent manner by the addition of retinol (Dixon & Goodman, 1987*a*; Marinari *et al.*, 1987). Recent studies have demonstrated that the movement of RBP from the endoplasmic reticulum to the Golgi apparatus is prevented in retinol-deficient animals (Rask *et al.*, 1983; Ronne *et al.*, 1983; Fries *et al.*, 1984). Hence, retinol regulates the secretion of RBP without influencing the synthesis of RBP or the level of RBP mRNA in the liver and the visceral yolk sac.

The effect of a number of hormones on the synthesis and secretion of RBP by the liver has been studied (Borek *et al.*, 1981; Dixon & Goodman, 1987*b*; Whitman *et al.*, 1990). Hormones such as estrogen, growth hormone, dexamethasone, thyroid hormone and testosterone have been examined and demonstrated to have little effect on liver RBP synthesis and secretion. In contrast to the liver, recent studies have shown that estrogen rapidly increases the level of RBP mRNA in the rat kidney (Whitman *et al.*, 1990). This suggests that the regulation of RBP gene expression in each tissue may be different. Much future work is necessary to understand the tissue-specific regulation of RBP gene expression.

The controlled delivery of retinol to target tissues is believed to occur via a specific RBP cell surface receptor. However, the biochemical purification and characterization of this putative receptor is yet to be reported. This is a question of great importance which has remained elusive to a number of investigators. Heller (1975) first suggested the possibility of a cell-surface receptor for RBP. Since that initial report there have been numerous reports by a large number of investigators using a number of experimental systems of the existence of a plasma membrane RBP-specific receptor (Rask & Peterson, 1976; Bok & Heller, 1976; Bhat & Cama, 1979; Eriksson *et al.*, 1986; Pfeffer *et al.*, 1986; Torma & Vahlquist, 1986; Sivaprasadarao & Findlay, 1988*a,b*). Other reports using model membrane systems have suggested that retinol uptake may not require a specific cell-surface RBP receptor (Creek *et al.*, 1988; Fex & Johannesson, 1987, 1988). Therefore, at the present time the exact mechanism of retinol uptake from RBP–retinol remains unclear.

In summary, the efficient delivery of retinol to target tissues is dependent on adequate synthesis and secretion of RBP. There are numerous clinical examples which demonstrate the very important role of RBP in the delivery of retinol from its storage site in the liver to target tissues of the body. Vitamin A deficiency symptoms such as impaired visual dark adaptation have been observed in clinical circumstances such as protein or calorie malnutrition and liver disease where plasma RBP levels are reduced (for review, see Goodman, 1984). Recently, the first case of familial hypo-RBP proteinuria was described (Matsuo *et al.*, 1988; Matsuo & Matsuo, 1988). This was detected in a 19-month, well-nourished child who developed keratomalacia after a measles infection. Analysis of serum RBP levels indicated that the child, a sister and the child's mother all had RBP levels about half that of normal individuals. Further studies of this family may provide interesting information concerning both the RBP gene and the metabolism of RBP.

Role of retinoid binding proteins in the intracellular transport of vitamin A

Retinol is delivered to target tissues of the body via its specific transport protein, RBP. As discussed above, the mechanism of uptake of retinol by the target tissue cells is still unclear. However, what is clear is that retinol enters the target cells. Within the target cells there are two types of proteins which are involved in the intracellular transport of vitamin A. These are the cellular retinol binding proteins (CRBPs) and the cellular retinoic acid binding proteins (CRABPs). To date two different CRBPs, termed CRBP-I and CRBP-II, have been identified along with two different CRABPs, termed CRABP-I and CRABP-II. The previous section dealing with the absorption of vitamin A by the intestines has already discussed CRBP-II in detail; hence this section will deal principally with CRBP-I, CRABP-I and CRABP-II.

In vivo, the endogenous ligand associated with CRBP-I is all-*trans* retinol and the endogenous ligand associated with both CRABP-I and CRABP-II is all-*trans* retinoic acid. Presumably, when all-*trans* retinol is taken up by target tissue cells from the RBP–retinol complex, CRBP-I functions as its first acceptor. The source of the retinoic acid which binds to CRABP-I and CRABP-II within cells is still an area of question. It would appear that the majority of retinoic acid arises from oxidation of retinol within target cells. This is supported by a number of observations, including: the fact that plasma concentrations of retinoic acid are low; natural food sources do not contain retinoic acid; and there appears to be no plasma transport system specific for the delivery of retinoic acid to target tissues.

CRBP-I (originally termed CRBP) was originally discovered in 1973 as a retinol-binding activity present in a large number of tissues (Bashor *et al.*, 1973). Since this initial observation CRBP-I has been purified from a number of tissues and species (for review, see Chytil & Ong, 1984). The complete amino acid sequence of rat CRBP-I was reported in 1985 by Sundelin *et al.* (1985*a*). Comparison of the amino acid sequence of rat

CRBP-I and rat CRBP-II has demonstrated that these proteins are collinear, with 75 of 134 amino acid residues identical. However, CRBP-I and CRBP-II are antigenically distinct (Ong, 1984).

CRABP was discovered as a retinoic acid-binding activity within the cytosol from a number of rat tissues (Ong & Chytil, 1975) and in chicken embryo skin (Sani & Hill, 1974). Since this initial discovery two CRABPs have been identified, which are termed CRABP-I and CRABP-II (Bailey & Siu, 1988; Giguère *et al.*, 1990*a*). Both CRABP-I and CRABP-II are a single polypeptide chain with a relative molecular mass of 15 500. The complete amino acid sequence of bovine CRABP-I has been determined (Sundelin *et al.*, 1985*b*) while only a partial amino terminal end sequence of rat pup CRABP-II has been reported (Bailey & Siu, 1988). Comparison of the deduced amino acid sequences obtained from mouse cDNA clones of CRABP-I and CRABP-II (see below) indicates that these two proteins are 73% homologous.

Within the past few years, cDNA clones have been isolated for CRBP-I, CRABP-I and CRABP-II. Both human (Colantuoni *et al.*, 1985; Wei *et al.*, 1987) and rat (Sherman *et al.*, 1987) CRBP-I cDNA clones have been isolated and the nucleotide sequence determined. In CRABP-I, bovine (Wei *et al.*, 1987; Shubeita *et al.*, 1987), mouse (Wei *et al.*, 1989; Stoner & Gudas, 1989) and human (Astrom *et al.*, 1991) cDNA clones have been isolated and characterized. Analysis of these cDNA clones has shown that CRABP-I is a highly conserved protein since bovine and mouse CRABP-I have identical deduced amino acid sequences while only one amino acid is different (amino acid 86, Ala instead of Pro) in human CRABP-I. Recently, cDNA clones for both mouse (Giguère *et al.*, 1990*a*) and human (Astrom *et al.*, 1991) CRABP-II have been isolated. Comparison of the deduced amino acid sequence of mouse and human CRABP-II demonstrates that CRABP-II is less highly conserved than CRABP-I since these two proteins demonstrate 93.5% identity at the amino acid level (nine amino acids are different).

On the basis of amino acid sequence homology obtained either by direct sequence analysis of purified protein or deduction of the amino acid sequence from cDNA clones, CRBP-I, CRBP-II, CRABP-I and CRABP-II have all been recognized as members of a family of small cytoplasmic, lipid-binding proteins (Takahashi *et al.*, 1982; Sundelin *et al.*, 1985*a,b,c*; Sacchettini *et al.*, 1986; Lowe *et al.*, 1985). Other members of this family include intestinal, heart and liver fatty acid binding proteins, aP2 and P2 myelin protein. The percentage of amino acid homology between these family members ranges from 42 to 52% with the greatest amount of homology at the amino terminal end.

Genomic clones have been isolated for both human CRBP-I (Nilsson *et al.*, 1988) and mouse CRBP-I (Smith *et al.*, 1991) along with genomic clones for bovine CRABP-I (Shubeita *et al.*, 1987) and mouse CRABP-I (Wei *et al.*, 1990). To date there has been no report of a genomic clone for CRABP-II. Both the human CRBP-I gene and the bovine CRABP-I gene are single copy genes composed of four exons and three introns. The second intron of human CRBP-I gene comprises 19 kilobases (kb) of the 21 kb gene. The positions of the introns in the human CRBP-I gene agree with those of CRBP-II. The promoter of the CRBP-I gene resembles many housekeeping genes in that it is highly G+C rich, with multiple copies of the Sp1 transcription factor binding site, and lacks a TATA box. Recently, a retinoic acid response element (RARE) which functions to modulate the response of this gene to retinoic acid has been identified in the promoter of CRBP-I approximately 1 kb upstream from the start site of transcription (Smith *et al.*, 1991). This element closely resembles RAREs found in the promoters of other genes that are retinoic acid responsive, such as RAR-β2 (Sucov *et al.*, 1990). The CRABP-I promoter also lacks a TATA box and contains putative cAMP-responsive elements and Sp1 transcription factor binding sites. The mouse CRBP-I gene is located on chromosome 9 (Demmer *et al.*, 1987) and the human CRBP-I gene is located on chromosome 3 (Demmer *et al.*, 1987; Colantuoni *et al.*, 1986; Levin *et al.*, 1988). In addition, the CRBP-I and CRBP-II genes in both humans and mice are closely linked.

Analysis of the retinoid binding site of CRBP-I *in vitro* has been performed using both pure protein isolated from tissue (MacDonald & Ong, 1987) and recombinant protein expressed in *Escherichia coli* (Levin *et al.*, 1988). Both studies have demonstrated that CRBP-I can bind all-*trans* retinol, 13-*cis*-retinol and 3-dehydroretinol with a similar affinity to CRBP-II. However, MacDonald & Ong (1987) have reported that, unlike CRBP-II, CRBP-I is unable to bind all-*trans* retinal. Levin and coworkers (Levin *et al.*, 1988) have reported that CRBP-II can bind all-*trans* retinal; however, all-*trans* retinal cannot displace all-*trans* retinol from the binding site in CRBP. Finally, neither CRBP-I nor CRBP-II was found to bind all-*trans* retinoic acid. It would appear from these studies that the CRBP-I and CRBP-II retinoid binding sites are distinct.

Both CRABP-I and CRABP-II specifically bind retinoic acid and do not bind either retinol or retinal (Bailey & Siu, 1988; Jetten & Jetten, 1979; Ong & Chytil, 1978). However, retinoic acid binding studies comparing rat testis CRABP-I and rat variant CRABP-II suggest that each of these proteins demonstrate different retinoid binding properties. For example, CRABP-II does not bind 13-*cis*-retinoic acid (Bailey & Siu,

1988) while CRABP-I has a high affinity for this retinoid (Jetten & Jetten, 1979). In addition, rat pup CRABP-II has been reported to display a dissociation constant of 65 nM for retinoic acid (Bailey & Siu, 1988) while rat testis CRABP-I has a reported dissociation constant of 4nM (Ong & Chytil, 1978). Taken together these initial observations suggest that the retinoic acid binding site of these two CRABPs may have unique features.

Both CRBP-I and CRABP have been quantitated using radioimmunoassay techniques in a large number of rat tissues (Adachi *et al.*, 1981; Ong *et al.*, 1982; Eriksson *et al.*, 1984, 1987; Kato *et al.*, 1985c) and cell lines (for review, see Chytil & Ong, 1984, 1987). It should be noted that the antibodies which were employed to quantitate CRABP concentrations were most likely prepared using CRABP-I as the immunogen; however, at the moment it is unclear whether these antibodies also cross-react with CRABP-II. Therefore I will refer to the binding protein studied in this work as CRABP and not designate it CRABP-I or CRABP-II. Unlike CRBP-II, CRBP-I and CRABP are detected in a large number of tissues, suggesting that each of these tissues is a target tissue for vitamin A action. Interestingly, CRBP-I and CRABP-I have been detected in all tissues studied; however, the two proteins differed greatly with regard to their tissue distribution. In addition, analysis of RNA isolated from various rat tissues using the CRBP-I cDNA as a probe has demonstrated that CRBP-I mRNA concentrations correlate highly with CRBP-I protein concentrations (Rajan *et al.*, 1990). In non-reproductive organs relatively high concentrations of CRBP-I are detected in the liver, kidney, lungs, lymph nodes, eye and spleen while CRABP concentrations are relatively high in the skin and the eyes. High concentrations of both CRBP-I and CRABP are detected in the male and female reproductive organs. The highest concentration of CRBP-I for any rat tissue was observed in the proximal portion of the epididymis; the highest concentration of CRABP for any rat tissue was observed in the seminal vesicles. Another interesting observation is that CRBP-I concentrations are highest in the gonads and the proximal portion of the male reproductive tract whereas CRABP concentrations are highest in the distal portion of the male reproductive tract. Finally, tissue localization of CRBP-I in human tissues (Ong & Page, 1986) suggests that the relative abundance of this protein in human tissues is significantly different from that observed in rat tissues.

In addition to measuring CRBP-I and CRABP concentrations within tissues, there have been many reports dealing with the localization of CRBP-I and CRABP within specific cell types of tissues (Kato *et al.*,

Table 1.3. *Summary of cellular localization of CRBP-I and CRABP in adult rats*

Tissue	CRBP-I	CRABP
Liver	Stellate cells and hepatocytes[ab]	
Kidney	Proximal convoluted tubule cells of renal cortex[c]	
Pancreas	Glucagon containing A cells[d]	
Skin		Epidermis[i]
Placenta	Trophoblastic layer of chorioallantoic placenta + endodermal layer of visceral yolk sac[a]	
Testes	Sertoli cells[befg]	Pachytene[befg] spermatocytes and spermatids
Epididymis	Caput[efg]	Stereocilia of principal cells[efg]

Sources: [a]Kato *et al.* (1985b); [b]Blaner *et al.* (1985); [c]Kato *et al.* (1984); [d]Kato *et al.* (1985c); [e]Porter *et al.* (1985); [f]Kato *et al.* (1985d); [g]Porter *et al.* (1983); [h]Astrom *et al.* (1991).

1984, 1985a,b,d; Blaner *et al.*, 1985, 1987; Porter *et al.*, 1983, 1985; Chytil, 1986; Perez-Castro *et al.*, 1990). Table 1.3 is a summary of a number of adult tissues with the specific cellular localization of CRBP-I and CRABP, determined by immunolocalization on tissue section or by radioimmunoassay of isolated cell types.

Since the cloning of the cDNA for these proteins, both CRBP-I mRNA and CRABP-I mRNA have been localized in developing mice fetuses using *in situ* hybridization techniques (Perez-Castro *et al.*, 1990). Both transcripts were detected at embryonic days 9.5 – 14.5. However, the cellular localization of each of these transcripts was unique. For example, in the craniofacial area CRABP-I mRNA was localized in the mesenchyme of the frontonasal mass and mandible, while CRBP-I mRNA was localized in the mesenchyme of the nasolachrymal duct and surrounding auditory surface. With the very recent discovery and cloning of CRABP-II cDNA, initial data indicate that CRABP-II mRNA is present at a high concentration in mid-gestation developing mouse fetuses (Giguère *et al.*, 1990a; Harnish *et al.*, 1992) and in adults is localized to the skin (Giguère *et al.*, 1990a; Astrom *et al.*, 1991).

In summary, it is clear from the tissue concentration and localization of CRBP-I, CRABP-I and CRABP-II (both protein concentration and mRNA concentration) that each of these retinoid binding proteins appears to display a unique, specific pattern of localization. These observations suggest that retinol and retinoic acid and their respective binding proteins may serve unique vital functions within different tissues and organs and at different times during development. It will be very interesting in the future to understand the basis of the tissue-specific expression of these genes.

An area of considerable interest has been the effect of nutritional retinoid status on tissue concentrations of CRBP-I, CRABP-I and CRABP-II and their respective mRNAs. Initial studies measured by radioimmunoassay the concentration of CRBP-I and CRABP in rats of different retinoid status. Reduced concentrations of CRBP-I were found in totally retinoid-deficient rats (Smith *et al.*, 1991); however, when rats were maintained on a retinol-deficient retinoic-acid-supplemented diet, only the CRBP-I concentration of the proximal epididymis was found to differ significantly from the concentration in the control epididymis (Blaner *et al.*, 1986). On the other hand, CRABP concentrations were not influenced by the retinoid status of rats (Blaner *et al.*, 1986). More recently the effect of retinoid status on CRBP-I mRNA concentrations has been examined. Sherman *et al.* (1987) have demonstrated that the concentration of lung CRBP-I mRNA was increased 4-fold when retinoid-deficient rats were re-fed with retinol; Haq & Chytil (1988) have demonstrated a similar response in lung CRBP-I mRNA concentrations when retinoid-deficient rats were re-fed with retinoic acid. Recently, Rajan and coworkers (Rajan *et al.*, 1990) have examined the effects of retinol status on CRBP-I mRNA concentrations in a number of rat tissues. Seven tissues were examined, four of which (lung, testis, spleen and small intestine) were found to have reduced tissue concentrations of CRBP-I mRNA in retinoid-deficient rats. Repletion of these rats with retinol resulted in a rapid rise in CRBP-I mRNA concentration near to or up to that of the control rats. In addition, studies with P19 embryonic carcinoma cells have demonstrated that CRBP-I mRNA concentrations are rapidly increased in a cycloheximide-independent fashion within 3 h of retinoic acid treatment, while CRABP-I mRNA concentrations respond in a cycloheximide-dependent fashion only 12–24 h after retinoic acid treatment (Wei *et al.*, 1989). Finally, teratogenic doses of retinoic acid have been found to result in a modest increase in the concentration of CRBP-I mRNA and CRABP-II mRNA in mouse fetuses, while CRABP-I mRNA concentrations were unaffected (Harnish *et al.*, 1992). Taken together it appears that CRBP-I protein and mRNA concentrations and CRABP-II mRNA concentrations in at least some tissues are retinol- or retinoic-acid-responsive. Studies of CRABP-I protein and mRNA concentrations suggest that CRABP-I may not be responsive to retinoids.

At present there is very little information concerning the actual role of these proteins within cells. The recent developments in the retinoid field indicate that at least retinoic acid, and potentially retinol, manifests its effects in a manner similar to that of steroid hormones. That is, retinoic acid (and potentially retinol) interacts in the nucleus with the promoter of specific genes via nuclear receptors. This interaction results in an alteration in the transcription of genes responsive to retinoic acid and possibly also to retinol. The net result is an alteration in the expression of specific genes and ultimately those proteins synthesized within cells. The prime candidates for the transport of retinol and retinoic acid to the nucleus are their respective retinoid-binding proteins. Results of *in vitro* experiments have suggested that there are saturable nuclear binding sites for retinol and retinoic acid (Takase *et al.*, 1979, 1986; Barkai & Sherman, 1987). These *in vitro* experiments have shown that retinol and retinoic acid recognize physically distinct sites within the chromatin. Furthermore, CRBP and CRABP do not appear to bind to these sites. Therefore, these retinoid binding proteins may function to shuttle their respective ligands to these nuclear binding sites. On the other hand these proteins may be an important way to control the concentration of free and bound retinoid within the cells and thereby regulate the responsiveness of cells to retinoic acid. This mechanism of action is supported by a recent report which has demonstrated that over-expression of CRABP-I in F9 teratocarcinoma cells reduces the potency of retinoic acid in this differentiating system (Boylan & Gudas, 1991). In other words, higher doses of retinoic acid were required to differentiate F9 cells which displayed elevated levels of CRABP-I compared with normal F9 cells.

It has been suggested that during development the retinoid binding proteins may play a role in spatially modulating the concentration of their respective ligand within cells. This idea comes from the suggestion that retinoic acid may act as a morphogen. Strong support for this idea has come from recent work which has demonstrated a concentration gradient of retinoic acid across the chicken limb bud with the highest portion on the posterior side (Thaller & Eichele, 1987). Using immunohistochemical techniques, CRABP has been localized with the highest concentration at the anterior margin of the chick limb bud (Maden *et al.*, 1988). This is in opposite polarity to retinoic acid. How such a gradient of retinoic acid and CRABP forms, and the role of this gradient, remains to be established.

Conclusions

Five retinoid-binding proteins have been described, which are extremely important for the intercellular and intracellular transport of retinoids. One should not consider this list complete since there have been several other novel retinoid-binding proteins which have yet to be characterized. Retinoid-binding proteins play important roles in the binding and transport of their hydrophobic ligands. These proteins provide a mechanism of transporting vitamin A from site to site between tissues and within cells. This eliminates the requirement of retinoids to diffuse through aqueous environments within the cytoplasm of cells and of plasma.

The author acknowledges the financial support of National Institute of Health Grants DK41089, EYO5640 and HD27556.

References

Adachi, N., Smith, J.E., Sklan, D. & Goodman, D.S. (1981) Radioimmunoassay studies of the tissue distribution and subcellular localization of cellular retinol-binding protein in rats. *J. Biol. Chem.* **256**, 9471–6.

Astrom, A., Tavakkol, A., Pettersson, V., Cromie, M., Elder, J.T. & Voorhees, J.J. (1991) Molecular cloning of two human cellular retinoic acid-binding proteins (CRABP). *J. Biol. Chem.* **266**, 17662–6.

Bailey, J.S. & Siu, C.-H. (1988) Purification and partial characterization of a novel binding protein for retinoic acid from neonatal rat. *J. Biol. Chem.* **263**, 9326–32.

Barkai, U. & Sherman, M.I. (1987) Analysis of the interactions between retinoid-binding proteins and embryonal carcinoma cells. *J. Cell Biol.* **104**, 671–80.

Bashor, M.M., Toft, D.O. & Chytil, F. (1973) In vitro binding of retinol to rat tissue components. *Proc. Natl. Acad Sci. USA* **70**, 3483—7.

Benbrook, D., Lernhardt, E. & Pfahl, M. (1988) A new retinoic acid receptor identified from a hepatocellular carcinoma. *Nature* **333**, 669–72.

Bhat, M.K. & Cama, H.R. (1979) Homeostatic regulation of free retinol-binding protein and free thyroxine pools of plasma by their plasma carrier protein in chicken. *Biochim. Biophys. Acta* **587**, 273–81.

Blaner, W.S., Das, K., Mertz, J.R., Das, S.R. & Goodman, D.S. (1986) Effects of dietary retinoic acid on cellular retinol- and retinoic acid-binding protein levels in various rat tissues. *J. Lipid Res.* **27**, 1084–8.

Blaner, W.S., Galdieri, M. & Goodman, D.S. (1987) Distribution and levels of cellular retinol- and cellular retinoic acid-binding protein in various types of rat testis cells. *Biol. Reprod.* **36**, 130–7.

Blaner, W.S., Hendricks, H.F.J., Brouwer, A., deLeeuw, A.M., Knook, D.L. & Goodman, D.S. (1985) Retinoids, retinoid-binding proteins, and retinyl palmitate hydrolase distributions in different types of rat liver cells. *J. Lipid Res.* **26**, 1241–51.

Blomhoff, R., Berg, T. & Norum, K.R. (1988) Transfer of retinol from parenchymal to stellate cells in liver is mediated by retinol-binding protein. *Proc. Natl. Acad. Sci. USA* **85**, 3455–8.

Blomhoff, R., Norum, K.R. & Berg, T. (1985) Hepatic uptake of [^{3}H]retinol bound to the serum retinol-binding protein involves both parenchymal and sinusoidal stellate cells. *J. Biol. Chem.* **260**, 13571–5.

Bok, D. & Heller, J. (1976) Transport of retinol from the blood to the retina: an autoradiographic study of the pigment epithelial cell surface receptors for plasma retinol binding protein. *Exp. Eye Res.* **22**, 395–402.

Borek, C., Smith, J.E., Soprano, D.R. & Goodman, D.S. (1981) Regulation of retinol-binding protein metabolism by glucocorticoid hormones in cultured H$_4$IIEC$_3$ liver cells. *Endrocrinology* **109**, 386–91.

Boylan, J.F. & Gudas, L.J. (1991) Overexpression of the cellular retinoic acid binding protein-I (CRABP-I) results in a reduction in differentiation-specific gene expression in F9 teratocarcinoma cells. *J. Cell. Biol.* **112**, 965–79.

Brand, N., Petkovich, M., Krust, A., Chambon, P., deThe, H. Marchio, A., Tiollais, P. & Dejean, A. (1988) Identification of a second human retinoic acid receptor. *Nature* **332**, 850–3.

Bridges, C.D.B. (1984) Retinoids in photo-sensitive system. In *The Retinoids* (ed. M.B. Sporn, A.B. Roberts & D.S. Goodman Vol. 2, pp. 125–76. Academic Press, Orlando, Florida.

Bridges, C.D., Alvarez, R.A., Fong, S.L., Gonzalez-Fernandez, F., Lam, D.M. & Liou, G.I. (1984) *Vision Res.* **24**, 1581–2594.

Chader, G.J. (1989) Interphotoreceptor retinoid-binding protein (IRBP): a model protein for molecular biological and clinically relevant studies. *Invest. Ophthalmol. Vis. Sci.* **30**, 7–22.

Chytil, F. (1986) Retinoic acid; biochemistry and metabolism. *J. Amer. Acad. Dermatol.* **15**, 741–7.

Chytil, F. & Ong, D.E. (1984) Cellular retinoid-binding proteins. In *The Retinoids* (ed. M.B. Sporn, A.B. Roberts & D.S. Goodman Vol. 2, pp. 89–123. Academic Press, Orlando, Florida.

Chytil, F. & Ong, D.E. (1987) Intracellular vitamin A-binding proteins. *Ann. Rev. Nutr.* **7**, 321–35.

Colantuoni, V., Cortese, R., Nilsson, M., Lundvall, J., Bavik, C.-O., Eriksson, U., Peterson, P.A. & Sundelin, J. (1985) Cloning and sequencing of a full length cDNA corresponding to human cellular retinol-binding protein. *Biochem. Biophys. Res. Commun.* **30**, 431–9.

Colantuoni, V., Rocchi, M., Roncuzzi, L. & Romeo, G. (1986) Mapping of human cellular retinol-binding protein to chromosome 3. *Cytogenet. Cell Res.* **43**, 221–2.

Colantuoni, V., Romano, V., Bensi, G., Santoro, C., Costanzo, F., Raugei, G. & Cortese, R. (1983) Cloning and sequencing of a full-length cDNA coding for human retinol-binding protein. *Nucleic Acids Res.* **11**, 7769–76.

Creek, K.E., Silverman-Jones, C.S. & DeLuca, L.M. (1988) Comparison of the uptake and metabolism of retinol delivered to primary mouse keratinocytes either free or bound to rat serum retinol-binding protein. *J. Invest. Dermatol.* **92**, 283–90.

Crow, J.A. & Ong, D.E. (1985) Cell-specific immunohistochemical localization of a cellular retinol-binding protein (type II) in the small intestine of rat. *Proc. Natl. Acad. Sci. USA* **82**, 4707–11.

Demmer, L.A., Birkenmeier, E.H., Sweetser, D.A., Levin, M.S., Zollman, S., Sparkes, R.S., Mohandas, T., Lusis, A.J. & Gordon, J.I. (1987) The cellular retinol-binding protein II gene. *J. Biol. Chem.* **262**, 2458–67.

Dixon, J.L. & Goodman, D.S. (1987*a*) Studies on the metabolism of retinol-binding protein by primary hepatocytes from retinol-deficient rats. *J. Cell Physiol.* **130**, 14–20.

Dixon, J.L. & Goodman, D.S. (1987*b*) Effects of nutritional and hormonal factors on the metabolism of retinol-binding protein by primary cultures of rat hepatocytes. *J. Cell Physiol.* **130**, 6–13.

D'Onofrio, C.D., Colantuoni, V. & Cortese, R. (1985) Structure and cell-specific expression of a cloned human retinol binding protein gene: the 5' flanking regions contains hepatoma specific transcriptional signals. *EMBO J.* **4**, 1981–9.

Eriksson, U., Das, K., Busch, C., Nordlinder, H., Rask, L. Sundelin, J., Sallstrom, J. & Peterson, P.A. (1984) *J. Biol. Chem.* **259**, 13464–70.

Eriksson, U., Hansson, E., Nilsson, M., Jonsson, K.H., Sundelin, J. & Peterson, P.A. (1986) Increased levels of several retinoid binding proteins resulting from retinoic acid-induced differentiation of F9 cells. *Cancer Res.* **46**, 717–25.

Eriksson, U., Hansson, E., Nordlinder, H., Busch, C., Sundelin, J. & Peterson, P.A. (1987) Quantitation and tissue localization of the cellular retinoic acid-binding protein. *J. Cell Physiol.* **133**, 482–90.

Fex, G. and Johannesson, G. (1987) Studies of the spontaneous transfer of retinol from the retinol: retinol-binding protein complex to unilamellar liposomes. *Biochim. Biophys. Acta* **901**, 255–64.

Fex, G. & Johannesson, G. (1988) Retinol transfer across and between phospholipid bilayer membranes. *Biochim. Biophys. Acta* **944**, 249–54.

Fries, E., Gustafsson, L. & Peterson, P.A. (1984) Four secretory proteins synthesized by hepatocytes are transported from endoplasmic reticulum to Golgi complex at different rates. *EMBO J.* **3**, 147—53.

Futterman, S., Saari, J.C. & Blair, S. (1977) Occurrence of a binding protein for 11-cis-retinal in retina. *J. Biol. Chem.* **252**, 3267–71.

Giguère, V., Lyn, S., Yip, P., Siu, C.-H. & Amin, S. (1990*a*) Molecular cloning of cDNA encoding a second cellular retinoic acid binding protein. *Proc. Natl. Acad. Sci. USA* **87**, 6233–327.

Giguère, V., Ong, E.S., Segui, P. & Evans, R.M. (1987) Identification of a receptor for the morphogen retinoic acid. *Nature* **330**, 624–9.

Giguere, V., Shago, M., Zirngibl, R., Tate, P., Rossant, J. & Varmuza, S. (1990*b*) Identification of a novel isoform of the retinoic acid receptor gamma expressed in the mouse embryo. *Mol. Cell Biol.* **10**, 2335–40.

Gjoen, T., Bjerkelund, T., Blomhoff, H.K., Norum, K.R., Berg, T. & Blomhoff, R. (1987) Liver takes up retinol-binding protein from plasma. *J. Biol. Chem.* **262**, 10926–30.

Goodman, D.S. (1965) Cholesterol ester metabolism. *Physiol. Rev.* **45**, 747–839.

Goodman, D.S. (1984) Plasma retinol-binding proteins. In *The Retinoids* (ed. M.B. Sporn, A.B. Roberts & D.S. Goodman), Vol. 2, pp. 41–88. Academic Press, Orlando, Florida.

Goodman, D.S. & Blaner, W.S. (1984) Biosynthesis, absorption and hepatic metabolism of retinol. IN *The Retinoids* (ed.) M.B. Sporn, A.B. Roberts & D.S. Goodman, Vol. 2, pp. 1.39. Academic Press, Orlando, Florida.

Haq, R. & Chytil, F. (1988) Retinoic acid rapidly induces lung cellular retinol-binding protein mRNA levels in retinol deficient rats. *Biochem. Biophys. Res. Commun.* **156**, 712–16.

Harnish, D.C., Soprano, K.J. & Soprano, D.R. (1992) Mouse conceptuses have a limited capacity to elevate the mRNA level of cellular retinoid binding proteins in response to teratogenic doses of retinoic acid. *Teratology* **46**, 137–46.

Heller, J. (1975) Interactions of plasma retinol-binding protein with its receptor. *J. Biol. Chem.* **250**, 3613–19.

Hendriks, H.F.J., Brouwer, A. & Knook, D.L. (1987) The role of hepatic fat-storing (stellate) cells in retinoid metabolism. *Hepatology* **7**, 1368–75.

Jetten, A.M. & Jetten, M.E.R. (1979) Possible role of retinoic acid binding protein in retinoid stimulation of embryonal carcinoma cell differentiation. *Nature* **278**, 180–2.

Kakkad, B.P. & Ong, D.E. (1988) Reduction of retinaldehyde bound and cellular retinol-binding protein (type II) by microsomes from rat small intestines. *J. Biol. Chem.* **263**, 12916–19.

Kanai, M., Raz, A. & Goodman, D.S. (1968) Retinol-binding protein: the transport protein for vitamin A in human plasma. *J. Clin. Invest.* **47**, 2025–44.

Kato, M., Kato, K. & Goodman, D.S. (1984) Immunohistochemical studies of the localization of plasma and of cellular retinol-binding proteins and of transthyretin (prealbumin) in rat liver and kidney. *J. Cell Biol.* **98**, 1696–704.

Kato, M., Kato, K. & Goodman, D.S. (1985*a*) Immunchemical studies on the localization and on the concentration of cellular retinol-binding protein in rat liver during perinatal development. *Lab. Invest.* **52**, 475–84.

Kato, M., Blaner, W.S., Mertz, J.R., Das, K., Kato, K. & Goodman, D.S. (1985*b*) Influence of retinol nutritional status on cellular retinol and cellular retinoic acid-binding protein concentrations in various rat tissues. *J. Biol. Chem.* **260**, 4832–8.

Kato, M., Kato, K., Blaner, W.S., Chertow, B.S. & Goodman, D.S. (1985*c*) Plasma and cellular retinoid-binding proteins and transthyretin (prealbumin) are all localized in the islets of Langerhans in the rat. *Proc. Natl. Acad. Sci. USA* **82**, 2488—92.

Kato, M., Sung, W.K., Kato, K. & Goodman, D.S. (1985*d*) Immunohistochemical studies on the localization of cellular retinol-binding protein in rat testis and epididymis. *Biol. Reprod.* **32**, 173–89.

Kopelman, M., Cogan, U., Mokady, S. & Shinitzky, M. (1976) The interaction between retinol-binding proteins and prealbumins studied by fluorescence polarization. *Biochim. Biophys. Acta* **439**, 449–60.

Krust, A., Kastner, P.H., Petkovich, M., Zelent, A. & Chambon, P. (1989) A third human retinoic acid receptor. *Proc. Natl. Acad. Sci. USA* **86**, 5310—14.

Laurent, B.C., Nilsson, M.H.L., Bavik, C.O., Jones, T.A. Sundelin, J. & Peterson, P.A. (1985) Characterization of the rat retinol-binding protein gene and its comparison to the three-dimensional structure of the protein. *J. Biol. Chem.* **260**, 11476–80.

Lee, S.Y., Ubels, J.L. & Soprano, D.R. (1992) The lacrimal gland synthesizes retinol-binding protein. *Exp. Eye Res.* **55**, 163–71.

Levin, M.S., Li, E., Ong, D.E. & Gordon, J.L. (1987) Comparison of the tissue-specific expression and developmental regulation of two closely linked rodent genes encoding cytosolic retinol binding protein. *J. Biol. Chem.* **262**, 7118–24.

Levin, M.S., Locke, B., Yang, N.C., Li, E. & Gordon, J. (1988) Comparison of the ligand binding properties of two homologous rat apocellular retinol-binding proteins expressed in *Escherichia coli. J. Biol. Chem.* **263**, 17715–23.

Lewis, K.C., Green, M.H. & Underwood, B.A. (1981) Vitamin A turnover in rats as influenced by vitamin A status. *J. Nutr.* **111**, 1135–44.

Li, E., Demmer, L.A., Sweetser, D.A., Ong, D.E. & Gordon, J.I. (1986) Rat cellular retinol-binding protein II: use of a cloned cDNA to define its primary structure, tissue-specific expression and developmental regulation. *Proc. Natl. Acad. Sci. USA* **83**, 5779–83.

Lowe, J.B., Boguski, M.S., Swetser, D.A., Elshourbagy, N.A., Taylor, J.M. & Gordon, J.I. (1985) Human liver fatty acid binding protein. Isolation of a full length cDNA and comparative sequence analysis of orthologous and paralogous proteins. *J. Biol. Chem.* **260**, 3413–17.

MacDonald, P.N. & Ong, D.E. (1987) Binding specificities of cellular retinol binding protein and cellular retinol binding protein type II. *J. Biol. Chem.* **262**, 10550–6.

Maden, M., Ong, D.E., Summerbell, D. & Chytil, F. (1988) Spatial distribution of cellular protein binding to retinoic acid in the chick limb bud. *Nature* **335**, 733–5.

Makover, A., Soprano, D.R., Wyatt, M.L. & Goodman, D.S. (1989*a*) Localization of retinol-binding protein messenger RNA in the rat kidney and in perinephric fat tissue. *J. Lipid Res.* **30**, 171–80.

Makover, A., Soprano, D.R., Wyatt, M.L. & Goodman, D.S. (1989*b*) An in situ-hybridization study of the localization of retinol-binding protein and transthyretin messenger RNA during fetal development. *Differentiation* **40**, 17–25.

Mangelsdorf, D.J., Ong, E.S., Dyck, J.A. & Evans, R.M. (1990) Nuclear receptor that identifies a novel retinoic acid response pathway. *Nature* **345**, 224–9.

Mangelsdorf, D.J., Umesono, K., Kliewer, S.A., Borgmeyer, U., Ong, E.S. & Evans, R.M. (1991) A direct repeat in the cellular retinol binding protein type II gene confers differential regulation by RXR and RAR. *Cell* **66**, 555–61.

Marinari, L., Lenich, C.M. & Ross, A.C. (1987) Production and secretion of retinol-binding protein by a human hepatoma cell line, HepG2. *J. Lipid Res.* **28**, 941–50.

Martone, R.L., Schon, E.A., Goodman, D.S., Soprano, D.R. & Herbert, J. (1988) Retinol-binding protein in synthesized in the mammalian eye. *Biochem. Biophys. Res. Commun.* **157**, 1078–84.

Matsuo, T. & Matsuo, N. (1988) Characterization of retinol-binding protein in familial hypo-retinol-binding proteinemia. *Jpn. J. Ophthalmol.* **32**, 379–84.

Matsuo, T., Matsuo, N., Shiraga, F. & Koide, N. (1988) Keratomalacia in a child with familial hypo-retinol-binding proteinemia. *Jpn. J. Ophthalmol.* **32**, 249–54.

Muto, Y. & Goodman, D.S. (1972) Vitamin A transport in rat plasma: Isolation and characterization of retinol-binding protein. *J. Biol. Chem.* **247**, 2533–41.

Muto, Y., Smith, J.E., Milch, P.O. & Goodman, D.S. (1972) Regulation of retinol-binding protein metabolism by vitamin A status in the rat. *J. Biol. Chem.* **247**, 2542–50.

Newcomer, M.E., Jones, T.A., Aqvist, J., Sundelin, J., Erikson, U., Rask, L. & Peterson, P.A. (1984) The three-dimensional structure of retinol-binding protein. *EMBO J.* **3**, 1451–4.

Nilsson, M.H., Spurr, N.K., Lundvall, J., Rask, L. & Peterson, P.A. (1988) Isolation and characterization of cDNA clone corresponding to bovine cellular retinoic acid binding protein and chromosomal localization of the corresponding human gene. *Eur. J. Biochem.* **173**, 35–44.

Ong, D.E. (1984) A novel retinol-binding protein from rat: Purification and partial characterization. *J. Biol. Chem.* **259**, 1476–82.

Ong, D.E. & Chytil, F. (1975) Retinol binding protein in rat tissues. *J. Biol. Chem.* **250**, 6113–17.

Ong, D.E. & Chytil, F. (1978) Cellular retinoic acid-binding protein from rat testis: Purification and characterization. *J. Biol. Chem.* **253**, 4551–4.

Ong, D.E., Crow, J.A. & Chytil, F. (1982) Radioimmunochemical determination of cellular retinol and cellular retinoic acid binding proteins in cytosols of rat tissues. *J. Biol. Chem.* **257**, 13385–9.

Ong, D.E., Kakkad, B.P. & MacDonald, P.N. (1987) Acyl-CoA independent esterification of retinol bound to cellular retinol binding protein (type II) by microsomes from rat small intestines. *J. Biol. Chem.* **262**, 2729–36.

Ong, D.E. & Page, D.L. (1986) Quantitation of cellular retinol-binding protein in human organs. *Am. J. Clin. Nutr.* **44**, 425–30.

Ong, D.E. & Page, D.L. (1987) Cellular retinol-binding protein (type two) is abundant in human small intestines. *J. Lipid Res.* **28**, 739–45.

Perez-Castro, A.V., Toth-Rogler, L.E., Wei, L.N. & Nguyen-Huu, M.C. (1989) Spatial and temporal pattern of expression of the cellular retinic acid-binding protein and the cellular retinol-binding protein during mouse embryogenesis. *Proc. Natl. Acad. Sci. USA* **86**, 8813–17.

Petkovich, M., Brand, N.J., Krust, A. & Chambon, P. (1987) A human retinoic acid receptor which belongs to the family of nuclear receptors. *Nature* **330**, 444–50.

Pfeffer, B.A., Clark, V.M., Flannery, J.G. & Bok, D. (1986) Membrane receptors for retinol binding protein in cultured human retinol pigment epithelium. *Invest. Ophthalmol. Vis. Sci.* **27**, 1031–40.

Porter, S.B., Fraker, L.D., Chytil, F. & Ong, D.E. (1983) Localization of cellular retinol-binding protein in several rat tissues. *Proc. Natl. Acad. Sci. USA* **80**, 6586–90.

Porter, S.B., Ong, D.E., Chytil, F. & Oregebin-Crist, M.C. (1985) Localization of cellular retinol-binding protein and cellular retinoic acid-binding protein in rat testes and epididymis. *J. Androl.* **6**, 197–212.

Quick, T.C. & Ong, D.E. (1989) Levels of cellular retinol-binding proteins in the small intestine of rats during pregnancy and lactation. *J. Lipid Res.* **30**, 1049–54.

Rajan, N., Blaner, W.S., Soprano, D.R., Suhara, A. & Goodman, D.S. (1990) Cellular retinol binding protein mRNA levels in normal and retinoic acid deficient rats. *J. Lipid Res.* **31**, 821–9.

Rask, L., Anundi, H. & Peterson, P.A. (1979) The cloning and characterization of human RBP cDNAs. *FEBS Lett.* **104**, 55–8.

Rask, L. & Peterson, P.A. (1976) In vitro uptake of vitamin A from the retinol-binding plasma protein to mucosal epithelial cells from monkey's small intestines. *J. Biol. Chem.* **251**, 6360–6.

Rask, L., Valtersson, C., Anundi, H., Kvist, S., Eriksson, U., Dallner, G. & Peterson, P.A. (1983) Subcellular localization in normal and vitamin A-deficient rat liver of vitamin A serum transport proteins, albumin, ceruloplasmin and class I major histocompatibility antigens. *Exp. Cell. Res.* **143**, 91–102.

Ronne, H., Ocklind, C., Wiman, K., Rask, L., Obrink, B. & Peterson, P.A. (1983) Ligand-dependent regulation of intracellular protein transport: Effect of vitamin A on the secretion of the retinol-binding protein. *J. Cell Biol.* **96**, 907–10.

Sacchettini, J.C., Said, B., Schulz, H. & Gordon, J.I. (1986) Rat heart fatty acid-binding protein is highly homologous to the murine adipocyte 422 protein and the P2 protein of peripheral nerve region. *J. Biol. Chem.* **261**, 8218–23.

Sani, B.P. & Hill, D.L. (1974) Retinoic acid: a binding protein in chick embryo metatarsal skin. *Biochem. Biophys. Res. Commun.* **61**, 1276–81.

Schaefer, W.H., Kakkad, B., Crow, J.A., Blair, I.A. & Ong, D.E. (1989) Purification, primary structure characterization, and cellular distribution of two forms of cellular retinol-binding protein, type II from adult rat small intestines. *J. Biol. Chem.* **264**, 4212–21.

Sherman, D.R., Lloyd, R.S. & Chytil, F. (1987) Rat cellular retinol binding protein: cDNA sequence and rapid retinol-dependent accumulation of mRNA. *Proc. Natl. Acad. Sci. USA* **84**, 3209–13.

Shubeita, H.E., Sambrook, J.F. & McCormick, A.M. (1987) Molecular cloning and analysis of functional cDNA and genomic clones encoding bovine cellular retinoic acid binding protein. *Proc. Natl. Acad. Sci. USA* **84**, 5645–9.

Sivaprasadarao, A. & Findley, J.B.C. (1988a) The interaction of retinol-binding protein with its plasma membrane receptor. *Biochim. J.* **255**, 561–70.

Sivaprasadarao, A. & Findlay, J.B.C. (1988b) The mechanism of uptake of retinol by plasma-membrane vesicles. *Biochem. J.* **255**, 571–9.

Sklan, D. & Ross, A.C. (1987) Synthesis of retinol-binding protein and transthyretin in yolk sac in the rat. *J. Nutr.* **117**, 436–5.

Smith, J.E., Muto, Y., Milch, P.O. & Goodman, D.S. (1973) The effects of chylomicron vitamin A on the metabolism of retinol-binding protein in the rat. *J. Biol. Chem.* **248**, 1554–9.

Smith, W.C., Nakshatri, H., Leroy, P., Rees, J. & Chambon, P. (1991) A retinoic acid response element is present in the mouse cellular retinol binding protein I (mCRBPI) promoter. *EMBO J.* **10**, 2223–30.

Soprano, D.R. & Blaner, W.S. (1993) Plasma retinol-binding protein. In *The Retinoids* (ed. M.B. Sporn, A.B. Roberts & D.S. Goodman). Raven Press, New York (in press).

Soprano, D.R., Pickett, C.B., Smith, J.E., & Goodman, D.S. (1981) Biosynthesis of plasma retinol-binding protein in liver as a larger molecular weight precursor. *J. Biol. Chem.* **256**, 8256–8.

Soprano, D.R., Smith, J.E. & Goodman, D.S. (1982) Effect of retinol status on retinol-binding protein biosynthesis rate and translatable messenger RNA levels in rat liver. *J. Biol. Chem.* **257**, 7693–7.

Soprano, D.R., Soprano, K.J. & Goodman, D.S. (1986a) Retinol-binding protein messenger RNA levels in the liver and in extrahepatic tissues of the rat. *J. Lipid Res.* **27**, 166–71.

Soprano, D.R., Soprano, K.J. & Goodman, D.S. (1986b) Retinol-binding protein and transthyretin mRNA levels in visceral yolk sac and liver during fetal development in the rat. *Proc. Natl. Acad. Sci. USA* **83**, 7330–4.

Soprano, D.R., Soprano, K.J., Wyatt, M.L. & Goodman, D.S. (1988a) Induction of the expression of retinol-binding protein and transthyretin in F9 embryonal carcinoma cells differentiated to embryoid bodies. *J. Biol. Chem.* **263**, 17897–900.

Soprano, D.R., Wyatt, M.L., Dixon, J.L. & Goodman, D.S. (1988*b*) Retinol-binding protein synthesis and secretion by the rat visceral yolk sac. *J. Biol. Chem.* **263**, 2934–8.

Steven, D. & Morris, G. (1975) *Comparative Placentation.* Academic Press, New York.

Stoner, C.M. & Gudas, L.J. (1989) Mouse cellular retinoic acid binding protein: Cloning, complementary DNA sequences and messenger RNA expression during the retinoic acid-induced differentiation of F9 wild type and RA-3-10 mutant teratocarcinoma cells. *Cancer Res.* **49**, 1497–504.

Sucov, H.M., Murakami, K.K. & Evans, R.M. (1990) Characterization of an autoregulated response element in the mouse retinoic acid receptor type beta gene. *Proc. Natl. Acad. Sci. USA* **87**, 5392–6.

Sundelin, J., Anundi, H., Tragardh, L., Eriksson, U., Lund, P., Ronne, H., Peterson, P.A. & Rask, L. (1985*a*) The primary structure of rat liver cellular retinol-binding protein. *J. Biol. Chem.* **260**, 6488–93.

Sundelin, J., Das, S.R., Eriksson, U., Rask, L. & Peterson, P.A. (1985*b*) The primary structure of bovine cellular retinoic acid binding protein. *J. Biol. Chem.* **260**, 6494–9.

Sundelin, J., Laurent, B.C., Anundi, H., Tragardh, L., Larhammar, D., Bjock, L., Eriksson, U., Akerstrom, B., Jones, A., Newcomer, M., Peterson, P.A. & Rask, L. (1985*c*) Amino acid sequence homologies between rabbit, rat and human serum retinol-binding proteins. *J. Biol. Chem.* **260**, 6472–87.

Takahashi, J., Odani, S. & Ono, T. (1982) A close structural relationship of rat liver Z-protein to cellular retinod binding proteins and peripheral myelin P2 protein. *Biochem. Biophys. Res. Commun.* **106**, 1099–05.

Takase, S., Ong, D.E. & Chytil, F. (1979) Cellular retinol-binding protein allows specific interaction of retinol with the nucleus in vitro. *Proc. Natl. Acad. Sci. USA* **76**, 2204–8.

Takase, S., Ong, D.E. & Chytil, F. (1986) Transfer of retinoic acid from its complex with cellular retinoic acid-binding protein to the nucleus. *Arch. Biochem. Biophys.* **247**, 328–33.

Thaller, C. & Eichele, G. (1987) Identification and spatial distribution of retinoids in the developing chick limb bud. *Nature* **327**, 625–8.

Thompson, J.N., Howell, J. & Pitt, G.A. (1964) Vitamin A and reproduction in rats. *Proc. R. Soc. Lond.* **159**, 510.

Torma, H. & Vahlquist, A. (1986) Uptake of vitamin A and retinol-binding protein by human placenta in vitro. *Placenta* **7**, 295–300.

Van Jaarsveld, P.O., Edelhoch, H., Goodman, D.S. & Robbins, J. (1973) The interaction of human plasma retinol binding protein with prealbumin. *J. Biol. Chem.* **248**, 4698–705.

Wald, G. (1968) Molecular basis of visual excitation. *Science* **162**, 230.

Wei, L.N., Mertz, J.R., Goodman, D.S. & Nguyen-Huu, M.C. (1987) Cellular retinoic acid and cellular retinol-binding proteins: Complementary deoxyribonucleic acid cloning, chromosomal assignment and tissue specific expression. *Molec. Endocrinol.* **1**, 526–34.

Wei, L., Blaner, W.S., Goodman, D.S. & Nguyen-Huu, M.C. (1989) Regulation of the cellular retinoid-binding proteins and their messenger ribonucleic acids during P19 embryonal carcinoma cell differentiation induced by retinol acid. *Molec. Endocrinol.* **3**, 454–63.

Wei, L., Tsao, J., Chu, Y., Jeannotte, L. & Nguyen-Huu, M. (1990) Molecular cloning and transcriptional mapping of the mouse cellular retinoic acid-binding protein gene. *DNA Cell Biol.* **9**, 471–8.

Whitman, M.M., Harnish, D.C., Soprano, K.J. & Soprano, D.R. (1990) Retinol-binding protein mRNA is induced by estrogen in the kidney but not in the liver. *J. Lipid Res.* **31**, 1483–90.

Wolback, S.B. & Howe, P.R. (1925) Tissue changes following deprivation of fat-soluble A vitamin. *J. Exp. Med.* **42**, 753–71.

Wolback, S.B. & Howe, P.R. (1933) Epithelial repair in recovery from vitamin A deficiency. *J. Exp. Med.* **57**, 511–27.

Yamada, M., Blaner, W.S., Soprano, D.R., Dixon, J.L., Kjeldbye, H.M. & Goodman, D.S. (1987) Biochemical characteristics of isolated rat liver stellate cells. *Hepatology* **7**, 1224–9.

Zelent, A., Krust, A., Petkovich, M., Kastner, P. & Chambon, P. (1989) Cloning murine alpha and beta retinoic acid receptors and a novel receptor gamma predominately expressed in the skin. *Nature* **339**, 714–17.

2

Retinoic Acid Receptors

Vincent Giguère

Introduction

The retinoids are a group of compounds that includes both natural and synthetic derivatives of retinol (vitamin A). Retinoic acid (RA), a member in this class of small lipophilic molecules, is known to exert profound effects on development and differentiation in a wide variety of systems. In the vitamin-A-deprived rat, exogenous RA is able to replace vitamin A for all its functions, with the exception of the requirement of vitamin A for vision. RA is therefore believed to act as the primary active metabolite of vitamin A in the regulation of development and homeostasis. The recent discovery of RA receptors which belong to the superfamily of steroid hormone receptors provided a model for the molecular mechanism of most, if not all, of the extravisual action of retinoids (Giguère et al., 1987; Petkovich et al., 1987). This family of nuclear receptors controls cell functions by directly regulating gene expression; by analogy, it is believed that the interaction between RA and one of its receptors may induce a cascade of regulatory events through the activation of specific gene networks by the ligand–receptor complex. This discovery not only offered the opportunity to analyze in great detail the structure, function and pattern of expression of the RA receptors, but also provided the necessary molecular tools to facilitate the identification and characterization of developmental control genes. In this review, I first give a brief overview of the numerous biological activities ascribed to natural retinoids, an exercise that emphasizes the need for the large family of RA receptors and the complex set of intracellular interactions involved in the transduction of the retinoid signal. I then describe the identification by molecular cloning of the two distinct families of RA receptors, referred to as the RARs and RXRs.

The analysis of the structure–function relations of the receptors is then discussed, followed by an overview of the interactions between RA receptors and their target genes.

Biological activities of retinoids

It is now more than 60 years ago that vitamin A was first shown to exert control over cell differentiation and proliferation. In 1925 Wolbach & Howe, in their classic studies on the effects of vitamin A deficiency in the rat, noted that vitamin A was essential for the normal differentiation of the epithelia (Wolbach & Howe, 1925). Vitamin A deficiency was leading to the conversion of secretory epithelium to a squamous epithelium while the reverse experiment, performed in 1953 by Fell & Mellanby, showed that an excess of the vitamin caused the disappearance of stratified squamous epithelium with replacement by a secretory epithelium (Fell & Mellanby, 1953). Following these pioneering experiments, organ and tissue culture systems were developed to investigate the role of retinoids in epithelial differentiation. In particular, hamster trachea organ cultures were extensively used to assay the ability of various retinoids to reverse *in vitro* the keratinization observed in vitamin-A-deficient animals (Sporn *et al.*, 1975). Using cultured human keratinocytes, Fuchs & Green (1981) showed that retinoids could regulate the terminal differentiation of epidermal cells, a change accompanied by marked alterations in the pattern of keratin genes expressed (Kopan *et al.*, 1987).

While early studies focused on the effects of retinoids on the various types of epithelium, their actions have been shown to be more widespread than previously suspected. A number of investigators have examined their effects on a variety of neoplastic cultured cell lines including neuroblastomas, melanomas and teratocarcinomas. For example, in the human myeloid leukemia cell line HL60, RA induces terminal granulocytic differentiation (Breitman *et al.*, 1980). In embryonic carcinoma (EC) cells such as the F9 EC line, RA will induce the differentiation of parietal endoderm (Strickland & Mahdavi, 1978), while the P19 EC cell line will differentiate into cardiac and smooth muscle cells, neurons or glia when treated with increasing concentrations of RA (Jones-Villeneuve *et al.*, 1982; Edwards & McBurney, 1983). Retinoic acid can also influence the proliferation of cells of mesenchymal origin, such as chick limb bud-derived mesenchymal cells often used to explore the cellular basis for RA-induced skeletal teratogenesis (see below) (Kisler, 1987). In contrast to the negative effect of RA observed in other systems, RA exerts a positive influence on the proliferation and chondrogenesis of chick limb bud cultures exposed to physiological levels of the retinoid (Ide & Aono, 1988; Paulsen *et al.*, 1988).

Retinoic acid has been shown to exert equally potent effects in development. Natural and synthetic retinoids are known as potent teratogens whose effects are characterized by a spectrum of severe limb and cranial defects and neuronal tube deformities in both mice (Rosa *et al.*, 1986) and humans (Lammer *et al.*, 1985). In *Xenopus*, administration of RA has been shown to disrupt the development of the nervous system by causing a marked loss of anterior structures (Durston *et al.*, 1989). Perhaps the most dramatic and studied developmental process influenced by RA is limb morphogenesis (reviewed by Tabin (1991)). For example, local application of RA can re-specify the positional memory of the blastema, the group of progenitor cells that arises at the amputation plane and constitutes the base on which regeneration proceeds, in the proximal–distal axis during limb regeneration in urodeles. Normally, if the amputation occurs at the level of the wrist, the blastema will regenerate a hand. However, treatment of a wrist blastema with high doses of RA will lead to the regeneration of the entire arm, producing a serial duplication of skeletal structures, while treatment with lower doses of RA will result in lesser proximal locations of the regenerate. In the developing chick limb bud, a zone at the posterior margin of the bud, referred to as the zone of polarizing activity (ZPA), is able to induce an extra set of digits when grafted to the anterior side of a limb bud (Saunders & Gasseling, 1968). On the basis of this observation, it was proposed that the ZPA might release a morphogen whose graded distribution in the developing field could specify digit patterning (Tickle *et al.*, 1985). The subsequent demonstration that implantation of a bead soaked in RA could induce digit duplications in a dose-dependent manner similar to those observed with the ZPA graft (Tickle *et al.*, 1982, 1985) and the measurement of graded concentrations of RA in the limb field (Thaller & Eichele, 1987) led to the suggestion that RA could be a natural morphogen produced and released by the ZPA. However, recent studies led to the suggestion that RA may act indirectly by inducing the formation of a ZPA, which then determines the anteroposterior axis of the limb bud by some undetermined mechanism not involving a gradient of RA (Noji *et al.*, 1991; Wanek *et al.*, 1991).

Finally, perhaps the most significant characteristic of retinoids is their inhibitory effect on tumor progression in animals and humans (Sporn & Roberts, 1983) and their ability to block the action of tumor promoter both *in vitro* (Yuspa *et al.*, 1981) and *in vivo* (Gendimenico *et al.*, 1990;

Leder *et al.*, 1990). The connection between retinoids and cancer was first made by Lasnitzki (1955) who showed that the pre-malignant phenotype of mouse prostate glands that has been treated with the carcinogen 3-methylcholantrene could be altered by retinoids. During the course of the next 35 years, a number of studies have demonstrated that treatment with retinoids can prevent mammary, skin, lung and urinary bladder cancer in experimental animals (Moon & Mehta, 1990). Recently, clinical trials have also shown that retinoids have excellent potential as therapeutic agents for the treatment of certain types of skin cancer (Kraemer *et al.*, 1988), head and neck cancer (Hong *et al.*, 1990) and acute promyelocytic leukemia (Huang *et al.*, 1988).

Nuclear receptors for retinoids

Identification of a retinoic acid receptor
Although various efforts were made to understand the role of retinoids in cellular differentiation and proliferation, the molecular mechanism underlying these effects eluded scientists for more than 50 years. The identification of cellular retinol and RA binding proteins (CRBP and CRABP, respectively) in the mid-1970s (Ong, 1987) led to the proposal that they might represent specific intracellular receptor systems. However, despite a detailed biochemical characterization of these proteins and the cloning of their cDNAs, no evidence was uncovered to establish a decisive role for CRBP and CRABP as direct mediators of retinoid action. In late 1987, two groups studying the superfamily of steroid hormone receptors identified a novel member of this family as a nuclear receptor for RA (Giguère *et al.*, 1987; Petkovich *et al.*, 1987). The studies on the structure–function relations of steroid and thyroid hormone receptors had revealed that these proteins share common structural and functional properties (reviewed in Evans (1988)), in particular a highly conserved cysteine-rich region which functions as the DNA-binding domain (discussed below). These findings led to the suggestion that this common segment might be used to scan the genome for related gene products (Arriza *et al.*, 1987; Giguère *et al.*, 1988). Both groups used a variation of this strategy to clone and subsequently identify RA as the ligand associated with a novel receptor. Giguère *et al.* (1987) used an oligonucleotide probe derived from the analysis of the integration site of a hepatitis B virus from a human hepatocellular carcinoma (Dejean *et al.*, 1986) to isolate full-length cDNAs encoding a novel member of the steroid hormone receptor family. Petkovich *et al.* (1987) used a consensus oligonucleotide derived from the nucleotide

sequences of a number of cloned steroid–thyroid hormone receptors as a hybridization probe to isolate a distinct cDNA clone encoding a large portion of this novel protein. As the ligand for this novel gene product was unknown, a quick and sensitive assay was developed to reveal its identity. The assay takes advantage of the modular structure of steroid hormone receptors (Giguère *et al.*, 1986) and the suggestion that functional domains may be interchangeable (Green & Chambon, 1987). In this procedure, the DNA-binding domain of the 'orphan' receptor is replaced by the well-characterized DNA-binding domain of either the human glucocorticoid or estrogen receptor to yield chimeric receptors bearing novel ligand specificity. After transfection of an expression vector carrying the chimeric receptor together with a glucocorticoid- or estrogen-responsive reporter gene, a battery of ligands can then be tested for their ability to activate transcription of the reporter gene. Using this assay, both groups found that RA was a potent activator of the chimeric receptors while other retinoids such as retinol and retinol palmitate were only poor inducers. To corroborate the identity of the novel gene product as the RA receptor, the putative receptor was expressed at high level in eucaryotic cells and shown to possess the ability to bind labeled RA with high affinity. Other ligands, such as steroid and thyroid hormones, vitamin D_3 and 25-hydroxycholesterol, were totally inactive in the transcriptional assay.

The identification of a RA receptor as a member of the superfamily of nuclear receptors provided a model for the molecular mechanism of action of RA, in principle if not in detail. As represented in Figure 2.1, RA receptors located in the nucleus are activated upon binding by RA, resulting in transcriptional stimulation or inhibition of the expression of RA-responsive genes. Although RA, like steroid and thyroid hormones, may regulate gene expression via other mechanisms such as increasing the stability of messenger RNA or the half-life of proteins, no evidence has yet been found that nuclear receptors are directly involved in these mechanisms. These so-called secondary effects are likely to result from the action of gene products whose expression is under the direct control of RA and its receptors. Thus, the activity of RARs as transcriptional factors may account for most, if not all, primary biological effects of RA.

The RAR family

Multiple genes, multiple isoforms

The first RAR, a polypeptide of 462 amino acids now referred to as RARα, was shown to be expressed in all tissues tested, a finding consistent with the broad spectrum of retinoid action. However, two sets of evidence suggested that RARα was probably not the only nuclear receptor implicated in the transduction of the retinoid signal. The first evidence was the demonstration, by low-stringency hybridization, of the existence of several loci in the human genome related to the RARα (Giguère *et al.*, 1987). The second evidence, consistent with the discovery of a family of RARα-related genes, was the close similarity observed between RARα and the gene product encoded by a cDNA cloned from the analysis of the integration site of the hepatitis B virus in a human hepatocellular carcinoma (de Thé *et al.*, 1987). Using the chimeric receptor assay, this putative receptor was also shown to respond to RA and consequently named RARβ (Benbrook *et al.*, 1988; Brand *et al.*, 1988). A year later, while attempting to clone the murine homologs of the human RARα and RARβ by screening a mouse embryonic cDNA library, Chambon and colleagues identified a third member of the RAR family, referred to as RARγ (Zelent *et al.*, 1989). The human RARγ homolog was cloned soon thereafter (Krust *et al.*, 1989; Ishikawa *et al.*,

Figure 2.1. Schematic representation of the mechanism by which retinoic acid controls gene expression. Retinol (R) enters the cell by active transport and/or passive diffusion. After cellular uptake, R is oxidized to its active metabolite retinoic acid (RA) in target cells. RA then diffuses to the nucleus and binds to its nuclear receptors. The activated RARs and RXRs recognize and bind specific retinoic acid response elements located in the promoter of RA-responsive genes, leading to positive or negative regulation of mRNA synthesis. Because their precise functions remain to be elucidated, the cellular retinoid-binding proteins CRBP and CRABP were omitted from this model.

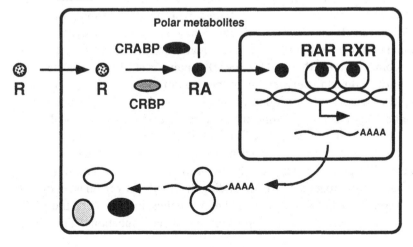

1990) and characterization of several partial cDNAs revealed the possible existence of many RARγ isoforms distinct at their amino terminus (Krust et al., 1989). The existence of two functional RARγ isoforms (RARγ1 and γ2) was confirmed by the cloning of their cDNAs from mouse embryonic cDNA libraries (Giguère et al., 1990; Kastner et al., 1990). As shown in Figure 2.2, it has now been established that all three RAR loci generate multiple isoforms, the RARα and RARβ genes producing two (RARα1 and α2) and three (RARβ1, β2 and β3) functional isoforms of each receptor subtype, respectively (Leroy et al., 1991; Zelent et al., 1991). Each isoform produced from the same RAR locus differs solely in its amino terminal region.

Because of their importance in the study of the action of retinoids in development, in particular limb morphogenesis, various RAR homologs were cloned from avian and amphibian species. Cloning of RARα has been reported in the newt (Ragsdale et al., 1989), RARβ was isolated in both chicken (Smith & Eichele, 1991) and newt (Giguère et al., 1989) while RARγ2 was identified in Xenopus laevis (Ellinger-Ziegelbauer & Dreyer, 1991). A possible fourth type of RAR, referred to as RARδ, was isolated in the newt (Ragsdale et al., 1989). However, its high overall homology with human, murine and Xenopus RARγ cast a doubt on the precise identity of the newt RARδ. Future experiments directed at the cloning of a true homolog of RARγ in the newt or a RARδ homolog from the human or mouse genome should provide a definitive classification of these receptors.

Genomic organization of the RAR genes

The molecular basis for the generation of alternative RAR transcripts was uncovered by the characterization of the genomic organization of each RAR locus. The elucidation of the human RARγ transcriptional unit reveals that it is composed of at least 11 exons, seven of which are common to RARγ1 and RARγ2 (Lehmann et al., 1991). Two distinct promoters direct the expression of the two isoforms. The 5' region of the RARγ1 transcript is encoded in three exons, the first two (Ia and Ib) containing most of the 5'-untranslated (UT) region while the third exon (Ic) encodes the remainder of the 5'UT and the first 61 amino acids, exclusive to RARγ1. This particular organization is relatively unusual: the 5'UT region of genes is rarely split into several exons. This organizational feature, together with the finding that the 5'UT region is relatively well conserved between human and mouse genes, indicates that the 5'UT region may perform some important and specific functions. The 5'UT

Figure 2.2. The large family of receptors for retinoic acid and other related members of the superfamily of nuclear receptors. All the RAR isoforms shown were characterized from the mouse while RXRα is from human, RXRβ is from mouse and USP referred to the gene product of the *ultraspiracle* locus in *Drosophila melanogaster*. The well conserved DNA- and ligand-binding domains are represented by large open boxes and less conserved regions by thin boxes. Numbers within these domains represent percentage amino acid identity when compared to the RARα for the RAR family and RXRα for the RXR family. Stippled and filled boxes indicate N-terminal domains generated by alternative RNA processing. Numbers above each box represent the boundaries of each functional domain as well as the total length of the protein.

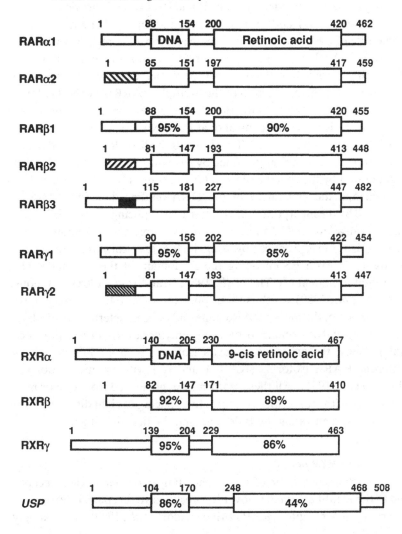

region of some yeast genes has been shown to be important for translational control and a similar function may provide a mechanism for differential expression of RAR isoforms. The exon encoding the first 50 amino acids of RARγ2 is located within the intron separating exon Ib and exon II, which encodes domain B and the first zinc finger of the DNA-binding domain. This genomic arrangement indicates that alternative promoter usage rather than differential splicing is responsible for the generation of the RARγ2 transcript. The genomic organization of the seven exons common to RARγ1 and RARγ2 shares many features with other members of the superfamily of nuclear receptors. The DNA-binding domain is encoded by exons II and III, exon IV encodes the hinge region (domain D) and part of the ligand-binding domain, while the remainder of this domain, the F domain and the 3'UT region are encoded by exons V–VIII. An even more complex genomic organization, involving both alternative splicing and differential usage of promoters, is responsible for the production of the multiple RARβ isoforms (Zelent *et al.*, 1991). As observed with the RARγ gene, the expression of the RARβ locus is under the control of at least two promoters. RARβ1 and RARβ3 share a common 5'UT region and the first 59 amino acids of the two isoforms, both regions encoded by an exon referred to as E1. RARβ1 is generated by the splicing of exon E1 to exon E4, which encodes the B domain and the first finger of the DNA-binding domain. RARβ3 is produced by the incorporation, via alternative splicing, of exon E2, which encodes an additional 27 amino acids that become part of domain A. The RARβ2 isoform is produced by using the RA-inducible promoter located upstream of exon E3 encoding the A domain of the original RARβ identified in human (de Thé *et al.*, 1987; Brand *et al.*, 1988) and mouse (Zelent *et al.*, 1989) which is then spliced to exon E4. The exon–intron organization of the mouse RARα gene and the characterization of a large number of cDNAs obtained by using polymerase chain reaction technology show the existence of seven isoforms with a minimum of two different RARα isoforms (RARα1 and α2) with unique A domains (Leroy *et al.*, 1991). All these isoforms are generated from one or more primary RARα transcripts by alternative splicing of eight different exons with the exon encoding the B domain and the first zinc finger.

The RXR family

Characterization of a gene encoding an orphan nuclear receptor led to the identification of a novel RA-responsive receptor, referred to as the retinoid X receptor (RXR) (Mangelsdorf *et al.*, 1990). Interestingly, amino acid sequence comparison between RXRα and RARα revealed only weak homology over the entire length of the proteins, with the highest degree of similarity found in the DNA-binding domain (61%) while the putative RA binding domain showed only 27% identity. Because of the weak homology between RARα and RXRα and the finding that RXRα responds to somewhat higher concentrations of RA than members of the RAR family, it was postulated that the natural high-affinity ligand for RXRα might not be all-*trans* RA. However, the ligand for RXR was likely to be closely related to RA, perhaps an unidentified natural retinoid, which suggested the existence of an alternative response pathway in retinoid physiology. Two teams of researchers, working independently and using different strategies, identified the RXR ligand as the 9-*cis* RA stereoisomer (Heyman *et al.*, 1992; Levin *et al.*, 1992). The 9-*cis* RA binds and *trans*-activates RXRα at physiological concentrations (ED_{50} ≈ 10 nM) and is present *in vivo* in significant levels in kidney and liver (Heyman *et al.*, 1992) which suggest that 9-*cis* RA is a natural hormone. It is also a specific ligand for RXRβ and RXRγ, two RXR isoforms closely related to RXRα and encoded by distinct genetic loci (Hamada *et al.*, 1989; Mangelsdorf *et al.*, 1992). However, 9-*cis* RA can also activate RARs with equal potency, which suggests that a more specific RXR ligand may exist. The finding of a RXR homolog in *Drosophila* encoded by the *utraspiracle* (USP) locus (Oro *et al.*, 1990) suggests that RXR may be the ancestral retinoid receptor. However, the *usp* gene product does not respond to all-*trans* or 9-*cis* RA, and *Drosophila* have no requirement for vitamin A other than for visual function.

Anatomy of the retinoic acid receptors

All nuclear receptors share a common structure and the RAR and RXR families constitute no exception. Domains can be divided according to the level of homology in the amino acid sequences of the same receptor subtype between different species (domains A to F) or according to their function (Figure 2.3). These proteins possess an amino terminus of variable length and homology (domains A and B), a short but highly conserved central region (domain C) and a relatively well-conserved carboxy terminal region (domain E) (Evans, 1988). These domains can be separated by hinge regions (such as domain D) and subdivided into smaller structural motifs. Mutational analysis of nuclear receptors showed that each structural domain performs specific functions. Because most of our knowledge on the structure–function relations of nuclear receptors has been gained from the study of steroid and thyroid hormone receptors, the reader should be aware that the functions associated with each of these regions within the RARs are often inferred from these studies.

The DNA-binding domain: interactions with RAREs

The central core region of nuclear receptors contains the DNA-binding domain that dictates specific recognition of hormone response element (HREs). The DNA-binding domain consists of two motifs known as 'zinc finger' (Evans & Hollenberg, 1988; Freedman *et al.*, 1988) and 'zinc twist' (Vallee *et al.*, 1991) and is represented schematically in Figure 2.4. Intensive *in vitro* mutational analysis showed that receptor mutants with changes in either motif were defective in DNA-binding activities (Hollenberg *et al.*, 1987; Hollenberg & Evans, 1988; Severne *et al.*, 1988). Unexpectedly, it was found that the finger loops themselves can be interchanged between heterologous receptors without affecting binding specificity. Instead, the target gene specificity resides in three amino acid residues clustered at the base or 'knuckle' of the first zinc finger; this subregion is now referred to as the P box (Green *et al.*, 1988;

Figure 2.3. Anatomy of a nuclear receptor. The prototype receptor can be divided according to the degree of homology of the same receptor subtype between species (Krust *et al.*, 1986) (upper panel) or according to function (Evans, 1988) (lower panel). The numbers represent the percentage amino acid identity between human and mouse RARα. The A/B domain resides in the N-terminal portion of the protein and has been shown to be necessary for maximal transactivation. The C domain is equivalent to the DNA-binding domain. The D domain functions as a hinge linking the DNA- and ligand-binding domains. The ligand-binding corresponds to the E domain while the F domain, absent in some receptors such as RXRα and RXRβ, has no assigned function. The bars indicate the location of each function within the protein. Functions represented by dashed lines have not been localized with precision or have not been well characterized in some receptors.

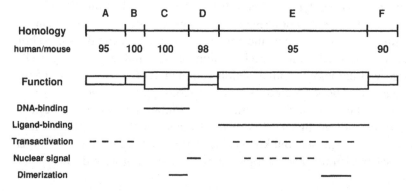

Danielsen *et al.*, 1989; Mader *et al.*, 1989; Umesono & Evans, 1989). The amino acid residues at the base of the second finger, referred to as the D box, are required for discrimination of HRE half-site spacing (Umesono & Evans, 1989). Consistent with these mutational studies, a three-dimensional structure analysis of the DNA-binding domain of the glucocorticoid receptor showed the domain to consist of a globular body from which the two finger regions extend. The P box forms a 'recognition helix' that positions itself into the major groove of the recognition site while the D box forms protein–protein contacts important for cooperative binding (Tsai *et al.*, 1989; Hard *et al.*, 1990).

Analysis of the DNA-binding domains of the RARs show them to be extremely well conserved across species: 97% amino acid identity between RARγ cloned from species as distantly related as *Xenopus* and human, for example. What may be more surprising is the high degree of similarity between RAR subtypes. Amino acid comparison of the mouse RARs reveals that RARα shares 63 out of 66 amino acid residues with

Figure 2.4. The human RARα DNA-binding domain. This model is based on mutational and protein structure analysis (see text for references). The domain contains two 'zinc fingers' represented by loops. Shaded residues represent the conserved cysteine residues implicated in zinc binding and known to affect DNA-binding specificity. The P box is believed to contact the DNA (HRE) within the major groove of the double helix and is responsible for sequence-specific recognition by the receptor. The P box makes protein:protein contact with the P box of a second receptor to form homo- or heterodimers and is responsible for recognition of the HRE half-site spacing.

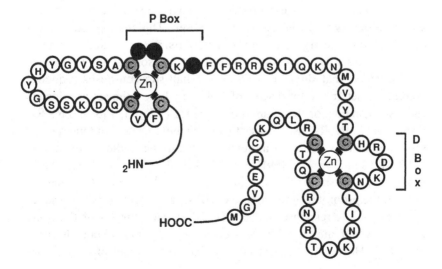

both RARβ and RARγ, or 94% homology. Because none of these mutations falls within the P or D boxes, it is expected that RARα, β and γ will recognize identical RARE, and therefore that RAR subtypes may not have the ability to discriminate between different RA-responsive genes. Even more challenging was the finding that RARs and thyroid hormone receptor (T3R) can induce gene expression through a common response element. RARα and RARβ expressed from cloned cDNAs or the endogenous RARs present in F9 EC cells were shown to activate gene expression from promoters fused to either the synthetic palindromic thyroid hormone response element (TREp) composed of an inverted repeat of the sequence TGACC or natural TREs extracted from the growth hormone gene or the myosin heavy chain gene (Umesono *et al.*, 1988; Graupner *et al.*, 1989). How then can specific control of gene expression by RARs be achieved? The recent identification of a variety of RA response elements (RARE) allowed this question to be investigated in detail. Analysis of a RARE located in the promoter of the mouse and human RARβ2 gene first revealed that the RAR recognition sequence was composed of two half-sites arranged in tandem repeat, not in an inverted repeat configuration as previously shown for most HREs (Figure 2.5) (deThé *et al.*, 1990*b*; Hoffman *et al.*, 1990; Sucov *et al.*, 1990). As expected, the βRARE, when linked to heterologous promoters, confers transcriptional activation via all three RAR subtypes. More importantly, however, the βRARE cannot be recognized efficiently by the T_3R, vitamin D_3 receptor and all steroid hormone receptors tested. Strong RAREs located in the promoters of the mouse complement factor H (Munoz-Cánoves *et al.*, 1990) and the human alcohol dehydrogenase (Duester *et al.*, 1991) genes were shown to share the same tandem repeat structure observed in the RARβ2 promoter. This finding led to a re-examination of the organization of other HREs, in particular TREs. In an elegant analysis, Umesono and colleagues (Umesono *et al.*, 1991) have demonstrated that direct repeats of the half-site consensus sequence AGGTCA separated by spacers of 3, 4 and 5 base pairs act as selective reponse elements for the vitamin D_3 receptor (VD_3R), T_3R and RARs, respectively. RARs also transactivate via binding to tandem repeats separated by two base pairs as demonstrated in the studies of the mouse CRBP-I (Smith *et al.*, 1991) and CRABP-II (Durand *et al.*, 1992) promoters. The 'spacer rule' also applies to the RXR family. Characterization of the promoter of the mouse CRBP-II gene revealed the presence of a retinoid X response element (RXRE) that consists of the same AGGTCA half-site but with a spacer of one nucleotide (Mangelsdorf *et al.*, 1991). However, characterization of a large number of RAREs has

Figure 2.5. Nucleotide sequences of a selection of natural and synthetic RA response elements (RARE and RXRE) that can confer retinoid-responsiveness on heterologous promoters. The core binding motif of the natural RARE and RXRE consists of repeats of the consensus half-site sequence A/GGGTCA, represented by arrows. Tandem repeats of the core motif can be arranged in a direct or inverted orientation with spacers of variable length. Elements composed of tandem repeats with a spacer of 1 bp (DR1) preferentially respond to RXRs, while elements with spacers of 2 or 5 bp (DR2 and DR5) are more specific for the RAR pathway. Composite elements contain multiple half-sites arranged in more than one configuration and therefore may be activated by more than one type of nuclear receptors. TREpal (or TREp) is a palindromic synthetic response element derived from the rat growth hormone TRE (Glass *et al.*, 1988). Other elements have been characterized from the following genes: CRBP-II, rat cellular retinol binding protein type II (Mangelsdorf *et al.*, 1991); CRBP-I, mouse cellular retinol binding protein type I (Smith *et al.*, 1991); RARβ2, mouse and human RARβ2 gene promoter (de Thé *et al.*, 1990*b*; Sucov *et al.*, 1990); ApoA1, human apolipoprotein A1 (Rottman *et al.*, 1991); LamB1, mouse laminin B1 (Vasios *et al.*, 1989).

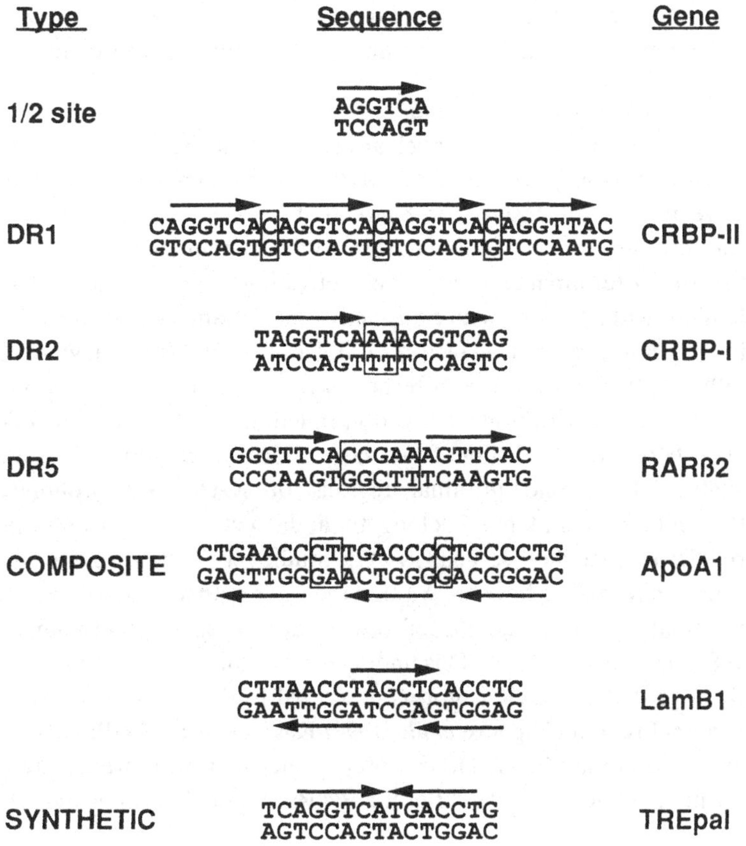

Type	Sequence	Gene
1/2 site	AGGTCA TCCAGT	
DR1	CAGGTCACAGGTCACAGGTCACAGGTTAC GTCCAGTGTCCAGTGTCCAGTGTCCAATG	CRBP-II
DR2	TAGGTCAAAAGGTCAG ATCCAGTTTTCCAGTC	CRBP-I
DR5	GGGTTCACCGAAAGTTCAC CCCAAGTGGCTTTCAAGTG	RARβ2
COMPOSITE	CTGAACCCTTGACCCCTGCCCTG GACTTGGGAACTGGGGACGGGAC	ApoA1
	CTTAACCTAGCTCACCTC GAATTGGATCGAGTGGAG	LamB1
SYNTHETIC	TCAGGTCATGACCTG AGTCCAGTACTGGAC	TREpal

demonstrated that flexibility is allowed in their composition. For example, HREs present in the promoter of the laminin B1 (Vasios *et al.*, 1989) and the phosphoenolpyruvate carboxykinase (Lucas *et al.*, 1991) genes composed of the repetition of the motif TGACC separated by spacers of various lengths can also function as RAREs, although they show weaker transcriptional activity than the DR-5 RAREs and generally require the expression of cloned RARs for maximal activation. While direct repeats such as the βRARE would confer selective hormonal response, promiscuous elements made of inverted repeats such as the TREp may provide a medium for cross-talk between different hormonal response systems.

Finally, the DNA-binding domain can also make protein–protein contacts with other transcription factors. Transcriptional interference between the proto-oncogene c-jun and RARα has recently been described (Schüle *et al.*, 1990) and shown to result in mutual inhibition of DNA binding properties owing to direct protein–protein interaction involving their respective DNA-binding domains. These findings not only revealed a cross-talk between two major signal transduction systems but also a novel molecular mechanism for transcriptional repression.

The N-terminal domain

This region of the nuclear receptors has been shown to contain *trans*-activation (τ) functions. *Trans*-activation domains can be described as regions of a transcription factor that increase effective initiation of transcription by RNA polymerases. In the human glucocorticoid receptor, the N-terminal transcriptional activating region is acidic in nature. Similar acidic regions with *trans*-activating capability have been located in the yeast transcription factors GAL4 and GCN4 (for review, see Sigler (1988)). In contrast, the N-terminus of the human estrogen receptor contains a transcriptional activation function that is non-acidic (Tora *et al.*, 1989) and thus distinct from those of acidic activators (Tasset *et al.*, 1990). The amino terminal regions of RARs are proline/serine/threonine-rich and may belong to a distinct class of transcriptional regulators (Mitchell & Tjian, 1989). Studies with the progesterone and estrogen receptors have shown that the *trans*-activating domain in the N-terminal region can dictate cell and target gene specificity (Kumar *et al.*, 1987; Tora *et al.*, 1988). This finding is particularly interesting in view of the fact that all RAR isoforms described to date possess a distinct N-terminal region (Giguère *et al.*, 1990; Kastner *et al.*, 1990; Leroy *et al.*, 1991; Zelent *et al.*, 1991). The precise mechanism by which the amino terminal regions in the different RARs might determine target gene specificity is unknown but is likely to involve differential interaction with other transcription factors which will be specific to the promoter of each target gene. Supporting this argument is the demonstration that steroid hormone receptors compete for functionally limiting transcription factors that mediate their enhancer function (Meyer *et al.*, 1989).

The ligand-binding domain

This region is large, complex and fulfills multiple functions. Primarily, it encodes the ligand-binding domain which encompasses a large portion of the C-terminus, approximately 220 amino acid residues. The degree of similarity of this domain between RAR subtypes is approximately 85%, which indicates that RAR subtypes might have different affinity for RA. Preliminary reports show a similar affinity of RA towards RARα and RARβ when expressed in HeLa cells or produced in *Escherichia coli* (Crettaz *et al.*, 1990; Hashimoto *et al.*, 1990), while certain synthetic retinoids displayed specific binding properties towards each isoform. The characterization of a series of synthetic retinoids that selectively activates RXR homodimers and not RAR–RXR heterodimers further demonstrates that both retinoid pathways can be independently activated by pharmacological agents (Lehmann *et al.*, 1992). Synthetic ligands that display an ability to selectively activate one retinoid signaling pathway are thus becoming invaluable tools for probing the physiological role of various RAR and RXR isoforms.

A ligand-dependent nuclear translocation signal that directs the ligand–receptor complex to the nucleus has been mapped within the hormone-binding domain of steroid hormone receptors (Picard & Yamamoto, 1987; Guiochon-Mantel *et al.*, 1989). This finding may not be relevant to the RARs since they have been shown to be localized in the nucleus in the absence of ligand (Gaub *et al.*, 1989) and able to bind DNA in a ligand-independent manner (Yang *et al.*, 1991). The hormone-binding domains of some steroid hormone receptors also contains *trans*-activation functions (Webster *et al.*, 1988; Tasset *et al.*, 1990), although this has not been tested with the RARs. Finally, a region within the hormone binding domain is required for both receptor dimerization and high affinity DNA-binding (Forman *et al.*, 1989; Fawell *et al.*, 1990). This dimerization domain has features in common with both the leucine zipper and helix–loop–helix motif which have been proposed as dimerization structures in other DNA binding proteins. This region of the RARα appears to be essential for the observed heterodimerization with T_3R (see below) (Forman *et al.*, 1989; Glass *et al.*, 1989). Such RAR–T_3R heterodimers have been shown to possess increased affinity for a subset of

TREs. This subdomain may also be responsible for mediating protein–protein interactions between RARα and cell-type-specific proteins shown to have the same properties as the T_3R, e.g. to increase the DNA-binding affinity of RARα for a variety of response elements (Glass *et al.*, 1990).

RAR and RXR: homodimers or heterodimers

Characterization of the mechanism of action of steroid hormone receptors had shown that these transcription factors bind DNA as homodimers (Kumar & Chambon, 1988; Tsai *et al.*, 1988). However, studies of the DNA binding properties of RAR and T_3R revealed that accessory factors were required for high-affinity binding of these receptors to their DNA recognition sites (Murray & Towle, 1989; Glass *et al.*, 1990). As mentioned above, it had been previously shown that RAR and T_3R could form heterodimers in solution when synthesized *in vitro* (Glass *et al.*, 1989), and together with data demonstrating that the accessory factors could recognize receptor binding sites, it was postulated that the accessory factors might be members of the superfamily of nuclear receptors (Glass *et al.*, 1990). This hypothesis was proven to be correct when a number of laboratories demonstrated that RXRs function as an accessory factor not only for RARs, but also for other receptors that include T_3R, VD_3R, the orphan nuclear receptor COUP and the peroxisome proliferator activated receptor (PPAR) (Yu *et al.*, 1991; Bugge *et al.*, 1992; Kliewer *et al.*, 1992*a,b,c*; Leid *et al.*, 1992; Marks *et al.*, 1992; Zhang *et al.*, 1992*a*). Thus, RXR appears to be used as a co-regulator of nuclear receptor function in a large number of hormonal signaling systems. Functionally, RXR is required for high-affinity binding of RAR to all types of RAREs and RXRE (DR-1, DR-2, DR-5 and inverted repeats) and enhances the activity of the RAR in transcriptional activation assays using reporter genes driven by promoter-containing RAREs. However, when similar transcriptional assays were performed using the RXRE derived from the promoter of the CRBP-II gene, co-transfection of RAR or COUP led to the repression of RXR function (Mangelsdorf *et al.*, 1991; Kliewer *et al.*, 1992*a*). The apparent paradox between the need for heterodimer formation between RAR and RXR for high-affinity binding to the CRBP-II RXRE and the repression of RXR activity by RAR on this response element was solved by Zhang *et al.* (1992*b*), who demonstrated that, in the presence of its specific ligand 9-*cis* RA, RXRs form and bind DNA with high affinity as homodimers. These observations thus defined two independent retinoid signaling pathways in which the ligands dictate whether RXR functions as a co-regulator of

RAR action or as an independent regulator of gene expression. Although it is relatively easy to envision a biological significance for a direct interaction between two retinoid receptor systems, the physiological significance of the convergence of the retinoid with other hormonal signaling systems is less apparent since the action of thyroid hormone and vitamin D, for example, does not appear to be retinoid-dependent.

Distribution and roles of the retinoic acid receptors

General distribution

The wide spectrum of biological activities affected by retinoids suggests that each RA receptor isoform may play a unique or combinatorial role as a regulator of mammalian development and homeostasis. It thus becomes crucial to identify the pattern of expression of each receptor isoform in biological systems regulated by the action of retinoids and to prove that a particular nuclear receptor is responsible for the transduction of the RA signal in a given system. Initial Northern blot analysis of RARα distribution using total RNA isolated from a variety of rat and human tissues revealed two RNA species of 3.2 and 2.3 kilobases (kb) expressed in an ubiquitous manner, the 3.2 kb transcript being the most abundant (Giguère *et al.*, 1987; de Thé *et al.*, 1989). Interestingly, the highest levels of expression were observed in specific areas of the brain such as the hippocampus and cerebellum, suggesting an unexpected role for RA in brain function (Giguère *et al.*, 1987). The two RARα transcripts are generated by alternative usage of polyadenylation sites; both encode a full-length protein. Analysis of the tissue distribution of the RARβ mRNAs showed two mRNA species of 3.0 and 2.5 kb in most tissues tested (de Thé *et al.*, 1989) as well as some other transcripts of various sizes (Benbrook *et al.*, 1988). Great variability in the transcription pattern of the RARβ gene expression was observed. High levels of expression were found in kidney, prostate, spinal cord, cerebral cortex, pituitary gland and adrenal gland; average levels were detected in liver, spleen, uterus, ovary, brain and testis while low levels were recorded in breast and eye (Benbrook *et al.*, 1988; de Thé *et al.*, 1989). The tissue distribution of RARγ transcripts was not as extensively studied. The RARγ mRNA can be found at high levels in the skin and lung but its general distribution is relatively unknown (Noji *et al.*, 1989*a*; Zelent *et al.*, 1989; Kastner *et al.*, 1990). RXR isoforms are widely distributed in adult tissues and display both unique and overlapping patterns of expression (Mangelsdorf *et al.*, 1992). In the mouse, RXRα mRNA consists of a single species of 5.6 kb and is abundant in visceral tissues such as the liver, spleen, kidney, lung and muscle. Two species of RXRβ mRNA (2.7 and 3.0 kb) are ubiquitously expressed at various levels in all tissues tested. The expression of RXRγ is more restricted; specific mRNA species are observed in distinct tissues, suggesting the existence of multiple RXRγ isoforms. For example, a 2.0 kb mRNA is present in adrenal, kidney and liver, whereas a 2.5 kb mRNA is observed in brain and lung. Overall, at least one RXR isoform is present in every tissue, an observation that is consistent with the role of RXR as a co-regulator essential for the action of a large number of nuclear receptors.

Although useful, these Northern blot analyses do not address the issue of the specific expression of each RAR isoform, and do not provide information on the cell type expressing the receptors in each tissue and the actual level of receptor protein present in these tissues. Clearly, *in situ* hybridization and immunohistochemistry studies using specific probes and antibodies directed at each receptor isoform will be necessary to obtain a complete picture of the distribution of these proteins.

RARs and RXRs in development

As discussed above, RA may act as a morphogenetic agent at various stages of vertebrate embryogenesis. In particular, the effects of exogenous RA treatment have implicated RA in anterior–posterior patterning both along the body axis and in the developing limb bud. Detailed analyses of the pattern of expression of the RARs by *in situ* hybridization in mouse embryo revealed that all RAR subtypes were expressed during development. While RARα transcripts are expressed in a ubiquitous manner, RARβ and RARγ show distinct spatial and temporal patterns of expression during morphogenesis and organogenesis (Dollé *et al.*, 1989, 1990; Noji *et al.*, 1989*b*; Osumi-Yamashita *et al.*, 1990; Ruberte *et al.*, 1990, 1991). For example, RARβ is the only receptor to show spatially restricted expression at presomatic stages, and its expression at later stages in tracheal, intestinal and genital tract epithelia suggests that the RARβ may play a major role in the differentiation of these epithelia during development. On the other hand, the distribution of RARγ in the late embryo suggests that this receptor could mediate the effects of retinoids on chondrogenesis and differentiation of squamous and mucous epithelia. Similar studies in the chick embryo showed that RARβ is abundantly expressed in the central nervous system (CNS) which indicates a potential role for retinoids in CNS development (Smith & Eichele, 1991).

The potential role of members of the RXR family in development has recently been addressed via the cloning of the avian RXRα (Rowe *et al.*,

1991) and murine RXRα, β and γ (Mangelsdorf *et al.*, 1992). Chicken RXRα is expressed in the chick embryo, predominantly in elements of the developing peripheral nervous system derived from the neural crest cells, and its general distribution in those cells provides a molecular marker for this specific subgroup of neural-crest-derived cells. In the mouse embryo, RXRα and RXRβ are abundant while levels of RXRγ are much lower. RXRα transcripts are found preferentially in the epithelia of the digestive tract, the skin and the liver. RXRβ expression during embryogenesis is less specific but the highest level of RXRβ transcript can be found in the CNS. In contrast, the pattern of expression of RXRγ is very specific, with the highest levels of transcripts found in the corpus striatum and the pituitary. These observations indicate that all three RXRs may play important and specific roles in development and that certain teratogenic effects of retinoids may result from the improper activation of gene network under the control of members of the RXR family.

Although the spatially restricted distribution patterns of RARs during embryogenesis can lead to speculation as to their potential roles in developmental processes, a more direct approach is needed to demonstrate their activity during morphogenesis. The level of RAR activity will be dictated by the presence of free endogenous RA in sufficient concentration and other auxiliary proteins involved in the transduction of the RA signal. By introducing a vector consisting of the βRARE linked to a minimal promoter-*lacZ* gene into transgenic mice, two independent groups (Rossant *et al.*, 1991; Balkan *et al.*, 1992) showed that the βRARE can direct specific spatial and temporal expression of the *lacZ* reporter gene during embryogenesis, thus providing direct evidence that some of the morphogenetic effects of RA could be mediated via localized transcriptional activity controlled by the various RAR subtypes. Definitive proof of an essential requirement for some or all RAR subtypes for normal embryonic development will require the generation of null mutations in each receptor gene by targeted mutagenesis via homologous recombination in transfected embryo-derived stem cells (Capecchi, 1989; Rossant & Joyner, 1989). The rapid pace of the development and improvement of the gene targeting technology and the great interest that large and well-funded laboratories show towards this issue should ensure that it is resolved in the very near future.

Another molecular approach to study receptor function is to create synthetic dominant-negative RAR mutants and investigate the effects of their expression in RA-responsive cell lines such as EC cells. The idea is to engineer a RAR mutant that will be not only non-functional but also able to interfere with the normal function of wild-type RAR present endogenously in EC cells. Such a mutant was made by fusing part of the RARα that included the N-terminal region and the DNA-binding domain to a bacterial protein, β-galactosidase (Espeseth *et al.*, 1989); β-galactosidase activity was used to monitor the expression of the chimeric protein in transfected F9 EC cells and its expression was found to correlate well with a RA-resistance phenotype. A dominant-negative RARα mutant was also characterized from a P19 EC cell line selected for its inability to differentiate in response to RA (Pratt *et al.*, 1990). This cell line was found to carry a rearrangement affecting one RARα allele, resulting in the production of a truncated RARα protein at its C-terminal end. This RARα mutant was shown to act as a dominant repressor of the endogenous RARs when expressed in normal P19 EC cells but its expression was insufficient to confer RA-resistance on F9 and P19 EC cells, suggesting that additional mutations might be needed to achieve complete RA non-responsiveness in P19 EC cells. Although the exact mechanism by which the dominant-negative effect is achieved remains to be demonstrated in both cases, this type of experiment certainly points to an active role for endogenous RARs in the RA-induced differentiation of EC cells.

Hematopoiesis and cancer

The ability of RA to induce terminal granulocytic differentiation of certain human myeloid leukemias (Breitman *et al.*, 1980, 1981) and to stimulate growth of normal hematopoietic progenitor cells in culture (Douer & Koeffler, 1982*a*,*b*) suggests that retinoids might regulate both normal and abnormal hematopoiesis. Northern blot analysis of the expression of RA receptors revealed that RARα transcripts were present in a wide variety of hematopoietic cells (Gallagher *et al.*, 1989; Largman *et al.*, 1989; Wang *et al.*, 1989) while expression of the RARβ appears to be repressed (de Thé *et al.*, 1989). No information is yet available on the expression of RARγ and RXRα in these cells. The level of expression of RARα in cultured myeloid leukocyte cell lines seems to correlate well with their level of RA-responsiveness (i.e. HL-60 > KG-1 > K562) (Gallagher *et al.*, 1989), while only a weak (Gallagher *et al.*, 1989) or no (Largman *et al.*, 1989; Wang *et al.*, 1989) correlation was found between RA-responsiveness and the level of RARα expression in freshly obtained myeloid leukemic cells. More direct evidence that RARα is the mediator of RA-induced granulocytic differentiation, at least in HL-60 cells, was the recent demonstration that retroviral vector-mediated transduction of a single copy of the RARα into an HL-60 RA-resistant subclone restored the sensitivity of these cells to RA (Collins *et al.*, 1990). This observation established a crucial role for RARα in myeloid cell differentiation. A second, more dramatic observation was the recent discovery that the t(15:17) translocation breakpoint of acute promyelocytic leukemia (APL) lies within the RARα locus (Borrow *et al.*, 1990; de Thé *et al.*, 1990*a*; Alcalay *et al.*, 1991). The translocation results in the creation of a novel transcription unit composed at the 5′ end of a portion of the *pml* gene located on chromosome 15 and, at the 3′ end, of a large portion of the RARα gene located on chromosome 17q21. Because the translocation occurs within the intron preceding the exon encoding the DNA-binding of RARα, the chimeric *pml*–RARα transcript, which is expressed in promyelocytic leukemic cells, is believed to encode a hybrid protein that will retain the DNA- and ligand-binding capacity of the RARα but possess a novel N-terminal domain provided by the *pml* locus. Given the important role played by the N-terminal region of nuclear receptors in transcriptional activation and in determining target gene specificity, the *pml*–RARα hybrid gene product may either (i) have acquired novel transcriptional function that could result in the activation of gene networks normally silent in hematopoietic cells, or (ii) be transcriptionally inactive and act as dominant negative RARα and/or *pml* mutants. In both cases, however, the observation that complete remission in APL can be attained by treatment with RA (Meng-er *et al.*, 1988) offers an interesting paradox and points to the fact that the *pml* gene might also play a crucial function in leukemic cell differentiation. The availability of molecular tools to study RAR functions combined with the molecular cloning and expression of the *pml*–RARα cDNA should provide new and exciting information not only on the molecular mechanism of RAR action but also on the physiological role of retinoids in hematopoiesis.

Conclusion

As members of the superfamily of nuclear receptors, RARs and RXRs appear to exert crucial functions as regulators of gene networks that control specific developmental programs and homeostasis in vertebrates. The large number of RAR and RXR isoforms expressed in a near-universal but subtype-specific fashion during development and in differentiated tissues is consistent with the diversity of retinoid actions. Evidence is now mounting that most of the physiological and pharmacological effects of retinoids can be associated with the presence of RARs or RXRs; in some cases, RARs have been clearly shown to be the mediators of RA action. However, the elucidation of the precise function of each receptor isoform, the identification of their target genes and the resolution of the molecular mechanisms by which cell-specific retinoid action is achieved represent complex challenges that, we hope, can now be addressed using the powerful tools of molecular genetics.

References

Alcalay, M., Zangrilli, D., Pandolfi, P.P., Longo, L., Mencarelli, A., Giacomicci, A., Rocchi, M., Biondi, A., Rambaldi, A., Coco, F.L., Divero, D., Donti, E., Grignani, F. & Pelicci, P.G. (1991) Translocation breakpoint of acute promyelocytic leukemia lies within the retinoic acid receptor α locus. *Proc. Natl. Acad. Sci. USA* **88**, 1977—81.

Arriza, J.L., Weinberger, C., Cerelli, G., Glaser, T.M., Handelin, B.L., Housman, D.E. & Evans, R.M. (1987) Cloning of human mineraloccorticoid receptor complementary DNA: structural and functional kinship with the glucocorticoid receptor. *Science* **237**, 268–75.

Balkan, W., Colbert, M., Bock, C. & Linney, E. (1992) Transgenic indicator mice for studying activated retinoic acid receptors during development. *Proc. Natl. Acad. Sci. USA* **89**, 3347—51.

Benbrook, D., Lernhardt, E. & Pfahl, M. (1988) A new retinoic acid receptor identified from a hepatocellular carcinoma. *Nature* **333**, 669–72.

Borrow, J., Goddard, A.D., Sheer, D. & Solomon, E. (1990) Molecular analysis of acute promyelocytic leukemia breakpoint cluster region on chromosome 17. *Science* **249**, 1577–80.

Brand, N., Petkovich, A., Krust, A., Chambon, P., de Thé, H., Marchio, A., Tiollais, P. & Dejean, A. (1988) Identification of a second human retinoic acid receptor. *Nature* **332**, 850–3.

Breitman, T.R., Collins, S.J. & Keene, B.R. (1981) Terminal differentiation of human promyelocytic leukemia cells in primary culture in response to retinoic acid. *Blood* **57**, 1000–4.

Breitman, T.R., Selonick, S.E. & Collins, S.J. (1980) Induction of differentiation of the human promyelocytic leukemia cell line (HL-60) by retinoic acid. *Proc. Natl. Acad. Sci. USA* **77**, 2936–40.

Bugge, T.H., Pohl, J., Lonnoy, O. & Stunnenberg, H.G. (1992) RXRα, a promiscuous partner of retinoic acid and thyroid hormone receptors. *EMBO J.* **11**, 1409—18.

Capecchi, M.R. (1989) Altering the Genome by Homologous Recombination. *Science* **244**, 1288–92.

Collins, S.J., Robertson, K.A. & Mueller, L. (1990) Retinoic acid-induced granulocytic differentiation of HL-60 myeloid leukemia cells is mediated directly through the retinoic acid receptor (RAR-α). *Molec. Cell. Biol.* **10**, 2154–63.

Crettaz, M., Baron, A., Siegenthaler, G. & Hunziker, W. (1990) Ligand specificities of recombinant retinoic acid receptors RAR alpha and beta. *Biochem. J.* **272**, 391–7.

Danielsen, M., Hinck, L. & Ringold, G.M. (1989) Two amino acids within the knuckle of the first zinc finger specify DNA response element activation by the glucocorticoid receptor. *Cell* **57**, 1131–8.

de Thé, H., Chomienne, C., Lanotte, M., Degos, L. & Dejean, A. (1990*a*). The t(15:17) translocation of acute promyelocytic leukemia fuses the retinoic acid receptor α gene to a novel transcribed locus. *Nature* **347**, 558–61.

de Thé, H., del Mar Vivanco-Ruiz, M., Tiollais, P., Stunneberg, H. & Dejean, A. (1990*b*) Identification of a retinoic acid responsive element in the retinoic acid receptor β gene. *Nature* **343**, 177–80.

de Thé, H., Marchio, A., Tiollais, P. & Dejean, A. (1987) A novel steroid thyroid hormone receptor-related gene inappropriately expressed in human hepatocellular carcinoma. *Nature* **330**, 667–70.

de Thé, H., Marchio, A., Tiollais, P. & Dejean, A. (1989) Differential expression and ligand regulation of retinoic acid receptor α and β genes. *EMBO J.* **8**, 429–33.

Dejean, A., Bougueleret, L., Grzeschik, K.H. & Tiollais, P. (1986) Hepatitis B virus DNA integration in a sequence homologous to v-erb-A and steroid receptor genes in a hepatocellular carcinoma. *Nature* **322**, 70–2.

Dollé, P., Ruberte, E., Kastner, P., Petkovich, M., Stoner, C.M., Gudas, L.J. & Chambon, P. (1989). Differential expression of genes encoding α, β and γ retinoic acid receptors and CRABP in the developing limb of the mouse. *Nature* **342**, 702–5.

Dollé, P., Ruberte, E., Leroy, P., Morris-Kay, G. & Chambon, P. (1990) Retinoic acid receptors and cellular retinoid binding proteins I. A systematic study of their differential pattern of transcription during mouse organogenesis. *Development* **110**, 1133–51.

Douer, D. & Koeffler, H.P. (1982*a*) Retinoic acid enhances growth of human early erythroid progenitor cells in vitro. *J. Clin. Invest.* **69**, 1039–41.

Douer, D. & Koeffler, H. P. (1982*b*) Retinoid enhances colomy-stimulating factor-induced growth of normal human myeloid progenitor cells in vitro. *Exp. Cell. Res.* **138**, 193–8.

Duester, G., Shean, M.L., McBride, M.S. & Stewart, M.J. (1991) Retinoic acid response element in the human alcohol dehydrogenase gene ADH3: implications for regulation of retinoic acid synthesis. *Molec. Cell. Biol.* **11**, 1638–46.

Durand, B., Saunders, M., Leroy, P., Leid, M. & Chambon, P. (1992) All-trans and 9-cis retinoic acid induction of CRABPII transcription is mediated by RAR-RXR heterodimers bound to DR1 and DR2 repeated motifs. *Cell* **71**, 73–85.

Durston, A., Timmermans, P., Hage, W., Hendriks, H., de Vries, N., Heideveid, M. & Nieuwkoop, P. (1989) Retinoic acid causes an anteroposterior transformation in the developing central nervous system. *Nature* **340**, 140–4.

Edwards, M.K.S. & McBurney, M.W. (1983) The concentration of retinoic acid determines the differentiated cell types formed by a teratocarcinoma cell line. *Dev. Biol.* **98**, 187–91.

Ellinger-Ziegelbauer, H. & Dreyer, C. (1991) A retinoic acid receptor expressed in the early development of *Xenopus laevis*. *Genes Dev.* **5**, 94–104.

Espeseth, A.S., Murphy, S.P. & Linney, E. (1989) Retinoic acid receptor expression vector inhibits differentiation of F9 embryonal carcinoma cells. *Genes Dev.* **3**, 1647–56.

Evans, R.M. (1988) The steroid and thyroid hormone receptor superfamily. *Science* **240**, 889–95.

Evans, R.M. & Hollenberg, S.M. (1988) Zinc fingers: gilt by association. *Cell* **52**, 1–3.

Fawell, S.E., Lees, J.A., White, R. & Parker, M.G. (1990) Characterization and colocalization of steroid binding and dimerization activities in the mouse estrogen receptor. *Cell* **60**, 953–62.

Fell, H.B. & Mellanby, E. (1953) Metaplasia produced in cultures of chick ectoderm by high vitamin A. *J. Physiol.* **119**, 470–88.

Forman, B. M., Yange, C.-r., Au, M., Casanova, J., Ghysdael, J. & Samuels, H.H. (1989) A domain containing leucine-zipper-like motifs mediate novel in vivo interactions between the thyroid hormone and retinoic acid receptors. *Mol. Endocrinol.* **3**, 1610–26.

Freedman, L. P., Luisi, B.F., Korszun, R., Basavappa, R., Sigler, P.B. & Yamamoto, K.R. (1988) The function and structure of the metal coordination sites within the glucocorticoid receptor DNA binding domain. *Nature* **334**, 543–6.

Fuchs, E. & Green, H. (1981) Regulation of terminal differentiation of cultured human keratinocytes by vitamin A. *Cell* **25**, 617–25.

Gallagher, R.E., Said, F., Pua, I., Papenhausen, P.R., Paietta, E. & Wiernik, P. (1989) Expression of Retinoic Acid Receptor-α m RNA in Human Leukemia Cells with Variable Responsiveness to Retinoic Acid. *Leukemia* **3**, 789–95.

Gaub, M.P., Lutz, Y., Ruberte, E., Petkovich, M., Brand, N. & Chambon, P. (1989) Antibodies specific to the retinoic acid human nuclear receptors α and β. *Proc. Natl. Acad. Sci. USA* **86**, 3089–93.

Gendimenico, G.J., Capetola, R.J., Rosenthale, M.E., McGuire, J.L. & Mezick, J.A. (1990) Retinoid modulation of phorbol ester effects in skin. In *Retinoids*, Part B (*Cell differentiation and clinical applications*) (ed. L. Packer), pp. 346–52. San Diego, CA: Academic Press.

Guiguère, V., Hollenberg, S., Rosenfeld, M. & Evans, R. (1986) Functional domains of the human glucocorticoid receptor. *Cell* **46**, 645–52.

Giguère, V., Ong, E., Evans, R. & Tabin, C. (1989) Spatial and temporal expression of the retinoic acid receptor in the regenerating amphibian limb. *Nature* **337**, 566–9.

Giguère, V., Ong, S., Segui, P. & Evans, R. (1987) Identification of a receptor for the morphogen retinoic acid. *Nature* **330**, 624–9.

Giguère, V., Shago, M., Zirngibl, R., Tate, P., Rossant, J. & Varmuza, S. (1990) Identification of a novel isoform of the retinoic acid receptor γ expressed in the mouse embryo. *Molec. Cell. Biol.* **10**, 2335–40.

Giguère, V., Yang, N., Segui, P. & Evans, R.M. (1988) Identification of a new class of steroid hormone receptors. *Nature* **331**, 91–4.

Glass, C.K., Devary, O.V. & Rosenfeld, M.G. (1990) Multiple cell type-specific proteins differentially regulate target sequence recognition by the α retinoic acid receptor. *Cell* **63**, 729–38.

Glass, C.K., Holloway, J.M., Devary, O.V. & Rosenfeld, M.G. (1988) The thyroid hormone receptor binds with opposite transcriptional effects to a common sequence motif in thyroid hormone and estrogen response elements. *Cell* **54**, 313–23.

Glass, C.K., Lipkin, S.M., Devary, O.V. & Rosenfeld, M.G. (1989) Positive and negative regulation of gene transcription by a retinoic acid-thyroid hormone receptor heterodimer. *Cell* **59**, 697–708.

Graupner, G., Wills, K.N., Tzukerman, M., Zhang, X.-k. & Pfahl, M. (1989) Dual regulatory role for thyroid-hormone receptors allows control of retinoic-acid receptor activity. *Nature* **340**, 653–6.

Green, S. & Chambon, P. (1987) Oestradiol induction of a glucocorticoid-responsive gene by a chimaeric receptor. *Nature* **325**, 75–8.

Green, S., Kumar, V., Theulaz, I., Wahli, W. & Chambon, P. (1988) The N-terminal DNA-binding 'zinc finger' of the oestrogen and glucocorticoid receptors determines target gene specificity. *EMBO J.* **7**, 3037–44.

Guiochon-Mantel, A., Loosfelt, H., Lescop, P., Sar, S., Atger, M., Perrot-Applanat, M. & Milgrom, E. (1989) Mechanisms of nuclear localization of the progesterone receptor: evidence for interactions between monomers. *Cell* **57**, 1147–54.

Hamada, K., Gleason, S.L., Levi, B.-Z, Hirschfeld, S., Appella, E. & Ozato, K. (1989) H-2RIIBP, a member of the nuclear hormone receptor superfamily that binds to both regulatory element of major histocompatibility class I genes and the estrogen response element. *Proc. Natl. Acad. Sci. USA* **86**, 8289—93.

Hard, T., Kellenbach, E., Boelens, R., Maler, B.A., Dahlman, K., Freedman, L.P., Carlstedt-Duke, J., Yamamoto, K.R., Gustafsonn, J.-A, & Kaptein, R. (1990) Solution structure of the glucocorticoid receptor DNA-binding domain. *Science* **249**, 157–60.

Hashimoto, Y., Kagechika, H. & Shudo, K. (1990) Expression of retinoic acid receptor genes and the ligand-binding selectivity of retinoic acid receptors (RAR's). *Biochem. Biophys. Res. Commun.* **166**, 1300–7.

Heyman, R.A., Mangelsdorf, D.J., Dyck, J.A., Stein, R.B., Eichele, G., Evans, R.M. & Thaller, C. (1992) 9-cis retinoic acid is a high affinity ligand for the retinoid X receptor. *Cell* **68**, 397–406.

Hoffman, B., Lehmann, J.M., Zhang, X.-k, Hermann, T., Husmann, M., Graupner, G. & Pfahl, M. (1990) A retinoic acid receptor-specific element controls the retinoic acid receptor-β promoter. *Molec. Endocrinol.* **4**, 1727–36.

Hollenberg, S.M. & Evans, R.M. (1988) Multiple and cooperative trans-activation domains of the human glucocorticoid receptor. *Cell* **55**, 899–906.

Hollenberg, S.M., Giguère V., Segui, P. & Evans, R.M. (1987) Colocalization of DNA-binding and transcriptional activation functions in the human glucocorticoid receptor. *Cell* **49**, 39–46.

Hong, W.K., Lippman, S.M., Itri, L.M., Karp, D.D., Lee, J.S., Byers, R.M., Schantz, S.P., Kramer, A.M., Lotan, R. & Peters, L.J. (1990) Prevention of second primary tumors with isotretinoin in squamous-cell carcinoma of the head and neck. *N. Engl. J. Med.* **323**, 825–7.

Huang, M.-H., Ye, Y.-C., Chen, S.R., Chai, J.-R., Lu, J.-X., Zhoa, L., Gu, L.-J. & Wang, Y. (1988) Use of all-trans retinoic acid in the treatment of acute promyelocytic leukemia. *Blood* **72**, 567–72.

Ide, H. & Aono, H. (1988) Retinoic acid promotes proliferation and chondrogenesis in the distal mesodermal cells of chick limb bud. *Developmental Biology* **130**, 767–73.

Ishikawa, T., Umesono, K., Mangelsdorf, D.J., Aburatani, H., Stanger, B.Z., Shibasaki, Y., Imawari, M., Evans, R.M. & Takaku, F. (1990). A functional retinoic acid receptor encoded by the gene on human chromosome 12. *Molec. Endocrinol.* **4**, 837–44.

Jones-Villeneuve, E.M.V., McBurney, M., Rogers, K.A. & Kalnins, V.I.J. (1982) Retinoic acid induces embryonal carcinoma cells to differentiate into neurons and glia cells. *J. Cell Biol.* **94**, 253–62.

Kastner, P., Krust, A., Mendelsohn, C., Garnier, J.M., Zelent, A., Leroy, P., Staub, A. & Chambon, P. (1990) Murine isoforms of retinoic acid receptor γ with specific patterns of expression. *Proc. Natl. Acad. Sci. USA* **87**, 2700—4.

Kisler, A. (1987) Limb bud cell cultures for estimating the teratogenic potential of compounds. Validation of the test system with retinoids. *Arch. Toxicol.* **60**, 403–14.

Kliewer, S.A., Umesono, K., Heyman, R.H., Mangelsdorf, D.J., Dyck, J.A. & Evans, R.M. (1992*a*) Retinoid X receptor-COUP-TF interactions modulate retinoic acid signaling. *Proc. Natl. Acad. Sci. USA* **89**, 1448–52.

Kliewer, S.A., Umesono, K., Mangelsdorf, D.J. & Evans, R.M. (1992*b*) Retinoid X receptor interacts with nuclear receptors in retinoic acid, thyroid hormone and vitamin D3 signaling. *Nature* **335**, 446–9.

Kliewer, S.A., Umesono, K., Noonan, D.J., Heyman, R.A. & Evans, R.M. (1992*c*) Convergence of 9-cis retinoic acid and peroxisome proliferator signalling pathways through heterodimer formation of their receptors. *Nature* **358**, 771–4.

Kopan, R., Traska, G. & Fuchs, E. (1987) Retinoids as important regulators of terminal differentiation: examining keratin expression in individual epidermal cells at various stages of keratinization. *J. Cell Biol.* **105**, 427–40.

Kraemer, K.H., DiGiovanna, J.J., Moshell, A.N., Tarone, R.E. & Peck, G.L. (1988) Prevention of skin cancer in xeroderma pigmentosum with the use of oral isotretinoin. *N. Engl. J. Med.* **318**, 633–7.

Krust, A., Green, S., Argos, P., Kumar, V., Walter, P., Bornert, J.-M. & Chambon, P. (1986) The chicken oestrogen receptor sequence: homology with v-erbA and the human oestrogen and glucocorticoid receptors. *EMBO J.* **5**, 891–7.

Krust, A., Kastner, P., Petkovich, M., Zelent, A. & Chambon, P. (1989) A third human retinoic acid receptor, hRAR-γ. *Proc. Natl. Acad. Sci. USA* **86**, 5310–14.

Kumar, V. & Chambon, P. (1988) The estrogen receptor binds tightly to its responsive element as a ligand-induced homodimer. *Cell* **55**, 145–56.

Kumar, V., Green, S., Staub, A. & Chambon, P. (1987) Functional domains of the human oestrogen receptor. *Cell* **51**, 941–51.

Lammer, E., Chen, D., Hoar, R., Agnish, N., Benke, P., Braun, J., Curry, C., Fernhoff, P., Grix, A., Lott, I., Richard, J. & Shyan, C. (1985) Retinoic acid embryopathy. *N. Engl. J. Med.* **313**, 837–41.

Largman, C., Detmer, K., Corral, J.C., Hack, F.M. & Lawrence, H.J. (1989) Expression of retinoic acid receptor alpha mRNA in human leukemia cells. *Blood* **74**, 99–102.

Lasnitzki, I. (1955) The influence of A hypervitaminosis on the effect of 20-methyl-cholanthrene on mouse prostate glands grown in vitro. *Br. J. Cancer* **9**, 434–41.

Leder, A., Kuo, A., Cardiff, R.D., Sinn, E. & Leder, P. (1990) v-Ha-ras transgene abrogates the initiation step in mouse skin tumorigenesis: effects of phorbol esters and retinoic acid. *Proc. Natl. Acad. Sci. USA* **87**, 9178–82.

Lehmann, J.M., Hoffmann, B. & Pfahl, M. (1991) Genomic organization of the retinoic acid receptor gamma gene. *Nucl. Acids. Res.* **19**, 573–8.

Lehmann, J.M., Jong, L., Fanjul, A., Cameron, J.F., Lu, X.P., Haefner, P., Dawson, M.I. & Pfahl, M. (1992) Retinoids selective for retinoid X receptor response pathways. *Science* **258**, 1944–6.

Leid, M., Kastner, P., Lyons, R., Nakshari, H., Saunders, M., Zacharewski, T., Chen, J.-Y., Staub, A., Garnier, J.-M., Mader, S. & Chambon, P. (1992) Purification, cloning, and RXR identity of the HeLa cell factor with which RAR or TR heterodimerizes to bind target sequences efficiently. *Cell* **68**, 377–95.

Leroy, P., Krust, A., Zelent, A., Mendelsohn, C., Garnier, J.M., Kastner, P., Dierich, A. & Chambon, P. (1991) Multiple isoforms of the mouse retinoic acid receptor α are generated by alternative splicing and differential induction by retinoic acid. *EMBO J.* **10**, 59–69.

Levin, A.A., Sturzenbecker, L.J., Kazmer, S., Bosakowski, T., Huselton, C., Allenby, G., Speck, J., Kratzeisen, C., Rosenberger, M., Lovey, A. & Grippo, J.F. (1992) 9-*Cis* retinoic acid stereoisomer binds and activates the nuclear receptor RXRα. *Nature* **355**, 359–61.

Lucas, P.C., O'Brien, R.M., Mitchell, J.A., Davis, C.M., Imai, E., Forman, B.C., Samuels, H.H. & Granner, D.K. (1991) A retinoic acid response element is part of a pleiotropic domain in the phosphoenolpyruvate carboxykinase gene. *Proc. Natl. Acad. Sci. USA* **88**, 2184–8.

Mader, S., Kumar, V., de Verneuil, H. & Chambon, P. (1989) Three amino acids of the oestrogen receptor are essential to its ability to distinguish an oestrogen from a glucocorticoid-responsive element. *Nature* **338**, 271–4.

Mangelsdorf, D.J., Borgmeyer, U., Heyman, R.A., Zhou, J.Y., Ong, E.S., Oro, A.E., Kakizuka, A. & Evans, R. M. (1992) Characterization of three RXR genes that mediate the action of 9-cis retinoic acid. *Genes Dev.* **6**, 329–44.

Mangelsdorf, D.J., Ong, E.S., Dyck, J.A. & Evans, R.M. (1990) Nuclear receptor that identifies a novel retinoic acid response pathway. *Nature* **345**, 224–9.

Mangelsdorf, D.J., Umesono, K., Kliewer, S.A., Borgmeyer, U., Ong, E.S. & Evans, R.M. (1991). A direct repeat in the cellular retinol binding protein type II gene confers differential regulation by RXR and RAR. *Cell,* **66**, 555–61.

Marks, M.S., Hallenbeck, P.L., Nagata, T., Segars, J.H., Appella, E., Nikodem, V.M. & Ozato, K. (1992) H-2 RIIBP (RXRβ) heterodimerization provides a mechanism for combinatorial diversity in the regulation of retinoic acid and thyroid hormone responsive genes. *EMBO J.* **11**, 1419—35.

Meng-er, H., Yu-chen, Y., Shu-rong, C., Jen-ren, C., Jia-Xiang, L., Long-jun, Z. & Zhen-yi, W. (1988) Use of all-trans retinoic acid in the treatment of acute promyelocytic leukemia. *Blood* **72**, 567–72.

Meyer, M.-E., Gronemeyer, H., Turcotte, B., Bocquel, M.-T, Tasset, D. & Chambon, P. (1989) Steroid hormone receptors compete for factors that mediate their enhancer function. *Cell* **57**, 433–42.

Mitchell, P.J. & Tjian, R. (1989) Transcriptional regulation in mammalian cells by sequence-specific DNA binding proteins. *Science* **245**, 371–8.

Moon, R.C. & Mehta, R.G. (1990) Cancer chemoprevention by retinoids: animal models. In *Retinoids,* part B (*Cell differentiation and clinical applications*) (ed. L. Packer), pp. 395–406. San Diego, CA: Academic Press.

Munoz-Cánoves, P., Vik, D.P. & Tack, B.F. (1990) Mapping of a retinoic acid-responsive element in the promoter region of the complement factor H gene. *J. Biol. Chem.* **265**, 20065–8.

Murray, M.B. & Towle, H.C. (1989) Identification of nuclear factors that enhance binding of the thyroid hormone receptor to a thyroid hormone response element. *Molec. Endocrinol.* **3**, 1434–42.

Noji, S., Nohno, T., Koyama, E., Muto, K., Ohyama, K., Aoki, Y., Tamura, K., Ohsugi, K., Ide, H., Taniguchi, S. & Saito, T. (1991) Retinoic acid induces polarizing activity but is unlikely to be a morphogen in the chick limb bud. *Nature* **350**, 83–6.

Noji, S., Yamaai, T., Koyama, E., Nohno, T., Fujimoto, W., Arata, J. & Taniguchi, S. (1989*a*) Expression of retinoic acid receptor genes in keratinizing front skin. *FEBS Lett.* **259**, 86–90.

Noji, S., Yamaai, T., Koyama, E., Nohno, T. & Taniguichi, S. (1989*b*) Spatial and temporal expression pattern of retinoic acid receptor genes during mouse bone development. *FEBS Lett.* **257**, 93–6.

Ong, D.E. (1987) Cellular retinoid-binding proteins. *Arch. Dermatol.* **123**, 1693–5.

Oro, A.E., McKeown, M. & Evans, R.M. (1990) Relationship between the product of the Drosophila ultraspiracle locus and the vertebrate retinoid X receptor. *Nature* **347**, 298–301.

Osumi-Yamashita, N., Noji, S., Nohno, T., Koyama, E., Doi, H., Eto, K. & Taniguchi, S. (1990) Expression of retinoic acid receptor genes in neural crest-derived cells during mouse facial development. *FEBS Lett.* **264**, 71–4.

Padanilam, B.J., Mcleod, L.B., Suzuki, H. & Solursh, M. (1991) Nucleotide sequence of an isoform of chicken retinoic acid binding protein-β varying in its A domain. *Nucleic Acids Res.* **19**, 395.

Paulsen, D.F., Langille, R.M., Dress, V. & Solursh, M. (1988) Selective stimulation of in vitro limb-bud chondrogenesis by retinoic acid. *Differentiation* **39**, 123–30.

Petkovich, M., Brand, N.J., Krust, A. & Chambon, P. (1987) A human retinoic acid receptor which belongs to the family of nuclear receptors. *Nature* **330**, 444–50.

Picard, D. & Yamamoto, K.R. (1987) Two signals mediate hormone-dependent nuclear localization of the glucocorticoid receptor. *EMBO J.* **6**, 3333–40.

Pratt, M.A.C., Kralova, J. & McBurney, M.W. (1990) A dominant negative mutation of the alpha retinoic acid receptor gene in a retinoic acid-nonresponsive embryonal carcinoma cell. *Molec. Cell. Biol.* **10**, 6445–53.

Ragsdale, C.W., Petkovich, M., Gates, P.B., Chambon, P. & Brockes, J.P. (1989) Identification of a novel retinoic acid receptor in regenerative tissues of the newt. *Nature* **341**, 654–7.

Rosa, F.W., Wilk, A.L., Kelsey, F.O. (1986) Teratogen update: vitamin A and congeners. *Teratology* **33**, 355–64.

Rossant, J. & Joyner, A. (1989) Towards a molecular genetic analysis of mammalian development. *Trends Genet.* **5**, 277–83.

Rossant, J., Zirngibl, R., Cado, D., Shago, M. & Giguère, V. (1991) Expression of a retinoic acid response element-*hsplacZ* transgene defines specific domains of transcriptional activity during mouse embryogenesis. *Genes Dev.* **5**, 1333–44.

Rottman, J.N., Widom, R.L., Nadal-Guinard, B., Mahdavi, V. & Karathanasis, S.K. (1991) A retinoic acid-responsive element in the apolipoprotein AI gene distinguishes between two different retinoic acid response pathways. *Molec. Cell. Biol.* **11**, 3814–20.

Rowe, A., Eager, N.S.C. & Brickell, P.M. (1991) A member of the RXR nuclear receptor family is expressed in neural-crest-derived cells of the developing chick peripheral nervous system. *Development* **111**, 771–8.

Ruberte, E., Dollé, P., Chambon, P. & Morris-Kay, G. (1991) Retinoic acid receptors and retinoid binding proteins II. Their differential pattern of transcription during early morphogenesis in mouse embryo. *Development* **111**, 45–60.

Ruberte, E., Dolle, P., Krust, A., Zelent, A., Morriss-Kay, G. & Chambon, P. (1990) Specific spatial and temporal distribution of retinoic acid receptor gamma transcripts during mouse embryogenesis. *Development* **108**, 213–22.

Saunders, J.W. & M.T. Gasseling, 1968. Ectodermal-mesenchymal interactions in the origins of limb symmetry. In *Epithelial-Mesenchymal Interactions* (ed. R. Fleischmajer & R.E. Billingham), pp. 78–97. Baltimore: Williams & Wilkins.

Schüle, R., Rangarajan, P., Kilewer, S., Ransone, L.J., Bolado, J., Yang, N., Verma, I. & Evans, R.M. (1990) Functional antagonism between oncoprotein c-Jun and the glucocorticoid receptor. *Cell* **62**, 1217–26.

Severne, Y., Wieland, S., Schaffner, W., & Rusconi, S. (1988) Metal binding 'finger' structures in the glucocorticoid receptor defined by site-directed mutagenesis. *EMBO J.* **7**, 2503–8.

Sigler, P.B. (1988) Acid blobs and negative noodles. *Nature* **333**, 210–12.

Smith, S.M. & Eichele, G. (1991) Temporal and regional differences in the expression pattern of distinct retinoic acid receptor-β transcripts in the chick embryo. *Development* **111**, 245–52.

Smith, W.C., Nakshatri, H., Leroy, P., Rees, J. & Chambon, P. (1991) A retinoic acid response element is present in the mouse cellular retinol binding protein I (mCRBPI) promoter. *EMBO J.* **10**, 2223–30.

Sporn, M.B. & Roberts, A.B. (1983) Roles of retinoids in differentiation and carcinogenesis. *Cancer Res.* **43**, 3034–40.

Sporn, M.B., Clamon, G.H., Dunlop, N.M., Newton, D.L., Smith, J.M. & Saffioti, U. (1975) Activity of vitamin A analogs in cell cultures of mouse epidermis and organ cultures of hamster trachea. *Nature* **253**, 47–50.

Strickland, S. & Mahdavi, V. (1978) The induction of differentiation in teratocarcinoma stem cells by retinoic acid. *Cell* **15**, 393–403.

Sucov, H.M., Murakami, K.K. & Evans, R.M. (1990) Characterization of an autoregulated response element in the mouse retinoic acid receptor type β gene. *Proc. Natl. Acad. Sci. USA* **87**, 5392–6.

Tabin, C.J. (1991) Retinoids, homeoboxes, and growth factors: towards molecular models for limb development. *Cell* **66**, 199–217.

Tasset, D., Tora, L., Fromental, C., Scheer, E. & Chambon, P. (1990) Distinct classes of transcriptional activating domains function by different mechanisms. *Cell* **62**, 1177–87.

Thaller, C. & Eichele, G. (1987) Identification and spatial distribution of retinoids in the developing chick limb bud. *Nature* **327**, 625–8.

Tickle, C., Alberts, B.M., Wolpert, L. & Lee, J. (1982) Local application of retinoic acid to the limb bud mimics the action of the polarizing region. *Nature* **296**, 564–5.

Tickle, C., Summerbell, D. & Wolpert, L. (1975) Positional signaling and specification of digits in chick limb morphogenesis. *Nature* **254**, 199–202.

Tickle, C., Lee, J. & Eichele, G. (1985) A quantitative analysis of the effect of all trans-retinoic acid on the pattern of chick wing development. *Dev. Biol.* **109**, 82–95.

Tora, L., Gronemeyer, H., Turcotte, B., Gaub, M.P. & Chambon, P. (1988) The N-terminal region of the chicken progesterone receptor specifies target gene activation. *Nature* **333**, 185–8.

Tora, L., White, J., Brou, C., Tasset, D., Webster, N., Scheer, E. & Chambon, P. (1989) The human estrogen receptor has two independent nonacidic transcriptional activation functions. *Cell* **59**, 477–87.

Tsai, S.Y., Carlstedt-Duke, J., Weigel, N.L., Dahlman, K., Gustafsson, J.-A, Tsai, M.-J & O'Malley, B.W. (1988) Molecular interactions of steroid receptor with its enhancer element: evidence for receptor dimer formation. *Cell* **55**, 361–9.

Tsai, S.Y., Tsai, M.J. & O'Malley, B.W. (1989) Cooperative binding of steroid hormone receptors contributes to transcriptional synergism at target enhancer elements. *Cell* **57**, 443–8.

Umesono, K. & Evans, R.M. (1989) Determinants of target gene specificity for steroid/thyroid hormone receptors. *Cell* **57**, 1139–46.

Umesono, K., Giguère, V., Glass, C., Rosenfeld, M. & Evans, R. (1988) Retinoic acid and thyroid hormone induce gene expression through a common responsive element. *Nature* **336**, 262–5.

Umesono, K., Murakami, K.K., Thompson, C.C. & Evans, R.M. (1991) Direct repeats as selective response elements for the thyroid hormone, retinoic acid, and vitamin D3 receptors. *Cell* **65**, 1255–66.

Vallee, B.L., Coleman, J.E. & Auld, D.S. (1991) Zinc fingers, zinc clusters, and zinc twists in DNA-binding protein domains. *Proc. Natl. Acad. Sci. USA* **88**, 999–1003.

Vasios, G., Gold, J.H., Petkovich, M., Chambon, P. & Gudas, L. (1989) A retinoic acid-response element is present in the 5'-flanking region of the laminin B1 gene. *Proc. Natl. Acad. Sci. USA* **86**, 9099–103.

Wanek, N., Gardiner, D.M., Muneoka, K. & Bryant, S.V. (1991) Conversion by retinoic acid of anterior cells into ZPA cells in the chick wing bud. *Nature* **350**, 81–3.

Wang, C., Curtis, J.E., Minden, M.D. & McCulloch, E.A. (1989) Expression of a retinoic acid receptor gene in myeloid leukemia cells. *Leukemia* **3**, 264–9.

Webster, N.J.G., Green, S., Jin, J.R. & Chambon, P. (1988) The hormone-binding domains of the estrogen and glucocorticoid receptors contain an inducible transcription activation function. *Cell* **54**, 199–207.

Wolbach, S.B. & Howe, P.R. (1925) Tissue changes following deprivation of fat soluble A vitamin. *J. Exp. Med.* **62**, 753–7.

Yang, N., Schüle, R., Mangelsdorf, D.J. & Evans, R.M. (1991) Characterization of DNA binding and retinoic acid binding properties of retinoic acid receptor. *Proc. Natl. Acad. Sci. USA* **88**, 3559–63.

Yu, V.C., Delsert, C., Andersen, B., Holloway, J.M., Devary, O.V., Näär, A.M., Kim, S.Y., Boutin, J.M., Glass, C.K. & Rosenfeld, M.G. (1991) RXRβ: a coregulator that enhances binding of retinoic acid, thyroid hormone, and vitamin D receptors to their cognate response elements. *Cell* **67**, 1251–66.

Yuspa, S.H., Lichti, U., Ben, T. & Hennings, H. (1981) Modulation of terminal differentiation and responses to tumor promoters by retinoids in mouse epidermal cell cultures. *Ann. NY Acad. Sci.* **359**, 260–73.

Zelent, A., Krust, A., Petkovich, M., Kastner, P. & Chambon, P. (1989) Cloning of murine α and β retinoic acid receptors and a novel receptor γ predominantly expressed in skin. *Nature* **339**, 714–17.

Zelent, A., Mendelsohn, C., Kastner, P., Garnier, J.M., Ruffenach, F., Leroy, P. & Chambon, P. (1991) Differentially expressed isoforms of the mouse retinoic acid receptor β are generated by usage of two promoters and alternative splicing. *EMBO J.* **10**, 71—81.

Zhang, X.-k., Hoffmann, B., Tran, P.B.-V., Graupner, G. & Pfahl, M. (1992*a*) Retinoid X receptor is an auxiliary protein for thyroid hormone and retinoic acid receptors. *Nature* **335**, 441–446.

Zhang, X.-K, Lehmann, J., Hoffmann, B., Dawson, M.I., Cameron, J., Graupner, G., Hermann, T., Tran, P. & Pfahl, M. (1992*b*) Homodimer formation of retinoid X receptor induced by 9-cis retinoic acid. *Nature* **358**, 587–91.

3

Vitamin D Receptors and the Mechanism of Action of 1,25-Dihydroxyvitamin D₃

J. Wesley Pike

Introduction

Steroid hormones exert profound regulatory control over complex gene networks through direct interactions at the level of the cellular genome (O'Malley, 1990; Yamamoto, 1985). The products of these modulated genes serve to control processes essential to cellular growth and differentiation as well as to influence mechanisms integral to the maintenance of intracellular and extracellular homeostasis. The actions of these blood-borne signals following their diffusion into distant cells are mediated by unique intracellular receptors (Jensen *et al.*, 1968; Jensen & De Sombre, 1972). Indeed, the presence of these receptors in cells and tissues represents the principal, although not the only, determinant of response to a particular hormone. These soluble signal-transducing proteins are members of a large gene family of latent transcription factors that acquire strong but unique gene-regulating capacities upon activation by their respective hormonal ligands (Evans, 1988). Recognition of the individual ligands is characterized by high affinity and specificity (Haussler, 1986). While hormone interactions have been well characterized, the events that follow association of the ligand with its receptor remain less well understood. These events include receptor activation, interaction with DNA, association with other transcription factors, and the eventual transactivation process itself. Despite the paucity of information about these mechanisms, the outcome of such events is modification of specific gene expression and in turn biological response.

The vitamin D hormone mechanism

The vitamin D hormone 1,25-dihydroxyvitamin D₃ (1,25(OH)₂-D₃) is believed to regulate a number of biological processes via a steroid hormone mechanism. These include cell growth, differentiation, immune function, and of course the numerous events in many target tissues that are associated with mineral metabolism (Haussler, 1986; DeLuca, 1988; Minghetti & Norman, 1988; Pike, 1991; Suda *et al.*, 1990; Manolagas *et al.*, 1985). As with other steroid hormone systems, the mechanism involves a specific receptor, that for $1,25(OH)_2D_3$ (VDR) (Haussler & Norman, 1969). While the exact details of how the receptor elicits transcriptional modulation is unclear, many of the protein products that participate in the ensuing biological responses are well known. They include such proteins as the calbindin gene family, osteocalcin (OC), matric gla protein, osteopontin (OP), and collagen to name just a few. A larger and more complete list has been compiled recently (Suda *et al.*, 1990). It seems likely that several mechanisms will eventually be identified to account for the effects of $1,25(OH)_2D_3$ on the levels of these biologically active proteins. The direct mechanism of action of vitamin D_3, however, involves a specific interaction of the VDR with DNA sequences within $1,25(OH)_2D_3$-responsive genes. The purpose of this chapter is to discuss recent advances that support this mechanism of action, particularly with regard to the structure of the receptor, its involvement in transactivation, the nature of DNA binding sites which mediate vitamin D_3 action, and the mechanism by which the protein interacts with these sites.

Structure of the vitamin D receptor

Current evidence suggests that the VDR may be involved in most if not all phases of vitamin D_3 action at the nuclear level. Considerable effort by many groups has been exerted over the years in characterizing this elusive protein (Brumbaugh & Haussler, 1975; Pike & Haussler, 1979; Dame *et al.*, 1986). Nevertheless, the recent cloning of the structural gene for the VDR from several species, including humans, has substantially increased our understanding of both its structure and its interrelations with the other steroid receptors (McDonnell *et al.*, 1987; Baker *et al.*, 1988; Burmester *et al.*, 1988). Inspection of the VDR's primary sequence has revealed that the protein belongs to the steroid, thyroid, and retinoic acid receptor gene family (Evans, 1988). Its basic structural organization, similar to that of other family members, consists of a distinct N-terminal DNA recognition domain and a large C-terminal $1,25(OH)_2D_3$ binding domain. The former region contains two finger structures each folded about a zinc atom tetrahedrally coordinated through the sulfhydryl moieties of four conserved cysteine residues (Hard *et al.*, 1990; Schwabe *et al.*, 1990). This motif appears to be reminiscent of the helix–loop–helix DNA binding motif found in a number of other proteins that carry out related transcriptional regulating functions (Johnston & McKnight, 1989). Perhaps more importantly, the cloning of this protein has provided a direct opportunity to determine its function. Indeed, studies on these structural regions of the VDR have confirmed their participation in both hormone and DNA binding functions as well as trans-activation (McDonnell *et al.*, 1989), and ongoing studies are aimed at further dissection of this interesting macromolecule. Thus, the concept that $1,25(OH)_2D_3$ likely functions through a steroid hormone-like mechanism has been strongly substantiated by our preliminary observations about the receptor.

Transcriptional response to 1,25-dihydroxyvitamin D₃

During the past decade a number of proteins have been identified whose genes represent potential transcriptional models for the study of both positive and negative regulation by $1,25(OH)_2D_3$. The genes for OC and OP, nevertheless, have emerged thus far as the most informative with respect to the inductive effects of this hormone. Determination of vitamin D inducibility at the level of transcriptional control was made initially using chimeric genes that contained the promoter and upstream 5' flanking region of the human and rat OC genes fused to the chloramphenicol acetyltransferase (CAT) structural gene containing an SV40 polyadenylation signal. These studies were made possible by the successful cloning of the chromosomal genes for both human (Celeste *et al.*, 1986) and rat (Yoon *et al.*, 1988; Demay *et al.*, 1989; Lian *et al.*, 1989) OC, as well as the availability of genomic DNA containing their potential control regions. Plasmids were introduced by DNA-mediated gene transfer into receptor-positive cultured cells (such as osteoblastic ROS 17/2.8) and the activity of the reporter function CAT determined following stimulation by $1,25(OH)_2D_3$. We and others observed that CAT activity was strongly elevated by physiological concentrations of $1,25(OH)_2D_3$ when compared with cells that received only vehicle (McDonnell *et al.*, 1989; Yoon *et al.*, 1988; Demay *et al.*, 1989; Lian *et al.*, 1989). Subsequent experiments using the cloned OP gene promoter have led to identical results (Noda *et al.*, 1990). Thus, $1,25(OH)_2D_3$ is indeed capable of direct action on transcription through these promoters, and activation likely requires the presence of *cis*-acting elements located within the promoters themselves. These actions in turn correlate well with the inducibility of both endogenous genes by $1,25(OH)_2D_3$ *in vivo* (Lian *et al.*, 1989; Butler, 1989; Price & Baukol, 1980).

Role of the receptor in trans-activation

The essential role of the VDR as a trans-activator in this $1,25(OH)_2D_3$-dependent response has been defined in related experiments (McDonnell *et al.*, 1989). We noted that while the OC promoter–CAT chimera was inducible by $1,25(OH)_2D_3$ in receptor-positive cells, $1,25(OH)_2D_3$ had no effect on induction in cells such as the CV-1 fibroblasts that do not contain receptor. In contrast, $1,25(OH)_2D_3$ response was promptly restored when the VDR was provided to these cells through introduction of a human VDR cDNA expression vector that directed the synthesis of full-length VDR (McDonnell *et al.*, 1989). This restoration of response was receptor-concentration-dependent and required synthesis of a normal receptor; receptors that were not full-length, did not bind $1,25(OH)_2D_3$, or contained mutations within the DNA binding domain (Sone *et al.*, 1989) were inactive. These studies demonstrate directly that the VDR is an essential component that acts in *trans* to mediate $1,25(OH)_2D_3$ action.

The nature of *cis*-acting DNA binding sites in vitamin D_3 responsive genes

Steroid receptors and their counterparts for ligands such as thyroid hormone and retinoic acid are known to interact directly with short *cis*-acting DNA sequences located within the vicinity of responsive gene promoters (Beato, 1989). These sequences are not unlike binding sites recognized by an extensive list of basal promoter activators that include CREB, JUN/FOS, SP1, and others. Hormone responsive elements (HRE) take a variety of motifs and have been characterized as either short single or repeated segments, inverted repeats, or combinations thereof. Examples include regions found in promoters for natural genes such as the vitellogenin (Klein-Hitpass *et al.*, 1986), tyrosine aminotransferase (Jantzen *et al.*, 1987), prolactin (Waterman *et al.*, 1988), and growth hormone (Glass *et al.*, 1987; Brent *et al.*, 1989). While initial examples of HREs seem to favor the palindromic motif, increased evidence has suggested that directly repeated segments may be a more common structural organization. We anticipated that an element related to the above type would likely mediate the action of vitamin D_3.

The location of VDREs

Precise locations of vitamin D_3 responsive elements (VDREs) within the OC and OP gene promoters were mapped functionally using unidirectional and/or internal analysis (Noda *et al.*, 1990; Kerner *et al.*, 1989; Morrison *et al.*, 1989; Demay *et al.*, 1990). An example of this strategy for the human OC promoter can be observed in Figure 3.1. Our studies and those of others revealed that the regions which confer vitamin D₃ response are located in the human OC promoter between −512 and −485 (relative to the start site of transcription) (Kerner *et al.*, 1989; Morrison *et al.*, 1989; Demay *et al.*, 1990), in the rat OC promoter between −458 and −433 (Denmay *et al.*, 1990), and in the murine OP promoter between −761 and −741 (Noda *et al.*, 1990). Deletion of this sequence in the human OC promoter leads to loss of vitamin D response, whereas fusion of a synthetic version of this sequence to downstream OC fragments in which the VDRE had been removed restores response (Kerner *et al.*, 1989). Most importantly, these regions were also capable

Figure 3.1. 1,25-Dihydroxyvitamin D₃ inducibility of deletion mutations of the human osteocalcin gene promoter. A series of human osteocalcin promoter deletion constructs (whose 5′ endpoints are indicated relative to the start site of transcription) were fused to the structural gene for chloramphenicol acetyltransferase in the plasmid vector pBLCAT3. The capacity of 1,25(OH)₂D₃ to induce these constructs following transfection into ROS 17/2.8 cells is indicated (+ or −). Plasmids designated −344 V, −193 V, and −75 V represent constructs that each contain a single copy of the VDRE sequence from −512 to −483 cloned at the 5′ end of the osteocalcin promoter at −344, −193, and −75. Plasmid −838 Δ 513/487 represents a construct in which the VDRE from −513 to −487 has been internally deleted using site-directed mutagenesis. B, BamHI; P, PstI; Bs, BspMII; Pv, PvuII; A, ApaI; N, NcoL; TATA, tata box.

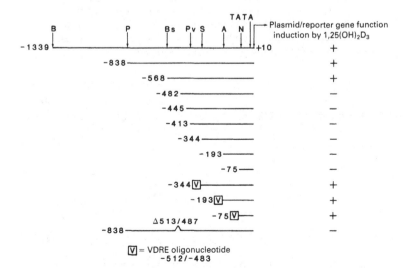

in either orientation of conferring novel vitamin D_3 response upon heterologous viral promoters such as those of the mouse mammary tumor virus LTR (Pike, 1990) or the herpes simplex virus thymidine kinase gene (Noda *et al.*, 1990; Kerner *et al.*, 1989; Morrison *et al.*, 1989; Demay *et al.*, 1990). These studies suggest that DNA sequences identified within the OC and OP genes are indeed capable of directing response to vitamin D_3, and that these sequences retain properties typical of hormone-responsive enhancers. Moreover, as in our previous studies, the presence of the VDR was required for activation when VDRE-containing promoters were introduced into receptor-negative cells (Pike, 1990).

Structure and sequence of natural VDREs

The sequences of the VDREs in the OC and OP genes are aligned in Figure 3.2. This alignment reveals that the overall structure of the VDRE comprises two direct but imperfect repeats separated by three nucleotide pairs. Four of six bases in the first repeat and only three of six bases in the second repeat are identical in all three sequences. These direct repeats are clearly reminiscent of sequences as well as the structural motif of elements that mediate thyroid hormone response in the growth hormone gene promoter (Glass *et al.*, 1987; Brent *et al.*, 1989) and an element that mediates retinoic acid activation in the growth hormone

Figure 3.2. Alignment of three naturally occurring vitamin D responsive regions. Sequences shown include the human (−512 to −483) and rat (−466 to −439) osteocalcin (OC) genes and the mouse (−761 to −741) osteopontin (OP) gene. The position of the VDRE in each gene is indicated where arrows mark the locations of the two direct repeats each spaced by a trinucleotide pair. The human osteocalcin AP-1 consensus sequence is also indicated.

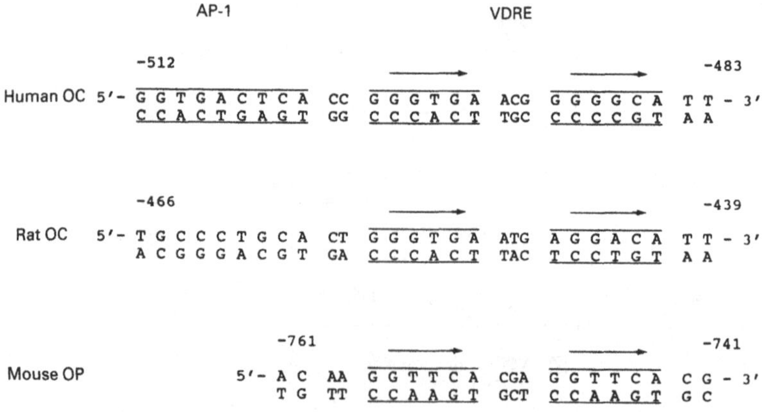

(Umesono *et al.*, 1988) and beta RAR (de Thé *et al.*, 1990) gene promoters. Perhaps not surprisingly, retinoic acid is capable of activating the OC VDRE in both ROS 17/2.8 cells and CV-1 cells supplemented with a retinoic acid receptor expression vector (Pike, 1990; Schule *et al.*, 1990). A degree of specificity is retained, however, as thyroid hormone in contrast is not able to activate this gene promoter.

Interrelationship between VDRE and AP-1 cis-*elements*

Also evident in the human sequence observed in Figure 3.2 is a unique and closely juxtaposed AP-1 consensus sequence that serves to bind members of the Jun/Fos proto-oncogene family (Lee *et al.*, 1987; Angel *et al.*, 1987). The action of the AP-1 family on this site, and the effect this action may have upon $1,25(OH)_2D_3$ induction, is controversial. We have found that an intact AP-1 site synergistically enhances $1,25(OH)_2D_3$ activation in ROS 17/2.8 and receptor cDNA-transfected CV-1 cells; point mutations in the AP-1 consensus sequence known to block AP-1 protein binding lead to loss of basal activity and reduce but do not eliminate $1,25(OH)_2D_3$ response (Ozono *et al.*, 1990). Moreover, treatment of ROS 17/2.8 cells with TPA leads to a modest induction (K. Ozono *et al.*, unpublished). In contrast, cotransfection of c-jun and c-fos expression vectors into ROS 17/2.8 cells suppresses the OC gene's basal activity and $1,25(OH)_2D_3$ response is reduced but not eliminated (Schule *et al.*, 1990). The latter studies have been used to support a hypothesis wherein this element might provide a paradigm for cross-communication between the actions of Jun and Fos heterodimers and steroid receptors; the former tending to promote cellular proliferation and the latter to promote cellular differentiation. This hypothesis was reiterated by Owen *et al.* (1990) who proposed that the rat OC VDRE also contained an AP-1-like element imbedded within the VDRE (not juxtaposed as in the human OC VDRE). Unfortunately, this report provided no direct evidence to suggest that this sequence represented a functional *cis* element and certainly no evidence to suggest that it mediated transcriptional suppression. Our unpublished studies reveal that this rat VDRE/AP-1 sequence does not compete significantly for ROS 17/2.8 cell-derived AP-1 binding to the human OC VDRE/AP-1 site *in vitro*, suggesting that this sequence is not comparable to the AP-1 site found in the human OC gene. The OP VDRE also does not contain an AP-1 sequence, nor is it induced by TPA (Noda *et al.*, 1990). Regardless, clarification of whether AP-1 proteins function to activate or suppress human OC gene promoter transcription, or direct a combination of the two depending upon the state of the cell remains for future endeavors, as does the identification of the specific AP-1 gene members in osteoblasts that are responsible for these effects.

A VDRE consensus sequence

Consensus sequences for the two hexanucleotide half-sites of these three elements are GG G/T G/C A and GG G/T G/T/A A. While additional natural VDREs will no doubt be identified to improve this consensus, it is clear from this derivation that the VDRE half-site is not unlike those of steroid responsive elements for other hormones (Beato, 1989). This finding is consistent with the observation that the nucleotide sequence of a response element for one receptor may resemble or be identical to that for another receptor, perhaps for several (Umesono et al., 1988; Stahle et al., 1987). Possibly more interesting than their sequence, however, is the emerging awareness that considerable plasticity exists within response elements for their cognate receptors. This observation is clearly evident from the divergent nucleotide sequence of each of the VDREs that have been identified. Taken together with the fact that sequences that serve as receptor binding sites in vitro may not represent functional sites in vivo, it seems likely that the specificity of interaction between a receptor and its DNA binding site and the subsequent functional response that ensues may be determined to a large degree by additional factors. These include the structural organization of the HRE, the transcriptional context of the promoter in which it resides, protein factors that participate in the induction phenomenon, chromatin structure, and finally the cellular context under which the response is recorded. All of these may in fact be more important than the nucleotide sequence of the HRE itself. These issues are becoming increasingly important to our understanding of the mechanisms of transcriptional regulation.

Interaction of the VDR with VDREs in vitro

Direct interaction of mammalian cell-derived VDR with these VDRE sequences in vitro has been demonstrated (Noda et al., 1990; Demay et al., 1990; Ozono et al., 1990; Liao et al., 1990). The VDR exhibits sufficiently high affinity for these elements that protein–DNA complexes can be identified following separation from unbound DNA by the stringent conditions of gel retardation. Interestingly, this interaction is $1,25(OH)_2D_3$-modulated (Liao et al., 1990). Thus, the addition of hormone during receptor incubation with the VDRE in vitro enhances dramatically, although not absolutely, the VDR's capacity to bind to a

VDRE. Difficulties have been encountered with other receptors in demonstrating hormone-dependent DNA binding (Umesono *et al.*, 1988; Bagchi *et al.*, 1990; Willman & Beato, 1986; Damm *et al.*, 1989).

The apparent structural organization of the three natural VDREs as direct repeats suggests that, as illustrated in Figure 3.3, the VDR may associate with these elements as dimers. This type of interaction forms the basis for the association of the progesterone, glucocorticoid, and estrogen receptors on their respective DNA elements (Kumar & Chambon, 1988; Fawell *et al.*, 1990). Surprisingly, and in contrast with mammalian cell-derived VDR, neither *in vitro* translated VDR nor crude or highly purified VDR derived from yeast transformed with a VDR expression plasmid demonstrates the capacity to interact directly with the OC VDRE in bandshift assays (Liao *et al.*, 1990; Sone *et al.*, 1990*b*). These data suggest

Figure 3.3. Model for the interaction of the VDR on a VDRE comprised of two direct repeats. VDR can interact as two independent monomers, as a cooperative homodimer or as a heterodimer. Evidence exists *in vitro* for the latter possibility.

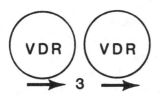

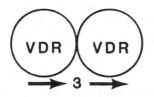

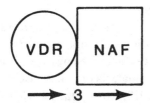

that perhaps mammalian cell extracts contain an additional protein factor that facilitates VDR–DNA interaction. Consistent with this hypothesis, the addition of VDR-free mammalian cell extracts to reticulocyte lysate-translated VDR, to yeast cytosols that contain recombinant VDR, or to highly purified VDR restores the ability of these receptors to bind the VDRE DNA in a fashion identical to that observed for their mammalian cell-derived counterparts (Sone et al., 1991). Thus, it is possible that the VDR is unable to form natural homodimers, and instead forms hetero-dimers with a nuclear accessory factor (NAF) that are capable of DNA binding affinities greater than that of the receptor monomer alone. Is this interaction of the VDR and NAF specific for the human OC VDRE? We have observed that mammalian cell extracts containing NAF are required for VDR interaction not only with the human OC VDRE, but also with the rat OC VDRE, the mouse OP VDRE and several synthetic versions of the VDRE that we have identified as well (K. Ozono et al., unpub-lished). These and additional experiments suggest that receptor–NAF heterodimer, and not receptor–receptor homodimer, formation may be essential for VDR interaction with VDREs.

The above observations are consistent with the DNA binding behavior of both retinoic acid and thyroid hormone receptors. The latter receptors form heterodimers in vitro both with each other (Glass et al., 1989; Forman et al., 1989) and with a set of nuclear protein factors whose identities currently remain unknown (Murray & Towle, 1989; Burnside et al., 1990; Glass et al., 1990; Lazar & Berrodin, 1990). In contrast to the VDR, however, thyroid and retinoic acid receptors are also capable of forming homodimers (Glass et al., 1989; Forman et al., 1989). While NAF heterodimer formation appears to be favored with regard to VDR interaction in the current in vitro studies, we cannot at present exclude the possibility that homodimer formation by this protein on novel VDRE sequences will be identified or indeed created under conditions different than those heretofore tested. Finally, it should be stressed that no direct evidence currently exists that either interaction mechanism is required for gene expression in vivo. Each of these important issues remains for further testing.

Hereditary 1,25(OH)$_2$D$_3$ resistance

The human syndrome of hereditary 1,25(OH)$_2$D$_3$ resistance (vitamin D$_3$ dependent rickets, type II) is a rare autosomal recessive disease that occurs early in infancy and is characterized by hypocalcemia, secondary hyperparathyroidism, and osteomalacia or rickets (Beer et al., 1981; Liberman et al., 1980; Marx et al., 1978; Rosen et al., 1979; Takeda *et al.*, 1987). These clinical features all persist in parallel with elevated levels of circulating $1,25(OH)_2D_3$, suggesting that they arise as a result of target organ resistance to the action of this hormone.

The vitamin D₃-resistant fibroblast model

Clinical cases of the syndrome of $1,25(OH)_2D_3$ resistant rickets were first documented in the late 1970s. However, the discovery that normal skin fibroblasts contain a functional effector system for $1,25(OH)_2D_3$ provided the key to our eventual understanding of the molecular basis for this disease (Eil & Marx, 1981; Feldman *et al.*, 1980). The observations that fibroblasts derived from patients resistant to the action of the vitamin D_3 hormone were either qualitatively or quantitatively unresponsive were not surprising (Balsan *et al.*, 1983; Chen *et al.*, 1984; Gamblin *et al.*, 1985; Hirst *et al.*, 1985; Liberman *et al.*, 1983, 1986). In the light of the foregoing discussion regarding the vitamin D_3 receptor, characterization of this protein in defective fibroblasts was initiated rapidly. The results of these efforts has been the definition of fibroblastic phenotypes in which defective hormone-binding or DNA-binding activities were evident (Takeda *et al.*, 1987; Eil & Marx, 1981; Feldman *et al.*, 1980; Balsan *et al.*, 1983; Chen *et al.*, 1984; Gamblin *et al.*, 1985; Hirst *et al.*, 1985; Liberman *et al.*, 1983, 1986) as well as a third defective phenotype normal with respect to hormone and DNA binding functions but incapable of facilitating the nuclear accumulation of $1,25(OH)_2D_3$ (Takeda *et al.*, 1987; Liberman *et al.*, 1983, 1986). Presently, these three abnormal receptor phenotypes have emerged as each potentially responsible for the disease phenotype of rickets (Marx & Barsony, 1988).

Genetic lesions in the vitamin D₃ receptor chromosomal locus

In order to determine the genetic basis for dysfunctional receptors, R.A. Kesterson & J.W. Pike (unpublished results) cloned the human chromosomal gene for the vitamin D_3 receptor, and determined its structural organization and nucleotide sequence. Hughes *et al.*(1988), Richie *et al.* (1989), and Sone *et al.* (1990a) then utilized short oligonucleotide sequences complementary to intron sequences to amplify selectively DNA from each of the coding exons of the VDR using polymerase chain reaction techniques. The incorporation of restriction sites at the 5' ends of each synthetic oligonucleotide allowed the subsequent cloning and sequencing of these individually amplified DNAs. In this manner, the presence and nucleotide sequence of each of the exons that encode the human vitamin D_3 receptor were determined rapidly for patients with $1,25(OH)_2D_3$ resistance.

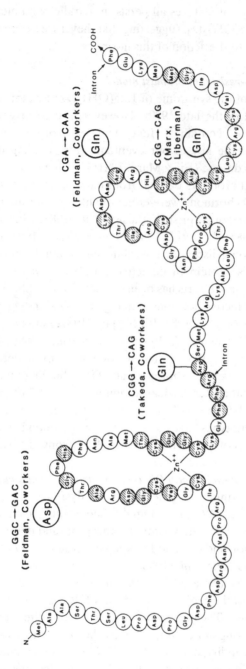

Figure 3.4. Natural mutations in the DNA binding domain of the vitamin D_3 receptor identified in chromosomal DNA from patients with hereditary resistance to 1,25-dihydroxyvitamin D_3. The dual DNA binding zinc finger motif at the amino terminus of the vitamin D receptor is illustrated. Residues which are shaded are conserved within the steroid receptor family. Four amino acid substitutions that have been identified in families with hereditary resistance are indicated. The point mutation in the natural codon that gives rise to the altered residue is summarized over each mutant.

The search for mutations within the VDR gene focused initially upon analysis of patients whose receptors exhibited defects in ability either to bind $1,25(OH)_2D_3$ or to recognize DNA. Three interrelated families were evaluated for genetic defects associated with an inability to bind hormonal $1,25(OH)_2D_3$ (Ritchie *et al.*, 1990). DNA amplification and sequence analysis of material from these patients revealed the presence in a single exon of a C to A nucleotide substitution that produced a premature termination codon. Parental DNA was heterozygotic for this mutation, and unaffected siblings were either homozygous or hetero- zygous for the normal gene. The apparent result of this mutation is the synthesis of a protein truncated at codon 291 within the $1,25(OH)_2D_3$ binding domain and unable to bind the ligand. Clearly, however, the loss of binding activity in cells can arise from many unrelated defects in the receptor gene and thus this mutation represents but one example.

Perhaps the most interesting category of receptor defect in patients with resistance to $1,25(OH)_2D_3$ is that which leads to altered ability to recognize DNA. As the DNA binding domain is restricted to two or three exons within the gene, the search for genetic mutations was initiated within those regions using DNA derived from six different families whose receptor proteins exhibited low affinity for immobilized DNA (Hughes et al., 1988; Ritchie *et al.*, 1989; Sone *et al.*, 1989). Four independent point mutations within these exons that together encode the DNA binding region of the VDR were identified (see Figure 3.4). These mutations result in aspartic acid or glutamine substitutions at different sites within the dual zinc finger motif that without exception result in alterations in the charge of the structure. While it is unclear precisely how these changes affect the DNA binding activity of the receptor, the preponder- ance of basic amino acids within this region as well as the conservation of these mutant amino acids within the receptor gene family underscore their importance (O'Malley, 1990; Evans, 1988).

Functional activity of receptors derived from patient genotypes

To gain additional evidence that the mutations identified within the mutated genes result in aberrant binding functions and inability to modify gene transcription, Hughes *et al.* (1988), Sone *et al.* (1989) and Richie *et al.* (1989) utilized the technique of site-directed mutagenesis to introduce each mutation independently into the normal VDR cDNA. Analysis of the mutant receptor products synthesized following transfec- tion into host mammalian fibroblasts revealed that each protein exhibited a functional phenotype identical to that in patient fibroblasts. Accord- ingly, introduction of a termination codon resulted in synthesis of an immunologically detected truncated protein of 32 kilodaltons that was unable to bind 1,25(OH)$_2$D$_3$ (Ritchie *et al.*, 1989). Likewise, the three point mutations introduced individually into the DNA binding domain of the receptor led to the synthesis of proteins of normal size and ability to bind 1,25(OH)$_2$D$_3$, but of proteins unable to interact appropriately with either non-specific DNA (Sone *et al.*, 1989, 1990*b*) or the osteocalcin VDRE (Sone *et al.*, 1991). Thus, the defects associated with these mutated receptors correspond identically to those found in their endogenously synthesized counterparts. Perhaps most important, these mutated receptors were also inactive in the transcriptional response assay outlined above capable of assessing the functional activity of the VDR. While the normal receptor was fully capable of activating osteocalcin promoter-mediated transcription in the presence of 1,25(OH)$_2$D$_3$, mutant forms of the protein did not function, even during receptor overexpression or under conditions of high hormone levels (Sone *et al.*, 1989, 1990*b*; Ritchie *et al.*, 1989). Thus, these defective receptors display not only abnormal binding functions but aberrant transcription-inducing actions as well. These data clearly establish the molecular basis for hormone unresponsiveness in at least a subset of patients with hereditary 1,25(OH)$_2$D$_3$ resistant rickets. They in turn strongly support the essential role of this receptor in mediating the biologic functions of hormonal 1,25(OH)$_2$D$_3$ that have been described in this chapter.

Summary

The purpose of this chapter has been to describe recently accumulated evidence that suggests that the hypothesis made in 1968 that 1,25(OH)$_2$D$_3$ might act as a steroid hormone was correct (Norman, 1968). This evidence includes the finding that a clear structural interrelationship exists between the VDR and other members of the steroid receptor gene family, the observation that the VDR is required for gene promoter transactivation, and the identification of VDREs which act in *cis* to mediate 1,25(OH)$_2$D$_3$ response. The VDR has been found to bind *in vitro* specifically to these functional DNA sites. Current evidence, however, indicates that the receptor may interact at these sites not as a monomer or homodimer but rather as a heterodimer with a protein whose identity remains unknown. Finally, the role of functional receptor in normal mineral metabolism is dramatically highlighted by the syndrome of 1,25-dihydroxyvitamin D$_3$ resistance, where the genetic basis of the disease lies in aberrant and non-functional receptors. Future studies with regard to the mechanism of vitamin D action must be aimed at gaining additional insight into the nature of VDREs, acquiring further detail about the interaction of the VDR with these elements, identifying factors that facilitate VDR DNA binding, and determining the biochemical mechanism whereby the binding of receptor to these elements leads to modulation of common transcriptional machinery. In addition, $1,25(OH)_2D_3$ acts to suppress a number of genes, for example collagen, calcitonin, and PTH. While efforts to elucidate these actions are currently underway, the mechanism by which attenuation of response occurs remains largely uncharacterized. Finally, it is possible that additional mechanisms of vitamin D action may exist, perhaps modes of action that involve metabolites of vitamin D other than $1,25(OH)_2D_3$, do not require the presence of the VDR, or require an as yet unidentified VDR gene product. Each of these areas offers a considerable challenge to future research.

References

Angel, P., Masayoshi, I., Chiu, R., Stein, B., Imbra, R.J., Rahmsdorf, H.J., Jonat, C., Herrlich, P. & Karin, M. (1987) Phorbol ester-inducible genes contain a common cis element recognized by a TPA-modulated trans-acting factor. *Cell* **49**, 729–39.

Bagchi, M.K., Tsai, S.Y., Tsai, M.-J. & O'Malley, B.W. (1990) Identification of a functional intermediate in receptor activation in progesterone-dependent cell-free transcription. *Nature* **345**, 547–50.

Baker, A.R., McDonnell, D.P., Hughes, M.R., Crisp, T.M., Mangelsdorf, D.J., Haussler, M.R., Pike, J.W., Shine, J. & O'Malley, B.W. (1988) Cloning and expression of full-length cDNA encoding human vitamin D receptor. *Proc. Natl. Acad. Sci. USA* **85**, 3294–8.

Balsan, S., Garabedian, M., Liberma, U.A., Eil, C., Bourdeau, A., Guillozo, H., Grimberg, R., Le Deunff, M.J., Lieberherr, M., Grimbaud, P., Broyer, M. & Marx, S.J. (1983) Rickets and alopecia with resistance to 1,25-dihydroxyvitamin D: two different clinical courses with two different cellular defects. *J. Clin. Endocrinol. Metab.* **57**, 803–11.

Beato, M. (1989) Gene regulation by steroid hormones. *Cell* **5**, 335–44.

Beer, S., Tieder, M., Kohelet, D., Liberman, U.A., Vure, G., Bar-Joseph, G., Gabizon, D., Borochowitz, Z.U., Varon, M. & Modai, D. (1981) Vitamin D resistant rickets with alopecia: a form of end organ resistance to 1,25 dihydroxyvitamin D. *Clin. Endocrinol.* **14**, 395–402.

Brent, G.A., Harney, J.W., Chen, Y., Warne, R.L., Moore, D.D. & Larsen, P.R. (1989) Mutations of the rat growth hormone promoter which increase and decrease response to thyroid hormone define a consensus thyroid hormone response element. *Molec. Endocrinol.* **3**, 1996–2004.

Brumbaugh, P.F. & Haussler, M.R. (1975) Specific binding of 1,25-Dihydroxychole-calciferol to nuclear components of chick intestine. *J. Biol. Chem.* **250**, 1588–94.

Burmester, J.K., Maeda, N. & DeLuca, H.F. (1988) Isolation and expression of rat 1,25-dihydroxyvitamin D_3 receptor cDNA. *Proc. Natl. Acad. Sci. USA* **85**, 1005–9.

Burnside, J., Darling, D.S. & Chin, W.W. (1990) A nuclear factor that enhances binding of thyroid hormone receptors to thyroid hormone response elements. *J. Biol. Chem.* **265**, 2500–4.

Butler, W.T. (1989) The nature and significance of osteopontin. *Connect. Tissue Res.* **23**, 123–36.

Celeste, A.J., Rosen, V., Buecker, J.L., Kriz, R., Wang, E.A. & Wozney, J.M. (1986) Isolation of the human gene for bone gla protein utilizing mouse and rat cDNA clones. *EMBO J.* **5**, 1885–90.

Chen, T.L., Hirst, M.A., Cone, C.M., Hochberg, A., Tietze, H. & Feldman, D. (1984) 1,25-Dihydroxyvitamin D resistance, rickets, and alopecia: analysis of receptors and bioresponses in cultured fibroblasts from patients and parents. *J. Clin. Endocrinol. Metab.* **59**, 383–8.

Dame, M.C., Pierce, E.A., Prahl, J.M., Hayes, C.E. & DeLuca, H.F. (1986) Monoclonal antibodies to the porcine intestinal receptor for 1,25-dihydroxyvitamin D_3: interaction with distinct receptor domains. *Biochemistry* **25**, 4523–34.

Damm, K., Thompson, C.C. & Evans, R.M. (1989) Protein encoded by v-erbA functions as a thyroid-hormone receptor antagonist. *Nature* **339**, 593–7.

de Thé, H., del Mar Vivanco-Ruiz, M., Tiollais, P., Stunnenberg, H. & Dejean, A. (1990) Identification of a retinoic acid responsive element in the retinoic acid receptor beta gene. *Nature* **343**, 177–80.

DeLuca, H.F., (1988) The vitamin D story: a collaborative effort of basic science and clinical medicine. *FASEB J.* **2**, 224–36.

Demay, M.B., Gerardi, J.M., DeLuca, H.F. & Kronenberg, H.M. (1990) DNA sequences in the rat osteocalcin gene that bind the 1,25-dihydroxyvitamin D_3 receptor and confer responsiveness to 1,25-dihydroxyvitamin D_3. *Proc. Natl. Acad. Sci. USA* **87**, 369–73.

Demay, M.B., Roth, D.A. & Kronenberg, H.M. (1989) Regions of the rat osteocalcin gene which mediate the effect of 1,25-dihydroxyvitamin D_3 on gene transcription. *J. Biol. Chem.* **264**, 2279–82.

Eil, C. & Marx, S.J. (1981) Nuclear uptake of 1,25-dihydroxy[^{3}H]cholecalciferol in dispersed fibroblasts cultured from normal human skin. *Proc. Natl. Acad. Sci. USA* **78**, 2562–6.

Evan, R.M. (1988) The steroid and thyroid hormone receptor superfamily. *Science* **240**, 889–95.

Fawell, S.E., Lees, J.A., White, R. & Parker, M.G. (1990) Characterization and colocalization of steroid binding and dimerization activities in the mouse estrogen receptor. *Cell* **60**, 953–62.

Feldman, D., Chen, T., Hirst, M., Colston, K., Karasek, M. & Cone, C. (1980) Demonstration of 1,25-dihydroxyvitamin D_3 receptors in human skin biopsies. *J. Clin. Endocrinol. Metab.* **51**, 1463–5.

Forman, B.M., Yang, C., Au, M., Casanova, J., Ghysdael, J. & Samuels, H.H. (1989) A domain containing leucine-zipper-like motifs mediate novel in vivo interactions between the thyroid hormone and retinoic acid receptors. *Molec. Endocrinol.* **3**, 1610–26.

Gamblin, G.T., Liberman, U.A., Eil, C., Downs, R.W., DeGrange, D.A. & Marx, S.J. (1985) Vitamin D-dependent rickets type II. *J. Clin. Invest.* **75**, 954–60.

Glass, C.K., Devary, O.V. & Rosenfeld, M.G. (1990) Multiple cell type-specific proteins differentially regulate target sequence recognition by the alpha retinoic acid receptor. *Cell* **63**, 729–36.

Glass, C.K., Franco, R., Weinberger, C., Albert, V.R., Evans, R.M. & Rosenfeld, M.G. (1987) A c-erb-A binding site in rat growth hormone gene mediates trans-activation by thyroid hormone. *Nature* **329**, 738–41.

Glass, C.K., Lipkin, S.M., Devary, O.V. & Rosenfeld, M.G. (1989) Positive and negative regulation of gene transcription by a retinoic acid-thyroid hormone receptor heterodimer. *Cell* **59**, 697–708.

Hard, T., Kellenbach, E., Boelens, R., Maler, B.A., Dahlman, K., Freedman, L.P., Carlstedt-Duke, J., Yamamoto, K.R., Gustafsson, J.-A. & Kaptein, R. (1990) Solution structure of the glucocorticoid receptor DNA-binding domain. *Science* **249**, 157–60.

Haussler, M.R. (1986) Vitamin D receptors: Nature and function. *Annu. Rev. Nutr.* **6**, 527–62.

Haussler, M.R. & Norman, A.W. (1969) Chromosomal receptor for a vitamin D metabolite. *Proc. Natl. Acad. Sci. USA* **62**, 155–62.

Hirst, M.A., Hochman, H.I. & Feldman, D. (1985) Vitamin D resistance and alopecia: a kindred with normal 1,25-dihydroxyvitamin D binding, but decreased receptor affinity for deoxyribonucleic acid. *J. Clin. Endocrinol. Metab.* **60**, 490–5.

Hughes, M.R., Malloy, P.F., Kieback, D.G., Kesterson, R.A., Pike, J.W., Feldman, D. & O'Malley, B.W. (1988) Point mutations in the human vitamin D receptor gene associated with hypocalcemic rickets. *Science* **242**, 1702–5.

Jantzen, H.-M, Strahle, U., Gloss, B., Stewart, F., Schmid, W., Boshart, M., Miksicek, R. & Schutz, G. (1987) Cooperativity of glucocorticoid response elements located far upstream of the tyrosine aminotransferase gene. *Cell* **49**, 29–38.

Jensen, E.V. & De Sombre, E.R. (1972) Mechanism of action of the female sex hormones. *Annu. Rev. Biochem.* **41**, 203–30.

Jensen, E.V., Suzuki, T., Kawashima, T., Stumpf, W.E., Jungblut, P.W. & DeSombre, E.R. (1968) A two step mechanism for the interaction of estradiol with rat uterus. *Proc. Natl. Acad. Sci. USA* **59**, 632–8.

Johnston, P.F. & McKnight, S.L. (1989) Eukaryotic transcriptional regulatory proteins. *Annu. Rev. Biochem.* **58**, 799–839.

Kerner, S.A., Scott, R.A. & Pike, J.W. (1989) Sequence elements in the human osteocalcin gene confer basal activation and inducible response to hormonal vitamin D_3. *Proc. Natl. Acad. Sci. USA* **86**, 4455–9.

Klein-Hitpass, L., Schorp, M., Wagner, U. & Ryffel, G.U. (1986) An estrogen-responsive element derived from the 5' flanking region of the Xenopus vitellogenin A2 gene functions in transfected human cells. *Cell* **4**, 1053–61.

Kumar, V. & Chambon, P. (1988) The estrogen receptor binds tightly to its responsive element as a ligand-induced homodimer. *Cell* **55**, 145–56.

Lazar, M.A. & Berrodin, T.J. (1990) Thyroid hormone receptors form distinct nuclear protein-dependent and independent complexes with a thyroid hormone response element. *Molec. Endocrinol.* **4**, 1627–35.

Lee, W., Mitchell, P. & Tjian, R. (1987) Purified transcription factor AP-1 interacts with TPA-inducible enhancer elements. *Cell* **49**, 741–52.

Lian, J.B., Stewart, C., Puchacz, E., Mackowiak, Shalhoub, V., Collart, D., Zambetti, G. & Stein, G. (1989) Structure of the rat osteocalcin gene and regulation of vitamin D-dependent expression. *Proc. Natl. Acad. Sci. USA* **86**, 1143–7.

Liao, J., Ozono, K., Sone, T., McDonnell, D.P. & Pike, J.W. (1990) Vitamin D receptor interaction with specific DNA requires a nuclear protein and 1,25-dihydroxyvitamin D_3. *Proc. Natl. Acad. Sci. USA* **87**, 9751–5.

Liberman, U.A., Eil, C. & Marx, S.J. (1983) Resistance to 1,25-dihydroxyitamin D. *J. Clin. Invest.* **71**, 192–200.

Liberman, U.A., Eil, C. & Marx, S.J. (1986) Receptor-positive hereditary resistance to 1,25-dihydroxyvitamin D: chromatography of hormone-receptor complexes on deoxyribonucleic acid-cellulose shows two classes of mutation. *J. Clin. Endocrinol. Metab.* **62**, 122–6.

Liberman, U.A., Halabe, A., Samuel, R., Kauli, R., Edelstein, S., Weisman, Y., Papapoulos, S.E., Clemens, T.L., Fraher, L.J. & O'Riordan, J.L.H. (1980) End-organ resistance to 1,25-dihydroxycholecalciferol. *Lancet* **1**, 504–7.

Manolagas, S.C., Provvedini, D.M. & Tsoukas, C.D. (1985) Interactions of 1,25-dihydroxyvitamin D_3 and the immune system. *Molec. Cell. Endocrinol.* **43**, 113–22.

Marx, S.J. & Barsony, J. (1988) Tissue-selective 1,25-dihydroxyvitamin D_3 resistance: Novel applications of calciferols. *J. Bone Min. Res.* **3**, 481–7.

Marx, S.J., Speigel, A.M., Brown, E.M., Gardner, D.G., Downs, R.W. Jr, Attie, M., Hamstra, A.J. & DeLuca, H.F. (1978) A familial syndrome of decreased insensitivity to 1,25-dihydroxycholecalciferol. *J. Clin. Endocrinol. Metab.* **47**, 1303–10.

McDonnell, D.P., Mangelsdorf, D.J., Pike, J.W., Haussler, M.R. & O'Malley, B.W. (1987) Molecular cloning of complementary DNA encoding the avian receptor for vitamin D. *Science* **235**, 1214–17.

McDonnell, D.P., Scott, R.A., Kerner, S.A., O'Malley, B.W. & Pike, J.W. (1989) Functional domains of the human vitamin D_3 receptor regulate osteocalcin gene expression. *Molec. Endocrinol.* **3**, 635–44.

Minghetti, P.P. & Norman, A.W. (1988) 1,25(OH)2-Vitamin D_3 receptors: gene regulation and genetic circuitry. *FASEB J.* **2**, 3043—53.

Morrison, N.A., Shine, J., Fragonas, J.-C., Verkest, V., McMenemy, M.L. & Eisman, J.A. (1989) 1,25-Dihydroxyvitamin D-responsive element and glucocorticoid repression in the osteocalcin gene. *Science* **246**, 1158–61.

Murray, M.B. & Towle, H.C. (1989) Identification of nuclear factors that enhance binding of the thyroid hormone receptor to a thyroid hormone response element. *Molec. Endocrinol.* **3**, 1434–42.

Noda, M., Vogel, R.L. Craig, A.M., Prahl, J., DeLuca, H.F. & Denhardt, D. (1990) Identification of a DNA sequence responsible for binding of the 1,25-dihydroxyvitamin D_3 receptor and 1,25-dihydroxyvitamin D_3 enhancement of mouse secreted phosphoprotein 1 (Spp-1 or osteopontin) gene expression. *Proc. Natl. Acad. Sci. USA* **87**, 9995–9.

Norman, A.W., (1968) The mode of action of vitamin D. *Biol. Rev.* **43**, 97–137.

O'Malley, B.W. (1990) The steroid receptor superfamily: more excitement predicted for the future. *Molec. Endocrinol.* **4**, 363–9.

Owen, T.A., Bortell, R., Yocum, S.A., Smock, S.L., Zhang, M., Abate, C., Shalhoub, V., Aronin, N., Wright, K.L., van Wunen, A.J., Stein, J.L., Curran, T., Lian, J.B. & Stein, G.S. (1990) Coordinate occupancy of AP-1 sites in the vitamin D-responsive and CCAAT box elements by Fos-Jun in the osteocalcin gene: Model for phenotype suppression of transcription. *Proc. Natl. Acad. Sci. USA* **87**, 9990–4.

Ozono, K., Liao, J., Kerner, S.A., Scott, R.A. & Pike, J.W. (1990) The vitamin D responsive element in the human osteocalcin gene. Association with a nuclear protooncogene enhancer. *J. Biol. Chem.* **265**, 21881–8.

Pike, J.W. (1990) Cis and trans regulation of osteocalcin gene expression by vitamin D and other hormones. In *Calcium regulation and bone metabolism, basic and clinical aspects* (ed. D.V. Cohn, F.H. Glorieux & T.J. Martin), vol. 10, pp. 127–136. Amsterdam: Elsevier Science.

Pike, J.W. (1991) Vitamin D_3 receptors: Structure and function in transcription. *Annu. Rev. Nutr.* **11**, 189–216.

Pike, J.W. & Haussler, M.R. (1979) Purification of chicken intestinal receptor for 1,25-dihydroxyvitamin D_3. *Proc. Natl. Acad. Sci. USA* **76**, 5488–9.

Price, P.A. & Baukol, S.A. (1980) 1,25-Dihydroxyvitamin D_3 increases synthesis of the vitamin K-dependent bone protein by osteosarcoma cell. *J. Biol. Chem.* **255**, 11660–3.

Ritchie, H.H., Hughes, M.R., Thompson, E.T., Malloy, P.F. Hochberg, Z, Feldman, D., Pike, J.W. & O'Malley, B.W. (1989) An ochre mutation in the vitamin D receptor gene causes hereditary 1,25-dihydroxyvitamin D_3-resistant rickets in three families. *Proc. Natl. Acad. Sci. USA* **86**, 9783–7.

Rosen, J.F., Fleischman, A.R., Finberg, L., Hamstra, A. & DeLuca, H.F. (1979) Rickets with alopecia: an inborn error of vitamin D metabolism. *J. Pediatr.* **94**, 729–35.

Schule, R., Umesono, K., Mangelsdorf, D.J., Bolado, J., Pike, J.W. & Evans, R.M. (1990) Jun-Fos and receptors for vitamins A and D recognize a common response element in the human osteocalcin gene. *Cell* **61**, 497–504.

Schwabe, J.W.R., Neuhaus, D. & Rhodes, D. (1990) Solution structure of the DNA-binding domain of the oestrogen receptor. *Nature* **348**, 458–61.

Sone, T., Kerner, S.A. & Pike, J.W. (1991) Vitamin D receptor interaction with specific DNA. *J. Biol. Chem.* **266**, 23296–305.

Sone, T., Marx, S.J., Liberman, U.A. & Pike, J.W. (1990*a*) A unique point mutation in the human vitamin D receptor chromosomal gene confers hereditary resistance to 1,25-dihydroxyvitamin D_3. *Molec. Endocrinol.* **4**, 623–31.

Sone, T., McDonnell, D.P., O'Malley, B.W. & Pike, J.W. (1990*b*) Expression of human vitamin D receptor in *Saccharomyces cerevisiae*. Purification, properties, and generation of polyclonal antibodies. *J. Biol. Chem.* **265**, 21997–2003.

Sone, T., Scott, R.A., Hughes, M.R., Malloy, P.J., Feldman, D. O'Malley, B.W. & Pike, J.W. (1989) Mutant vitamin D receptors which confer hereditary resistance to 1,25-dihydroxyvitamin D_3 in humans are transcriptionally inactive in vitro. *J. Biol. Chem.* **264**, 20230–4.

Stahle, U., Kloch, G. & Schutz, G. (1987) A 15 bp oligomer is sufficient to mediate both glucocorticoid and progesterone induction. *Proc. Natl. Acad. Sci. USA* **84**, 7871–5.

Suda, T., Shinki, T. & Takahashi, N. (1990) The role of vitamin D in bone and intestinal cell differentiation. *Annu. Rev. Nutr.* **10**, 195–211.

Takeda, E., Kuroda, Y., Saijo, T., Naito, E., Kobashi, H., Iwakuni, Y. & Miyao, M. (1987) 1-Hydroxyvitamin D_3 treatment of three patients with 1,25-dihydroxyvitamin D-receptor-defect rickets and alopecia. *Pediatrics* **80**, 97–101.

Umesono, K., Giguère, V., Glass, C.K., Rosenfeld, M.G. & Evans, R.M. (1988) Retinoic acid and thyroid hormone induce gene expression through a common responsive element. *Nature* **336**, 262–5.

Waterman, M., Adler, S. Nelson, C., Greene, G.L., Evans, R.M. & Rosenfeld, M.G. (1988) A single domain of the estrogen receptor confers DNA binding and transcriptional activation of the rat prolactin gene. *Molec. Endocrinol.* **2**, 14–21.

Willman, T. & Beato, M. (1986) Steroid-free glucocorticoid receptor binds specifically to mouse mammary tumour virus DNA. *Nature* **324**, 688–91.

Yamamoto, K.R. (1985) Steroid receptor regulated transcription of specific genes and gene networks. *Annu. Rev. Genet.* **19**, 209–52.

Yoon, K., Rutledge, S.J.C., Buenaga, R.F. & Rodan, G.A. (1988) Characterization of the rat osteocalcin gene: stimulation of promoter activity by 1,25-dihydroxyvitamin D_3. *Biochemistry* **27**, 8521–6.

4

Cobalamin Binding Proteins and their Receptors

B. Seetharam and D.H. Alpers

Introduction

Mammalian cobalamin binding proteins include those proteins that bind cobinamides. All molecules containing four reduced pyrrole rings linked together (called 'corrin' because they are at the core of the molecule) are called corrinoids. The prefix 'cob' is used when a cobalt atom is present, and such corrinoids are termed cobinamides. The cobinamides that show biological activity in microorganisms and/or mammals are called vitamin B_{12} and those that are active only in human and mammalian metabolism are called cobalamins. The two naturally occurring forms of cobalamin (cbl) in man are methyl-cobalamin (found mostly in the cytoplasm) and adenosylcobalamin (found in mitochondria). Cobinamides used by bacteria lack the ribosyl, aminopropanol, or nucleotide groups of cobalamin, or contain substitutions on these moieties (Hogenkamp, 1975).

The cobalamin binding proteins are three in number, each with a different membrane receptor that identifies the cell that subserves the function of the binding protein. *Intrinsic factor* (IF) refers to a protein secreted by the stomach which mediates the transepithelial absorption of cbl in the ileum (Allen, 1975). Haptocorrins comprise a group of immunologically identical proteins produced in a variety of body fluids (plasma, milk, amniotic fluid, saliva) from many cell types (granulocytes, mammary gland, salivary duct and acinar cells) (Nexo *et al.*, 1985; Grasbeck *et al.*, 1962). These proteins were formerly referred to as R protein (for rapid electrophoresis) or non-intrinsic factor (Stenman, 1975*a*). Specific haptocorrins are characterized by different glycosylation, and include transcobalamin I and III (TCI, TCIII) (Stenman, 1975*b*). Although these proteins account for the majority of the total cobalamin bound in serum, their function is unclear. *Transcobalamin II*

Table 4.1. Properties of cobalamin binding protein ligands

	Intrinsic factor	Haptocorrin	Transcobalamin II
Location	Gastric juice and accessory digestive organs	Plasma, granulocytes, secretions and accessory digestive organs	Plasma
Immunological specificity	Distinct	Distinct	Distinct
Amino acid composition	Distinct	Distinct	Distinct
Carbohydrate	15%	33–40%	0
Molecular mass (actual)	45–48 kDa	60–66 kDa	43 kDa
Subunit	1	1	1
B_{12} binding site	1	1	1
Spectral maximum for bound cbl	362	361 or 363	364

(TCII) is present in small amounts in plasma (0.5–1.5 μg l^{-1}), but accounts for most of the unsaturated cbl binding capacity in serum. TCII mediates the uptake of cbl into all somatic cells of the body.

IF, when complexed with cbl, is bound to the *IF–cbl receptor* localized to the ileum (Kouvonen & Grasbeck, 1979). Haptocorrin–cbl complexes are recognized by the *asialoglycoprotein receptor* found largely in the liver (Allen, 1975). TCII–cbl is bound to a *TCII receptor* found in all tissues (Seligman & Allen, 1978). The stoichiometry of all cbl binding to the specialized proteins is one binding site per molecule of protein, with affinity constants ranging from $10^{10}M^{-1}$ to $11^{10}M^{-1}$. The protein–cbl complexes also bind to their receptors with 1:1 stoichiometry, and the affinity of these binding reactions is of the same magnitude as for the ligand–binding protein interactions (Sennett *et al.*, 1981). The characteristics of the cbl binding proteins are outlined in Table 4.1.

Cobalamin binding proteins

Occurrence and distribution

IF and haptocorrin are produced primarily by tissues originating from foregut Anlage (Lee *et al.*, 1989). The primary source of IF is the gastric mucosa in most species, with the oxyntic (parietal) cell as the most common cell of origin (Allen, 1975). In the rat and mouse IF is produced in chief cells (Hoedemaeker *et al.*, 1964, 1966), while in the rat, at least, IF is found in both cell types. In the dog, extragastric locations of IF appear to be the major source of the luminal IF. The pancreas is an important source of an IF that closely resembles gastric IF (Batt & Horadagoda, 1989). This pancreatic IF promotes the absorption of cbl by receptor-mediated endocytosis in the dog (Batt & Horadagoda, 1989). Similarly, an active IF has been identified in pure human pancreatic juice obtained by endoscopic drainage (Carmel *et al.*, 1985).

Cells containing haptocorrin have been reported to be more widespread, in some cases including mucin-producing cells of the intestinal tract (Kudo *et al.*, 1987). Haptocorrin location was found by immunohistochemistry in another study to occur in parietal cells in the stomach of rat, human, and dog, but in mucous cells of the hog (Lee *et al.*, 1989). Accessory digestive organs, such as salivary glands and pancreas, contained haptocorrin (and IF) in the ductal cells, and in the rat, in the acini as well for haptocorrin only (Lee *et al.*, 1989). The haptocorrin in gastric juice comes presumably from both salivary glands and gastric mucosa. Leukocytes also produce haptocorrins (Simons & Weber, 1966), and mammary glands and amnion lining cells must also be sources to account for the presence of the protein in milk and amniotic fluid, although direct demonstration of the cell of origin in these tissues has not been made.

Unlike IF and haptocorrin, whose origin is restricted to certain cells, TCII has its origin in many cells. These cells include hepatocytes (Cooksley *et al.*, 1974; Savage & Green, 1976), heart, spleen, and kidney (Hall & Rappazzo, 1975), fibroblasts (Green & Eastwood, 1963), macrophages (Rachmilewitz *et al.*, 1977), and intestine (Chanarin *et al.*, 1978). It seems likely that TCII production will be found in nearly all cell types, but the low rate of production and insignificant storage has limited our ability to demonstrate TCII cytochemically in cells.

Cbl binding proteins are synthesized by the human fetus from at least 16–19 gestational weeks (Hansen *et al.*, 1989a). In the amniotic fluid IF, haptocorrin, and TCII were identical in size and pI with their adult counterparts. Another study confirmed the presence of all three binding proteins in amniotic fluid and showed that the percentage of cbl binding capacity due to each protein changed with gestational age (Gueant *et al.*, 1989). IF and TCII declined with age, while the alkaline isomer of haptocorrin increased. Moreover, this study found that amniotic fluid IF and haptocorrin had less acidic isoproteins compared with adult proteins (Gueant *et al.*, 1989). IF was localized in parietal cells in human fetuses from 11 weeks of gestation onwards (Aitchison & Brown, 1988). These cells were found mainly at the base of the developing gastric glands. Developmental regulation of rat IF was studied post-natally, a situation which mimics the *in utero* development of higher mammals. IF mRNA increased from 5 to 20 days after birth, owing both to increased numbers of chief cells making IF and to increased IF mRNA per cell (Dieckgraefe *et al.*, 1988*a*). Cortisol was found to regulate IF mRNA content, but thyroxine had no effect, unlike the situation in developing small intestine.

Cbl absorption in the elderly has been thought to be impaired, but recent studies show that such is not the case (Nilsson-Ehls *et al.*, 1986), suggesting that IF production is unchanged in most elderly humans. Likewise, the unsaturated cbl-binding capacity in plasma was unchanged in individuals from age 21 to 87, demonstrating that production of haptocorrin and TCII was not modified significantly with aging (Gimsing & Nexo, 1989). Although cell type and storage content has not been directly assessed in aged animals or humans, these data would indicate that such parameters do not change with increasing age.

Synthesis and secretion

Compared with the large amount of information concerning pepsinogen synthesis and secretion in the stomach, the data regarding IF are slight. Blocking vagal stimulation with atropine (Vatn *et al.*, 1975*b*) or vagotomy (Meikle *et al.*, 1977) decreases IF output. All substances that stimulate acid secretion also increase IF secretion, including histamine (Jeffries & Sleisenger, 1965), gastrin (Irvine, 1965), and methacholine (Vatn *et al.*, 1975*a*). The secretory rate is maximal in the first 20 min after stimulation and declines thereafter. This pattern was considered to be due to 'wash-out' of intracellular IF by the increased fluid movement accompanying acid secretion.

Although heterogeneity of cell types and loss of blood flow make cell and organ culture different from the intact stomach, such systems have extended the information obtained in whole animals or man. In stomach explants IF secretion was confirmed to be due to release of preformed protein, and the synthesis rate was found to be constant, regardless of the state of stimulation of secretion (Serfilippi & Donaldson, 1986). This conclusion was supported by the finding in intact rats that potent secretagogues did not alter the IF mRNA concentration (Dieckgraefe *et al.*, 1988*a*). Drugs that increase intracellular cAMP stimulate IF secretion (Kapadia *et al.*, 1979), but it is not clear whether this is the most important mediator. Recent studies have suggested that calcium-mediated mechanisms may also be important in regulating IF release (Ballantyne *et al.*, 1989). Inhibitors of IF secretion have been discovered. Prostaglandin E analogs block IF release potently in histamine-stimulated parietal cells (Zucker *et al.*, 1988). Moreover, somatostatin inhibits histamine-stimulated IF secretion from isolated guinea pig gastric glands (Oddsdotter *et al.*, 1987). Finally, inhibitors of histamine$_2$ receptors, such as cimetidine (Steinberg *et al.*, 1980) and ranitidine (Belaiche *et al.*, 1983), impair IF secretion, but only partly. Omeprazole, a substituted benzimidazole, is a potent inhibitor of the H^+/K^+ ATPase in man, but has no effect on IF output (Kitting *et al.*, 1985).

During acid secretion the tubulovesicular membranes present in the resting parietal cell proliferate, and long apical microvilli appear (Ito, 1981). IF is located in the tubulovesicles, ER, and multivesicular bodies of the resting parietal cell, and is transferred to the microvilli upon stimulation of secretion (Levine *et al.*, 1981). The regulation of secretion of IF differs in rodents in whom the major storage cell is the chief cell, not the parietal cell. In the rat, carbachol and dibutyryl cAMP, but not histamine or gastrin, stimulate IF secretion (Schepp *et al.*, 1983). It is assumed that IF is located in the granules of chief cells, but this has not been demonstrated conclusively.

In vivo in man some of these control mechanisms have received confirmation. When the stomach is bypassed for obesity surgery, IF is found in the mucosal parietal cells, but is absent from the gastric lumen, presumably owing to lack of vagal and hormonal stimulation (Marcuard *et al.*, 1989). Inhibitors of acid secretion such as cimetidine (Binder & Donaldson, 1978) and secretin (Vatn *et al.*, 1974) suppress IF secretion.

In contrast with the data for IF, the control of haptocorrin synthesis and secretion is poorly understood. The percentage of cbl binding in gastric juice of any species that is due to haptocorrin will depend upon the synthetic and secretory rate of the protein from salivary and gastric sources. It is assumed at present that the secretion of haptocorrin is the regulated step, as with IF, and that synthesis is a constitutive phenomenon, also as it is with IF. Haptocorrin is localized in granules within leukocytes, and release from these granules is inhibited by sulfhydryl reagents (Corcino *et al.*, 1970), and by sodium fluoride (Scott *et al.*, 1974), and was stimulated by lithium (Scott *et al.*, 1974) and the calcium ionophore A23187 (Simon *et al.*, 1976).

In the adult animal, haptocorrin in the intestinal lumen is degraded by proteases, largely pancreatic in origin (Marcoullis *et al.*, 1980). In the newborn animal, pancreatic secretion of enzymes is limited, and the haptocorrin from milk, saliva, gastric, or pancreatic sources survives intact. This haptocorrin may mediate uptake of cbl, either via non-specific ileal endocytosis, or by a specific mechanism present in the brush border membrane (Trugo *et al.*, 1985). Because haptocorrin from salivary glands and stomach is largely unsialylated, the asialoglycoprotein receptor (ASGP-R) was examined as a possible mediator for this uptake. The minor components of the rat ASGP-R (RHL 2/3) were found on the apical brush border of 14-day-old suckling animals, and low binding activity of haptocorrin–R complex was detected (Hu *et al.*, 1991). The binding capacity seems rather low to be of physiological significance for large amounts of cbl uptake and the receptor may have additional functions.

Studies using liver perfusion (Cooksley *et al.*, 1974; Tan & Hansen, 1968) and cell culture (Savage & Green, 1976) have shown that the liver is a source of TCII. Production from all cells used in culture suggests that the synthetic and/or secretion rate is constant (Hall & Rappazzo, 1975; Rachmilewitz *et al.*, 1977), a phenomenon akin to that seen with IF, and most likely also with haptocorrin. TCII accumulates in the culture medium of guinea pig ileal mucosal explants (Rothenberg *et al.*, 1978), suggesting also an intestinal origin for TCII. Such production would be consistent with the need for TCII to pick up cbl that has been translocated from the intestinal lumen into the enterocyte interior. Recent studies (Ramanujam *et al.*, 1991*b*) on cbl transcytosis using polarized human epithelial cells derived from human colon adenocarcinoma (Caco-2) have shown *de novo* synthesis and basolaterally directed secretion of TCII. However, observations of the direction of secretion of TCII in polarized cells derived from opossum kidney (Ramanujam *et al.*, 1991*a*) have shown that TCII is secreted in equal amounts in both directions (apical and basolateral). Secretion of TCII appears to be constitutive: secretion was independent of cbl entry into these cells. Various cell lines secrete and internalize varying amounts of TCII; the amount internalized is proportional to the amount secreted (Hall & Colligan, 1989). Thus, TCII produced locally by many cells may be used to mediate tissue uptake in all cells, and in some cells such as enterocytes and renal proximal tubule cells it may mediate its exit as well.

Structure and function

The major advance in understanding the structure of the cbl binding proteins occurred when Allen & Majerus prepared a monocarboxylic acid derivative of cbl and coupled it to Sepharose beads (Allen & Majerus, 1972*a*). The amide structure of cbl was regenerated during coupling, thus providing a native cbl structure on a solid-state matrix. Purified human IF was found to have a molecular mass of 45–47 kDa on equilibrium dialysis, and 59–66 kDa on gel filtration (Allen & Mehlman,

1973*a*). This difference was probably due to the 15% carbohydrate content. IF from hog gastric mucosa was 5–10 kDa larger than the human IF (Allen & Mehlman, 1973*b*), whereas the molecular mass of non-glycosylated rat IF deduced from a cDNA clone was 46 kDa (Dieckgraefe *et al.*, 1988*b*). The primary translation product of IF produced in a cell-free system displayed cbl binding activity, despite the lack of glycosylation (Dieckgraefe *et al.*, 1988*b*). Recent studies have shown that non-glycosylated IF expressed in COS-1 cells by transfection and tunicamycin treatment bind equally well to cbl and to the IF–cbl receptor when coupled with glycosylated IF (Gordon *et al.*, 1991). Based on the predicted sequence of the full length IF-cDNA and crystallographic data for malate dehydrogenase, another nucleotide binding protein, there was some similarity in certain crucial residues which were structural requirements for folding and binding to the nucleotide region. The cbl binding region appears, however, to require folding of the entire molecule, as transfection into COS-1 cells of IF constructs lacking 25% of the molecule produce a truncated IF without cbl binding activity (Gordon *et al.*, 1991).

Even though the entire IF molecule seems to be important for cbl binding, the structural element on IF important for receptor binding is not known. The existence of two domains on IF, one for cbl binding and the other for receptor binding, was recognized a long time ago using antisera to IF from patients with pernicious anemia (Ardeman & Chanarin, 1963). Recent work demonstrates that the IF–cbl receptor binding region is restricted to the amino-terminus of IG (Tang *et al.*, 1992).

The serum of many, if not all, pernicious anemia patients have two types of antibodies directed against IF: a) 'blocking' or 'type I' IgG antibody that blocks cbl binding to IF and b) 'binding' or type II IgG (or sometimes IgM) that blocks the binding of IF-cbl to the receptor. Nearly 70% of the pernicious anemia patients have the type I antibody. About half of these patients have also type II, which rarely exists alone. The existence of these domains on IF has recently been confirmed using monoclonal antibodies to pure human IF (Smolka & Donaldson, 1990).

One fourth of the human IF has been sequenced and is 67% identical with the rat sequence (Hansen *et al.*, 1989*b*). Human IF cDNA now has been cloned, and the deduced sequence is 80% identical with that of the rat (Hewitt *et al.*, 1991). Juvenile pernicious anemia describes a heterogenous group of disorders which might include lack of synthesis of IF, a block in IF secretion, production of an IF that does not bind cbl, or of an IF that is rapidly degraded (Levine & Allen, 1985; Yang *et al.*, 1985). Southern blotting of DNA obtained from three families with congenital IF deficiency and absent IF reveals an EcoR1 pattern identical with normal DNA (Hewitt *et al.*, 1991), suggesting either a limited mutation and creation of a rapidly degraded IF, or a mutation in a regulatory sequence for IF. The gene for IF has been localized to chromosome 11 (Hewitt *et al.*, 1991).

Haptocorrins have been purified from hog gastric mucosa (Allen & Mehlman, 1973*b*), human milk and saliva (Burger & Allen, 1974), human granulocytes (Allen & Majerus, 1972*b*), normal plasma (Burger *et al.*, 1975*a*), plasma from hepatoma patients (Burger *et al.*, 1975*c*), and amniotic fluid (Stenman, 1974). These proteins have been remarkably similar in amino acid and sugar content, molecular mass, and amino-terminal sequences. Thus, it is likely that these proteins are identical or very closely related, at least within the same species (Stenman, 1974; Jacob *et al.*, 1980). Human transcobalamin I, a major constituent of granules in neutrophils, has been sequenced by isolation of a cDNA clone (Johnson *et al.*, 1989). TCI and IF were found to have extensive areas of homology (not identity), which suggested regions potentially important for cbl binding. A cDNA clone encoding hog haptocorrin has also been isolated, and its structure is only 45% identical with human TCI (Hewitt *et al.*, 1990). Interestingly, the regions of homology between human TCI and rat IF were not the same as those between hog haptocorrin and human TCI. Thus, it does not appear that primary structure will provide clues to regions needed for cbl binding. Strikingly, the signal peptides of human TCI and hog haptocorrin show 80% identity, much higher than anywhere else in the molecules.

Both human IF and haptocorrin were treated sequentially with glycosidases and proteases. It was concluded that the carbohydrate core of IF protects the whole protein whereas the carbohydrate core of haptocorrin protects only half of the molecule (Gueant *et al.*, 1988), and that the carbohydrates were implicated in the cbl binding site on both molecules. Recent studies (Ramasamy *et al.*, 1990) using various lectins have shown that carbohydrate on IF was not important for the binding of IF and [^{57}Co]cbl. Moreover, glycosylation of IF is under developmental control, and the IF present in neonatal rat stomach bound [^{57}Co]cbl and the complex bound to the ileal receptor equally well as the adult IF that was differently glycosylated (Dieckgraefe *et al.*, 1988*a*). These studies have indicated that the carbohydrate on IF is not important for either cbl binding to IF or for IF–cbl binding the receptor. Furthermore, studies with non-glycosylated IF, produced in tunicamycin-treated cells transfected with the IF gene, show that carbohydrate is important for protease resistance, but not for cbl binding (Gordon *et al.*, 1991). This latter result is consistent with the earlier finding that loss of unsaturated haptocorrin

Table 4.2 *Cell surface receptors for cobalamin* (cbl)

Protein ligands	Cellular receptor expression	Estimated affinity for ligand	Function mediated	Binding specificity and requirements
Intrinsic factor	Intestine, kidney, placenta	0.25 nM (for IF–cbl)	Intestinal uptake of cbl	Only IF–cbl Ca^{2+}, neutral pH
Transcobalamin II	Many cells	0.56 nM (for TCII–cbl)	Delivery of cbl into cells	TCII bound mainly to cbl and a few analogs Ca^{2+}, neutral pH
Haptocorrin	Liver	0.1 nM	Clearance of cbl analogs via bile	cbl and many analogs bound to haptocorrin Ca^{2+}, neutral pH

in suckling pig intestine was larger than with binder saturated with cbl (Trugo & Newport, 1985).

TCII has been purified by three groups (Allen & Majerus, 1972c; Puutula & Grasbeck, 1972; Quadros *et al.*, 1986). Because it is present at very low levels in plasma, several hundred liters of human plasma were needed for its isolation. It has a molecular mass of 38–43 kDa and is not a glycoprotein. It binds 27 μg of cbl per milligram of protein, consistent with a mole : mole binding ratio (Allen & Majerus, 1972c). A 1.9 kb cDNA for human TCII has been recently isolated (Platica *et al.*, 1991). The cDNA predicted amino acid sequence of TCII and other cbl binders (Dieckgraefe *et al.*, 1988b; Johnson *et al.*, 1989) have revealed seven hydrophobic domains in which considerable homology exists. Although not proven, the involvement of one or more of these domains in cbl binding has been suggested (Platica *et al.*, 1991). A genetic polymorphism for TCII has been found in men, rabbits, and mice (Frater-Schroder, 1983). In man there are four polymorphic and several rare alleles. Skin fibroblasts and cultured bone marrow cells synthesize and secrete iso-types corresponding with those observed in the plasma. On complex formation of cbl with TCII the major bands shift to longer wavelengths, although the electron density around the cobalt atom is held fairly constant, in contrast with the observations for the same cbl isomers attached to human IF and haptocorrin (Nexo *et al.*, 1977). Some of the properties of the three classes of cbl binding proteins are summarized in Table 4.1.

Cellular uptake and utilization of cobalamins

Many higher animals including man have developed a complex, yet a very efficient mechanism to transport cbl into and out of cells and in between to store and utilize cbl for intracellular metabolic reactions. Even though cbl, a water-soluble molecule, can slowly diffuse across mammalian cell membranes, protein-mediated translocation of cbl appears to be faster and more efficient.

The average daily intake of cbl in Western countries is about 4–5 μg; however, the physiological daily requirement is only between 0.5 and 1 μg. The absorption and the transport mechanisms effectively handle this tiny amount and distinguish cbl from its non-physiologic analogs, which can enter the system orally or be produced by microorganisms. Cobalamin bound to IF, haptocorrin, or TCII is transported across cell membranes via receptors which are specific for each ligand. For IF and TCII the receptors are expressed in more than one tissue (Table 4.2). It is important to recognize at the outset that several important binding and transfer reactions occur before the eventual utilization of dietary cbl by cells. Much of the knowledge gained in this area has evolved from studies aimed at understanding the pathophysiology of cbl malabsorption in man. This condition arises from defective synthesis or structure of transport proteins or from congenital or acquired disorders in the transport or cellular utilization of cbl. A more detailed discussion of the pathophysiology of cbl malabsorption and the resulting hematological and neurological disorders can be found in other excellent reviews (Davis, 1985; Kapadia & Donaldson, 1985).

Intestinal transport of cbl

Dietary cobalamin is bound to protein from which it is released in the stomach by acid and pepsin. The affinity of IF for cbl is low at pH < 3. *In vivo* at pH 2.0, cbl binds exclusively to haptocorrin when presented with an equimolar mixture of IF and haptocorrin, since the association constant (K_a) is 50 times higher for haptocorrin than for IF (Grasbeck *et al.*, 1962). Cobalamin is released from haptocorrin following treatment with pancreatic proteases. The exact site of cleavage of the haptocorrin–cbl complex is not known, but the apparent M_r of 150 000 for haptocorrin–cbl is reduced to 75 000 with the liberation of cbl. The degraded haptocorrin has very little affinity for cbl.

These *in vitro* reactions have been shown to have physiological consequences *in vivo*. Allen *et al.* (1978a,b) first proposed that dietary cbl may first bind to salivary or gastric haptocorrin in the stomach, later to be released by pancreatic proteases. This concept is supported by several lines of evidence (Allen et al., 1978a,b). First, patients with pancreatic insufficiency malabsorb crystalline cbl, a condition which can be reversed by ingestion of pancreatic extract or trypsin (Allen *et al.*, 1978a). Second, luminal cbl in such untreated patients is exclusively bound to haptocorrin while in normal subjects cbl is bound to IF (Marcoullis *et al.*, 1980). Finally, when cobinamide, an analog of cbl, was first administered to saturate all cbl binding sites in haptocorrin, subsequent absorption of cbl in these patients improved (Allen *et al.*, 1978b). Pancreatic proteases have no effect on IF or IF–cbl attachment to the IF–cbl receptor, but some studies (Teo *et al.*, 1980) have suggested a role for luminal bile in the ileal absorption of cbl. Although the mechanism by which bile influences cbl transport is not clear, some studies (Seetharam *et al.*, 1983, 1991b) have suggested that bile salts are needed to optimize the binding of IF–cbl to the membrane receptor.

Analogs of cbl that enter the intestinal lumen are present in either the diet or the bile and are poorly absorbed. These analogs are tightly bound to haptocorrin and, even if released by pancreatic proteases, rarely gain entry to enterocytes, owing to poor binding of analogs by IF and to not being recognized by the IF–cbl receptor. There is some evidence (Shaw *et al.*, 1989) to suggest that the analogs of cbl may gain entry to the enterocytes via an IF-independent mechanism, thus accounting for the presence of analogs in serum and tissues. Further studies are needed to understand the mechanism by which the cbl analogs enter the circulation and the effect that structure of these analogs have on their entry into and exit out of the enterocytes because some analogs may enter, but do not exit (Allen, 1982).

Following the transfer of cbl from haptocorrin to IF the resulting IF–cbl complex attaches to a specific ileal receptor located in the microvillus pits (Levine *et al.*, 1984) of the apical brush border membrane (MacKenzie & Donaldson, 1972). The receptor has been purified to homogeneity from various mammalian intestines (Kouvonen & Grasbeck, 1979; Seetharam *et al.*, 1981*a*). It is a 220–230 kDa protein whose ligand binding site is exposed on the luminal side of brush border membrane (Seetharam *et al.*, 1982). The primary structure of the receptor or the ligand binding domain is not fully known. The receptor, however, contains two subunits (Kouvonen & Grasbeck, 1979; Seetharam *et al.*, 1981*a*) in all species tested except rat; in dog, hog or human, ligand binding can be demonstrated with receptor peptides of M_r between 75 000 and 80 000 (Seetharam *et al.*, 1982). In addition to ileal enterocytes, an IF–cbl receptor with high activity has been detected in mammalian kidney and has been purified from rat and canine kidney (Seetharam *et al.*, 1988; Fyfe *et al.*, 1991*b*). The rat receptor is unique and differs from canine and human receptor immunologically and structurally. It is synthesized as a single chain precursor of 215–220 kDa by an mRNA of 5–7 (Seetharam *et al.*, 1981*a*).

The canine kidney receptor, like the rat kidney receptor, is a glycoprotein of M_r 230 000. It has 5–7 N-linked oligosaccharide chains of the complex type (Fyfe *et al.*, 1991*b*). However, the kidney IF–cbl receptor purified using kidney from a family of dogs that exhibited selective intestinal malabsorption of cbl contained 5–7 N-linked oligosaccharide chains that were sensitive to Endoglycosidase H, and thus were of the high-mannose type (Seetharam *et al.*, 1991*a*). In these dogs, ileal receptor was localized only in the intracellular membranes (Fyfe *et al.*, 1991*a*) and immunoblotting of the brush border membrane revealed the absence of IF–cbl receptor in the apical membranes isolated from both the kidney and the ileal mucosa. These results have suggested that IF–cbl receptor expressed in both the kidney and the ileal mucosa are products of the same gene. Furthermore, the apical expression of IF–cbl receptor is important for IF-mediated cbl transcytosis. Although the details of post-translation processing of IF–cbl receptor either in the kidney or in the ileal mucosa are not known, it appears that processing of the oligosaccharide chains in the Golgi is important for the intracellular targeting of the receptor to the apical membrane. Even though the molecular basis for the noted defect of trafficking of the IF–cbl receptor in these dogs is not clear, the available evidence such as (i) identical M_r of 230 000; (ii) very subtle changes in amino acid composition; and (iii) increased sensitivity to pancreatic proteases suggest that the IF–cbl receptor might have undergone a subtle mutation in the coding region. The disease noted in this dog family resembles the human disease known as Immerslund–Grasbeck Syndrome (McKusick Catalogue Number 26110) (McKusick, 1992). Any conclusions regarding the nature of defect which results in selective cbl malabsorption in humans must await further study.

The physiological importance of the presence of the receptor in renal brush borders is not clear at the present time, as the ligand IF has never been shown to enter the circulation (Cooper & White, 1968). Minuscule amounts of IF, however, have been detected in the human urine (Wahlstedt & Grasbeck, 1985). An immunoreactive IF that binds to the IF–cbl receptor has also been detected in renal brush borders (Ramanujam et al., 1990). The presence of IF–cbl receptor in the brush border membrane of the kidney of many species (Seetharam et al., 1988) and the ability to transport [^{57}Co]cbl bound only to IF by polarized proximal tubule-derived cells (Ramanujam et al., 1991a) suggests that the renal IF–cbl receptor may have a role in the readsorption of cbl that is filtered during cbl overload. Some of the properties of various IF–cbl receptors are shown in Table 4.3.

IF–cbl receptor activity is known to be under developmental (Ramasamy et al., 1989) and hormonal regulation (Robertson & Gallagher, 1979). One remarkable feature of the IF–cbl receptor is its ability to bind more IF–cbl during pregnancy (Robertson & Gallagher, 1979). Physiologically it would make sense that more cbl is transported across the intestine as the demand for cbl increases during pregnancy, owing to the requirement for cbl by the fetus. The exact mechanism by which the receptor meets the increased demand is poorly understood, but it has been proposed to involve recruitment of intracellular receptors to the apical membrane (Robertson & Gallagher, 1983).

Once the IF–cbl is attached to the membrane, there is a delay of 3–4 h before cbl exits the enterocytes bound to the plasma transporter, TCII. Even though not proven conclusively, available evidence suggests that

Table 4.3. *Comparison of the properties of mammalian receptor for intrinsic factor–cobalamin*

Source	Intestine		Kidney	
	Canine[a]	Porcine[b]	Human[b]	Rat[c]
Specific activity, IF–cbl bound (nmol mg^{-1} protein)	2.5	0.24	ND[d]	4.3
Molecular mass (SDS)	220 kDa	200 kDa	230 kDa	230 kDa
Subunits	62 kDa, 48 kDa	70 kDa, 130 kDA	90 kDa, 130 kDa	None
Carbohydrate	Trace	+	+	+
Reaction with anti-IF	−	+	+	−
Reaction with anti-canine receptor	+	+	+	−
Binding to:				
Human IF–cbl	+	ND	ND	−
Canine IF–cbl	+	ND	ND	−
Rat IF–cbl	+	ND	ND	+
Hog IF–cbl	+	+	+	−
Abnormal IF–cbl	−	ND	ND	ND
Free IF	−	ND	ND	+
Free cbl	−	ND	ND	−

[d]ND, not determined.
Sources of data: [a]Kapadia & Donaldson (1985); [b]Kouvonen & Grasbeck (1979) and Davis (1985); [c]Teo *et al.* (1980).

IF–cbl bound to the receptor is internalized (Kapadia *et al.*, 1983; Seetharam *et al.*, 1985) via receptor-mediated endocytosis (Robertson & Gallagher, 1985a). The intracellular release of cbl can be blocked by chloroquine (Ramasamy *et al.*, 1989; Robertson & Gallagher, 1985b) but not by leupeptin, an inhibitor of lysosomal cathepsins. These latter observations have argued for involvement of an acidic vesicle in the release of cbl inside the cell. Because lysosomal degradation of IF is not needed for the release of cbl from IF or the formation of TCII–cbl complex, the vesicles involved could be prelysosomal acidic vesicles (endosomes?). It is not known where and how the transfer of cbl to TCII occurs and how TCII–cbl complex exits from the enterocyte. The existence of regulatory mechanisms for the exit of cbl may exist, as the amount of cbl transported to the other tissues is not always proportional to the amount of cbl that enters the enterocytes (Ramasamy *et al.*, 1989). Irrespective of how cbl gains entry to the enterocytes (either free or bound to IF), its exit occurs always bound to TCII (Ramanujam *et al.*, 1991b; Ramasamy *et al.*, 1989). Within the enterocytes cbl is never free and is bound either to IF or to TCII (Ramanujam *et al.*, 1991b).

Intracellular distribution of [^{57}Co]cbl during cellular transcytosis has shown that even after 8 h nearly 16% of the internalized [57Co]cbl was still within the cell, equally distributed between IF and TCII (Ramanujam *et al.*, 1991*b*).

Other aspects of ileal transport of cbl which are poorly understood include the cause for the 3–4 h delay in the appearance of [^{57}Co]cbl in the circulation. Subcellular fractionation studies using animal models have identified [^{57}Co]cbl in the lysosomal fraction (Horadagoda & Batt, 1985) and in mitochondria (Peters & Hoffbrand, 1970). It is possible that the delay could simply be due to the time required for cbl to enter and exit these organelles, or in other organelles in these fractions only enriched for each of the marker organelles. The transfer of cbl from IF to TCII must occur prior to its exit. The $T_{1/2}$ for this transfer is about 4 h in human epithelial cells (Ramanujam *et al.*, 1991*b*), although transcytosis of cbl bound to IF through renal polarized cells is relatively faster (Ramanujam *et al.*, 1991*a*). Despite these studies it is not known where in the cell this transfer occurs and whether receptor and/or IF is degraded intracellularly. Thus, it is obvious from studies carried out so far that the transcytosis of cbl through polarized epithelial cells is complex and needs further investigation.

Plasma transport of cbl

Transcobalamin II represents the bulk of the unsaturated cbl binding activity (1–1.5 ng ml^{-1}) of human plasma. TCII delivers absorbed cbl to cells that need them and does so via the TCII receptor located on many cells. The cell surface TCII receptor activity appears to be dependent on the cell cycle, being highest during the active dividing phase of the cell (Hall *et al.*, 1987). Furthermore, in some cells, such as Leukemia L1210 cells, the TCII receptor is segregated and expressed in individual microvilli that fail to express receptors for transferrin (Kishimoto *et al.*, 1987). It is clear from these studies that very little information is available on the factors which regulate the synthesis, vectorial migration, and plasma membrane expression of the TCII receptor. Receptor for TCII–cbl has been purified to homogeneity from human placenta (Seligman & Allen, 1978). The placental receptor contains 33% carbohydrate and has an apparent M_r around 55 000, which is considerably less than the M_r of the IF–cbl receptor. One important feature of the TCII receptor is the remarkable similarity of its amino acid composition to that of TCII, which is not a glycoprotein. On a mole : mole basis, both TCII and its receptor have nearly identical amino acid composition, leading to the speculation that TCII and its receptor have evolved from a common structural gene, the former being non-glycosylated and secreted while the latter is glycosylated and anchored in the plasma membrane of cells (Seligman & Allen, 1978). This hypothesis, though proposed more than ten years ago, has not been proved.

Following binding, TCII–cbl complex is internalized via absorptive endocytosis (Youngdhal-Turner *et al.*, 1979). TCII is degraded in the lysosomes and cbl exits the lysosomes to be converted to adenosyl-cbl in the mitochondria or to methyl-cbl in the cytosol. Even though the lysosomal degradation of TCII–cbl has been demonstrated quite convincingly (Youngdhal-Turner *et al.*, 1979), the details of internalization, such as internalization in coated pits, and the fate of TCII receptor is not known. Further morphological and biochemical studies are warranted to understand fully this important cbl uptake system. Furthermore, how cbl exits from acid vesicles is not known, but a specific carrier may exist. Recent studies (Idriss & Jonas, 1991) have found evidence for the existence of a specific transport system for cbl in rat liver lysosomal membranes. The uptake of [^{57}Co]cbl was pH- and Mg^{2+}-dependent, saturable with a K_m of 3.5 μM, and was dependent on either a proton gradient or proton movement across the lysosomal membranes. Failure to exit lysosomes or acid vesicles and accumulation of free cbl within these structures has been noted in a patient who, as a consequence, developed cbl deficiency (Rosenblatt *et al.*, 1985). Whether this defect is related to the synthesis of a defective carrier needs further investigation. Given the fact that the sole function of this uptake system is to deliver cbl to tissues to make coenzyme forms of cbl (Kolhouse & Allen, 1977) and that TCII receptor is not subjected to metabolic regulation by cbl or TCII concentration (Youngdhal-Turner *et al.*, 1979) it is possible that TCII-mediated cbl uptake is constitutive, and not a regulated process.

Transport of cbl bound to haptocorrin

Within the intestinal lumen, cbl bound to haptocorrin is not transported across the gut of an adult. However, when the IF–cbl receptor activity is low, as in infancy, cbl bound to haptocorrin in milk or in bile is taken up by non-specific pinocytosis (Boass & Wilson, 1963) or by a receptor for haptocorrin–cbl which is expressed only by suckling animals (Trugo *et al.*, 1985; Hu *et al.*, 1991). It is not known whether this physiological haptocorrin–cbl receptor is similar to the asialoglycoprotein receptor which is known to bind and internalize cbl bound to plasma haptocorrins (TCI and TCIII), but components of the asialoglycoprotein receptor are found in suckling rat intestine (Hu *et al.*, 1991).

In addition to the delivery of absorbed cbl by TCII to the tissues, cbl can also be recruited from the body stores to the tissues. Hepatocytes have a receptor for asialoglycoproteins which binds galactose-terminated glycoproteins and clears these glycoproteins from plasma (Ashwell & Morell, 1974). This receptor binds primarily TCIII–cbl, and binds TCI–cbl after TCI is deglycosylated. Following binding the complexes are internalized rapidly ($T_{1/2}$ = 3 min) via pinocytosis and cbl is released intracellularly after lysosomal degradation of the haptocorrins (Burger et al., 1975b).

The physiological significance of this uptake system, though a major one, is not clear. It was suggested by Allen (1982) that this system might exist to clear harmful analogs of cbl, which bind tightly to these haptocorrins, and clear them rapidly via the bile. These analogs are not reabsorbed by the intestine and are excreted in the feces. This mechanism may have some importance, because patients with congenital lack of haptocorrins develop neurological disorders, which may be related to the incorporation of the cbl analogs into cbl coenzymes. By this hepatic clearance system 3–9 µg of cbl is excreted in human bile per day. This amount of cbl is easily handled since granulocytes produce enough TCIII to bind 150 µg of cbl per day.

Uptake of free cbl

Even though under normal physiological conditions cbl is translocated across membranes always bound to transport proteins, there is evidence to suggest that there is a place for transport of free cbl across membranes. Patients with congenital TCII deficiency improve remarkably when large doses of cbl are administered parenterally (Hall et al., 1979). Furthermore, free cbl has been shown to be taken up by many human cells, although the amount internalized was only a fraction of that taken up when bound to TCII (Hall & Colligan, 1989). There is some evidence to suggest that uptake of free cbl by human fibroblasts (Berliner & Rosenberg, 1981) is due to a protein-mediated, facilitated diffusion system. The physiological importance and the nature of that carrier remain to be clarified.

It is clear from the above discussion that cbl can be transported across cellular plasma and intracellular organelle membranes. Some of the binding proteins and their receptors have been isolated and their roles studied. Much more molecular, cellular and morphological studies are needed to understand the complexities involved in the cellular sorting of cbl in health and disease. An overall schematic diagram of cbl transport into and utilization by cells is shown in Figure 4.1.

Figure 4.1 Schematic representation of transport and cellular utilization of dietary cbl. (*a*) Release of cbl from food proteins, formation of IF–cbl in the lumen and attachment of IF–cbl to the ileal receptor. (*b*) Ileal transport of cbl. (*c*) Cellular utilization of cbl internalized bound to TCII (Tc II) or asialo R protein or via a carrier protein. Pathways which are incompletely understood in (*b*) and (*c*) are represented by dashed lines. Numbers (1)–(13) represent the sites at which inherited disorders (refer to Table 4.5) are localized. The receptors involved are for IF–cbl (circles), TCII–cbl (squares) and asialo R proteins (haptocorrins) (diamonds).

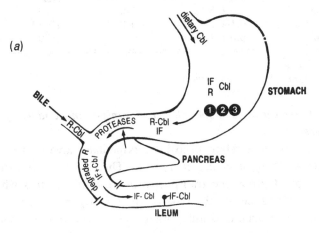

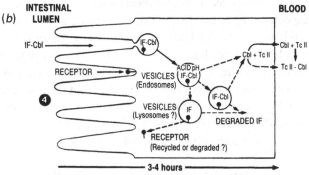

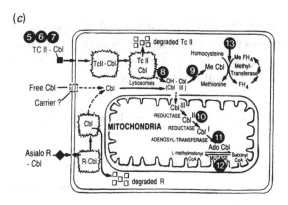

Table 4.4. *Endogenous cbl uptake by mammalian cells*

cbl transporter	Endogenous cbl presented (total, μg)	$T_{1/2}$ for cbl clearance (h)	cbl cleared (μg per 24 h)
Intrinsic factor	8–10	2–3	ND[a]
Transcobalamin I	0.6–0.8	240	<0.1
Transcobalamin II	0.1–0.2	0.1	15–30
Transcobalamin III	0.0–0.1	0.5	0–30

[a]ND, not determined.
Source: Data adapted from Allen (1982).

Quantitative aspects of cbl uptake and transport into and out of tissues

The three classes of human cbl binders mediate movement across membranes of 60–65 μg of cbl in a 24 h period. The $T_{1/2}$ for the transport across cells depends upon the transport protein. Other factors which could influence clearance are the type of vitamin B_{12} transported and interactions of cbl within intracellular organelles. The maximum amount of cbl transported by IF across the gut following a single oral dose is about 1–1.5 μg, irrespective of the size of the dose presented to the intestinal lumen. It is difficult to predict how much cbl is cleared through the enterocytes in 24 h, as not much is known about the turnover of the ileal receptor. However, it is safe to assume that transport across the gut will be just enough to balance the daily loss of cbl from the total body stores of about 4 mg. The daily turnover of cbl is about 0.03% of the body stores, that is about 1.5 μg (Seetharam & Alpers, 1982). A quantitative aspect of cbl transport across cells is shown in Table 4.4.

Inherited disorders of cobalamin transport and cellular metabolism

Cobalamin from the diet is absorbed and transported to the tissues for its utilization as coenzymes. The two forms of cbl that function as cofactors for the enzymatic reactions are methyl-cbl and adenosyl-cbl. These two forms mediate the following reactions.

$$\text{Homocysteine} + CH_3\,\text{cbl} \xrightarrow{\text{Methionine synthetase}} \text{Methionine} + \text{cbl.}$$

$$\text{Methylmalonyl-coenzyme A} + \text{Adenosyl cbl} \xrightarrow{\text{Mutase}}$$
$$\text{succinyl CoA} + \text{cbl.}$$

Table 4.5. *Inherited disorders of cbl transport and intracellular metabolism*

Site[a]	Mutant class	Nature of the defect
(1)	—	Congenital absence of IF
(2)	—	Abnormal IF, unable to bind receptor
(3)	—	Abnormal IF, sensitive to acid/pepsin
(4)	—	Defective ileal transport
(5)	—	Congenital lack of functional YCII
(6)	—	Abnormal TCII, does not bind cbl
(7)	—	Abnormal TCII, does not promote tissue uptake
(8)	cbl F	Failure to exit lysosomes
(9)	cbl C, cbl D	Cytosolic cbl metabolism
(10)	cbl A	Mitochondrial cbl reduction
(11)	cbl B	cob^{1+} alamin: ATP adenosyltransferase
(12)	Mut^0 Mut^-	Lack of apomutase
(13)	cbl E, cbl G	Methyltransferase: associated cbl utilization

[a]For locations of sites, see Figure 4.1.

The major intracellular cbl binding proteins are methionine synthetase and methylmalonyl CoA mutase (Kolhouse & Allen, 1977). Defective synthesis of either the cofactors or the apoenzyme has been implicated in the hematological or neurological disorders that are noted with congenital cbl deficiency (Fenton & Rosenberg, 1978). To a large extent these disorders are autosomal recessive traits and involve defective transport proteins or defective metabolism in the cell. Signs of these defects occur early in life and should be diagnosed promptly so that the patient be treated with cbl at an early stage. Failure to diagnose early can lead to severe mental retardation and in some cases even death. Some well documented inherited defects in cobalamin transport and metabolism are summarized in Table 4.5. A detailed discussion of these defects can be found in several review articles (Cooper & Rosenblatt, 1987; Fenton & Rosenberg, 1989; Matthews & Linell, 1982; Kano *et al.*, 1985). Defects at many sites (1–7) involve abnormalities in structure or synthesis of cbl-binding proteins or their receptors. These disorders, although rare, will provide excellent models for discovering some of the missing information in the transcellular movement of cbl.

The authors' work in this chapter was supported by NIH grants DK-26638 and DK-33487.

References

Aitchison, M. & Brown, I.L. (1988) Intrinsic factor in the human fetal stomach. An immunocytochemical study. *J. Anat.* **160**, 211–17.

Allen, R.H. (1982) Cobalamin absorption and malabsorption. In: *Viewpoints on digestive disease*, vol. 14, pp. 17–20. Thorofare, NJ: American Gastroenterological Association.

Allen, R.H., (1975) Human vitamin B12 binding proteins. *Prog. Hematol.* **9**, 57–84.

Allen, R.H. & Majerus, P.W. (1972a) Isolation of vitamin B12 binding proteins using affinity chromatography. I. Preparation and properties of vitamin B12-Sepharose. *J. Biol. Chem.* **247**, 7695–701.

Allen, R.H. & Majerus, P.W. (1972b) Isolation of vitamin B12 binding proteins using affinity chromatography. II. Purification and properties of a human granulocyte vitamin B12 binding protein. *J. Biol. Chem.* **247**, 7702–8.

Allen, R.H. & Majerus, P.W. (1972c) Isolation of vitamin B12 binding proteins using affinity chromatography. III. Purification and properties of human plasma transcobalamin II. *J. Biol. Chem.* **247**, 7709–17.

Allen, R.H. & Mehlman, C.S. (1973a) Isolation of gastric vitamin B12 binding proteins using affinity chromatography. I. Purification and properties of human intrinsic factor. *J. Biol. Chem.* **248**, 3660–9.

Allen, R.H. & Mehlman, C.S. (1973b) Isolation of gastric vitamin B12 binding proteins using affinity chromatography. II. Purification and properties hog intrinsic factor and hog non-intrinsic factor. *J. Biol. Chem.* **248**, 3670–80.

Allen, R.H., Seetharam, B., Allen, N.C., Podell, E. & Alpers, D.H. (1978a) Correction of cobalamin malabsorption in pancreatic insufficiency with a cobalamin analogue that binds with high affinity to a protein but not to intrinsic factor. *J. Clin. Invest.* **61**, 1628–34.

Allen, R.H., Seetharam, B., Podell, E. & Alpers, D.H. (1978b) Effect of proteolytic enzymes in the binding of cobalamin to R protein and intrinsic factor. *J. Clin. Invest.* **61**, 47–54.

Ardeman, S. & Chanarin, I. (1963) A method for the assay of human gastric intrinsic factor for the detection and titration of antibodies against intrinsic factor. *Lancet* **ii**, 1350–4.

Ashwell, G. & Morell, A.G. (1974) The role of surface carbohydrates in the hepatic recognition and transport of circulating glycoproteins. *Adv. Enzymol.* **41**, 99–128.

Ballantyne, G.H., Zdon, M.J., Zucker, K.A. & Modlin, I.M. (1989) cAMP and calcium-mediated modulation of intrinsic factor secretion in isolated rabbit gastric glands. *J. Surg. Res.* **46**, 241–5.

Batt, R.M. & Horadagoda, N.U. (1989) Gastric and pancreatic intrinsic factor-mediated absorption of cobalamin in the dog. *Am. J. Physiol.* **257**, G344–9.

Batt, R.M., Horadagoda, N.U., McLean, L., Morton, D.B. & Simpson, K.W. (1989) Identification and characterization of a pancreatic intrinsic factor in the dog. *Am. J. Physiol.* **256**, G517–23.

Belaiche, J., Zittoun, J., Marquet, J., Nurit, Y. & Yvart, J. (1983) Effect of ranitidine on secretion of gastric intrinsic factor and absorption of vitamin B12. *Gastroenterol. Clin. Biol.* **7**, 381–4.

Berliner, N. & Rosenberg, L.E. (1981) Uptake and metabolism of free cyanocobalamin by cultured fibroblasts from control and a patient with transcobalamin II deficiency. *Metabolism* **30**, 230–6.

Binder, H.J. & Donaldson, R.M. (1978) Effect of cimetidine on intrinsic factor and pepsin secretion in man. *Gastroenterology* **74**, 371–5.

Boass, A. & Wilson, T.H. (1963) Development of transport mechanisms for intestinal absorption of vitamin B12 in growing rats. *Am. J. Physiol.* **204**, 101–4.

Burger, R.L. & Allen, R.H. (1974) Characterization of vitamin B12 binding proteins isolated from human milk and saliva by affinity chromatography. *J. Biol. Chem.* **249**, 7220–7.

Burger, R.L., Mehlman, C.S. & Allen, R.H. (1975a) Human plasma R-type vitamin B12 binding proteins. I. Isolation and characterization of transcobalamin I, transcobalamin III, and the normal granulocyte vitamin B12 binding protein. *J. Biol. Chem.* **250**, 7700–6.

Burger, R.L., Schnider, R.J., Mehlman, C.S. & Allen, R.H. (1975b) Human plasma R-type vitamin B12 binding proteins. Role of TCI and TCII and normal granulocyte vitamin B12 binding proteins in plasma transport of vitamin B12. *J. Biol. Chem.* **250**, 7707–13.

Burger, R.L., Waxman, L.S., Gilbert, H.S., Mehlman, C.S. & Allen, R.H. (1975c) Isolation and characterization of a novel vitamin B12 binding protein associated with hepatocellular carcinoma. *J. Clin. Invest.* **56**, 1262–70.

Carmel, R., Hollander, D., Gergely, A.M., Renner, I.G. & Abramson, S.B. (1985) Pure human pancreatic juice directly enhances uptake of cobalamin by guinea pig ileum in vivo. *Proc. Soc. Exp. Biol. Med.* **178**, 143–50.

Chanarin, I., Muir, M., Hughes, A. & Hoffbrand, A.V. (1978) Evidence for intestinal origin of transcobalamin II during vitamin B12 absorption. *Br. Med. J.* **1**, 1453–5.

Cooksley, W.G.E., England, J.M., Louis, L., Down, M.C. & Tavill, A.S. (1974). Hepatic vitamin B12 release and transcobalamin II synthesis in the rat. *Clin. Sci. Mol. Med.* **47**, 531–45.

Cooper, B.A. & Rosenblatt, D.S. (1987) Inherited defects of vitamin B12 metabolism. *Ann. Rev. Nutr.* **7**, 291–320.

Cooper, B.A. & White, J.J. (1968) Absence of intrinsic factor from human portal plasma during ^{57}Co-B12 absorption in man. *Br. J. Haematol.* **14**, 73–8.

Corcino, J., Krauss, S., Waxman, S. & Herbert, V. (1970) Release of vitamin B12-binding protein by human leucocytes in vitro. *J. Clin. Invest.* **49**, 2250–5.

Davis, R.E. (1985) Clinical chemistry of vitamin B12. In: *Advances in Clinical Chemistry* (ed. H.E. Spiegel), vol. 24, pp.163–216. New York: Academic Press.

Dieckgraefe, B.K., Seetharam, B. & Alpers, D.H. (1988a) Developmental regulation of rat intrinsic factor mRNA. *Am. J. Physiol.* **254**, 913–19.

Dieckgraefe, B.K., Seetharam, B., Banaszak, L., Leykam, J.F. & Alpers, D.H. (1988b) Isolation and structural characterization of a cDNA clone encoding rat gastric intrinsic factor. *Proc. Natl. Acad. Sci. USA* **85**, 46–50.

Fenton, W.A. & Rosenberg, L.E. (1978) Genetic and biochemical analysis of human cobalamin mutants in cell culture. *Ann. Rev. Genet.* **12**, 223–48.

Fenton, W.A. & Rosenberg, L.E. (1989) Inherited disorders of cobalamin transport and metabolism. In: *Metabolic Basis of Inherited Disease,* vol. 2 (ed. C.R. Scriver, A.L. Beaudet, W.S. Sly & D. Valley), pp. 2065–82. New York: McGraw Hill.

Frater-Schroder, M. (1983) Genetic patterns of transcobalamin II and the relationships with congenital defects. *Molec. Cell. Biochem.* **56**, 5–31.

Fyfe, J.C., Giger, U., Hall, C.A., Jezyk, P.F., Klumpp, S.A., Levine, J.S. & Patterson, D.F. (1991a) Inherited selective intestinal cobalamin malabsorption and cobalamin deficiency in dogs. *Pediat. Res.* **29**, 24–31.

Fyfe, J.C., Ramanujam, K.S., Ramaswamy, K., Patterson, D.F. & Seetharam, B. (1991b) Defective brush border expression of intrinsic factor-cobalamin receptor in canine inherited intestinal cobalamin malabsorption. *J. Biol. Chem.* **266**, 4489–94.

Gimsing, P. & Nexo, E. (1989) Cobalamin-binding capacity of haptocorrin and transcobalamin: age-correlated reference intervals and values from patients. *Clin. Chem.* **35**, 1447–51.

Gordon, M., Hu, C., Chokshi, H., Hewitt, J.E. & Alpers, D.H. (1991) Glycosylation is not required for ligand or receptor binding by expressed rat intrinsic factor. *Am. J. Physiol.* **260** (*Gastrointest. & Liver Physiol.* 23), G736–42.

Grasbeck, R., Simons, K. & Sinkkonen, I. (1962) Purification of intrinsic factor and vitamin B12 binders from human gastric juice. *Ann. Med. Exp. Biol. Fenn.* **40** (suppl. 6), 1–24.

Green, C.D. & Eastwood, D.W. (1963) Effects of nitrous oxide inhalation on hemopoiesis in rats. *Anesthesiology* **24**, 341–5.

Gueant, J.L., Hambaba, L., Vidailhet, M., Schaefer, C., Wahlstedt, V. & Nicholas, J.P. (1989) Concentration and physicochemical characterization of unsaturated cobalamin binding proteins in amniotic fluid. *Clin. Chem. Acta* **181**, 151–61.

Gueant, J.L., Monin, B., Djalali, M., Wahlstedt, V., Bois, F. & Nicholas, J.P. (1988) Effect of glycosidases and proteinases on cobalamin binding and physiochemical properties of purified saturated haptocorrin and intrinsic factor. *Biochim: Biophys. Acta* **957**, 390–5.

Hall, C.A. & Colligan, P.D. (1989) The function of cellular transcobalamin II in cultured human cells. *Exp. Cell Res.* **183**, 159–67.

Hall, C.A., Colligan, P.D. & Begley, J.D. (1987) Cyclic activity of the receptors of cobalamin bound to transcobalamin II. *J. Cell Physiol.* **133**, 187–91.

Hall, C.A., Hitzig, W.H., Green, P.D. & Begley, J.A. (1979) Transport of therapeutic cobalamin in congenital deficiency of transcobalamin II (TCII). *Blood* **53**, 251–63.

Hall, C.A. & Rappazzo, M.E. (1975) Release of transcobalamin II by canine organs. *Proc. Soc. Exp. Biol. Med.* **148**, 1202–5.

Hansen, M., Hogdell, D.K. & Kryger-Baggasen, N. (1989a) Cobalamin binding proteins in the human fetus. *Scand. J. Clin. Lab. Invest.* (suppl.) **194**, 23–6.

Hansen, M.R., Nexo, E., Svendsen, I., Bucher, D., and Olesen, H. (1989b) Human intrinsic factor. Its primary structure compared to the primary structure of rat intrinsic factor. *Scand. J. Clin. Lab. Invest.* (suppl.) **194**, 19–22.

Hewitt, J.E., Gordon, M., Taggart, R.T., Mohandas, T.K. & Alpers, D.H. (1991) Human gastric intrinsic factor: Characterization of cDNA and genomic clones and localization to human chromosome 11. *Genomics* **10**, 432–40.

Hewitt, J.E., Seetharam, B., Leykam, J. & Alpers, D.H. (1990) Isolation and characterization of a cDNA encoding porcine gastric R protein. *Eur. J. Biochem.* **189**, 125–30.

Hoedemaeker, P.J., Abels, J., Watchers, J.J., Arends, A. & Nieweg, H.O. (1964) Investigations about the site of production of Castle's gastric intrinsic factor. *Lab. Invest.* **13**, 1394–9.

Hoedemaeker, P.J., Abels, J., Watchters, J.J., Arends, A. & Nieweg, H.O. (1966) Further investigations about the site of production of Castle's intrinsic factor. *Lab. Invest.* **15**, 1163–73.

Hogenkamp, H.P.C. (1975) The chemistry of cobalamins and related compounds. In: *Cobalamin: Biochemistry and Pathophysiology* (ed. B.M. Babior), pp. 21–73. New York: Wiley.

Horadagoda, N.U. & Batt, R.M. (1985) Lysosomal localization of cobalamin during absorption by the ileum of the guinea pig. *Biochim. Biophys. Acta* **838**, 206–10.

Hu, C., Lee, E.Y., Hewitt, J.E., Baenziger, J.U., DeSchryver-Kecskemeti, K. & Alpers, D.H. (1991) The minor components of the rat asialoglycoprotein receptor are apically located in neonatal enterocytes. *Gastroenterology* **101**, 1477–87.

Idriss, J.M. & Jonas, A.J. (1991) Vitamin B12 transport by rat liver lysosomal membrane vesicles. *J. Biol. Chem.* **266**, 9438–41.

Irvine, W.J. (1965) Effect of gastrin I and II on secretion of intrinsic factor. *Lancet* **i**, 736–7.

Ito, S. (1981) Functional gastric morphology. In: *Physiology of Gastrointestinal Tract* (ed. L.R. Johnson), p. 517. New York: Raven Press.

Jacob, E., Baker, S.J. & Herbert, V. (1980) Vitamin B12 binding proteins. *Physiol. Rev.* **60**, 918–60.

Jeffries, G.H. & Sleisenger, M.H. (1965) The pharmacology of intrinsic factor secretion in man. *Gastroenterology* **48**, 444–8.

Johnson, J., Bollekens, J., Allen, R.H. & Berliner, N. (1989) Structure of the cDNA encoding transcobalamin I, a neutrophil granule protein. *J. Biol. Chem.* **25**, 15754–7.

Kano, Y., Sakamoto, S., Miura, Y. & Takaku, F. (1985) Disorders of cobalamin malabsorption. *CRC Crit. Rev. Oncology/Hematology* **3**, 1–34.

Kapadia, C.R. & Donaldson, R.M. (1985) Disorders of cobalamin absorption and transport. *Annu. Rev. Med.* **36**, 98–110.

Kapadia, C.R., Schafer, D.E., Donaldson, R.M. & Ebersole, E.R. (1979) Evidence for involvement of cyclic nucleotides in intrinsic factor secretion by isolated rabbit gastric mucosa. *J. Clin. Invest* **64**, 1044–9.

Kapadia, C.R., Serfilippi, D., Voloshin, K. & Donaldson, R.M. (1983) Intrinsic factor mediated absorption of cobalamin by guinea pig cells. *J. Clin. Invest.* **71**, 440–8.

Kishimoto, T., Tavassoli, M., Green, R. & Jacobson, D.W. (1987) Receptors for transferrin and transcobalamin II display segregated distribution on microvilli of leukemia L1210 cells. *Biochem. Biophys. Res. Commun.* **146**, 1102–8.

Kitting, E., Aadland, E. & Schjonsby, H. (1985) Effect of omeprazole on the secretion of intrinsic factor gastric acid and pepsin in man. *Gut* **26**, 594–8.

Kolhouse, J.F. & Allen, R.H. (1977) Recognition of two intracellular cobalamin binding proteins and their identification as methyl malonyl CoA mutase and methionine synthetase. *Proc. Natl. Acad. Sci. USA* **74**, 921–5.

Kouvonen, I. & Grasbeck, R. (1979) A simplified technique to isolate the porcine and human ileal intrinsic factor and studies on their subunit structures. *Biochem. Biophys. Res. Commn.* **86**, 358–64.

Kudo, H., Inada, M., Ohshio, G., Wakatsuki, Y., Ogawa, K., Nawashima, Y. & Miyake, T. (1987) Immunohistochemical localization of vitamin B12 R-binder in the human digestive tract. *Gut* **28**, 339–45.

Lee, E.Y., Seetharam, B., Alpers, D.H. & DeSchryver-Kecskemeti, K. (1989) Cobalamin binding proteins (IF and R): an immunohistochemical survey. *Gastroenterology* **97**, 1171–80.

Levine, J.S. & Allen, R.H. (1985) Intrinsic factor within parietal cells of patients with juvenile pernicious anemia. *Gastroenterology* **88**, 1132–6.

Levine, J.S., Allen, R.H., Alpers, D.H. & Seetharam, B. (1984) Immunocytochemical localization of the intrinsic factor-cobalamin receptor in canine ileum: distribution of intracellular receptor during cell maturation. *J. Cell Biol.* **98**, 1111–18.

Levine, J.S., Nakane, P.K. & Allen, R.H. (1981) Human intrinsic factor secretion: Immunocytochemical demonstration of membrane associated vesicular transport in parietal cells. *J. Cell Biol.* **90**, 644–55.

MacKenzie, I.L. & Donaldson, R.M. (1972) Effect of divalent cations and pH on intrinsic factor-mediated attachment of vitamin B12 to intestinal microvillus membranes. *J. Clin. Invest.* **51**, 2464–71.

Marcoullis, G., Parmentier, Y., Nicholas, J.P., Jimenez, M. & Gerard, P. (1980) Cobalamin malabsorption due to nondegradation of R proteins in the human intestine. Inhibited cobalamin absorption in exocrine pancreatic dysfunction. *J. Clin. Invest.* **66**, 430–40.

Marcuard, S.P., Sinar, D.R., Swanson, M.S., Silverman, J.F. & Levine, J.S. (1989) Absence of luminal intrinsic factor after gastric bypass surgery for morbid obesity. *Dig. Dis. Sci.* **34**, 1238–42.

Matthews, D.M. & Linell, J.C. (1982) Cobalamin deficiency and related disorders in infancy and childhood. *Eur. J. Pediatr.* **138**, 6–16.

McKusick, V.A. (1992) *Mendelian Inheritance in Man.* (Eighth edition.) Baltimore, MD: The Johns Hopkins University Press.

Meikle, D.D., Bull, J., Callender, S.T. & Truelove, S.C. (1977) Intrinsic factor secretion after vagotomy. *Br. J. Surg.* **64**, 795–7.

Nexo, E., Olesen, H., Bucher, D. & Thomsen, J. (1977) Purification and characterization of rabbit transcobalamin II. *Biochim. Biophys. Acta* **494**, 395–402.

Nexo, E., Hansen, M., Poulsen, S. & Olsen, P. (1985) Characterization and immunohistochemical localization of rat salivary cobalamin-binding protein and comparison with human salivary haptocorrin. *Biochim. Biophys. Acta* **838**, 264–9.

Nilsson-Ehls, H., Jagenburg, R., Landahl, S., Lundstedt, G., Swoler, B. & Westin, J. (1986) Cyanocobalamin absorption in the elderly: results for healthy subjects and for subjects with low serum cobalamin concentration. *Clin. Chem.* **32**, 1368–71.

Oddsdotter, M., Ballantyne, G.H., Adrian, T.E., Zdor, M.J., Zucker, K.A. & Modlin, I.M. (1987) Somatostatin inhibition of intrinsic factor secretion from isolated guinea pig gastric glands. *Scand. J. Gastroenterol.* **22**, 233–8.

Peters, T.J. & Hoffbrand, A.V. (1970) Absorption of vitamin B12 by the guinea pig. 1. Subcellular localization of vitamin B12 in the ileal enterocyte during absorption. *Br. J. Haematol.* **19**, 369–82.

Platica, O., Janeczko, R., Quadros, E., Regee, A., Romain, R. & Rothenberg, S.P. (1991) The cDNA sequence and deduced amino acid sequence of human transcobalamin II show homology with rat intrinsic factor and human transcobalamin I. *J. Biol. Chem.* **266**, 7860–3.

Puutula, L. & Grasbeck, R. (1972) One millionfold purification of transcobalamin II from human plasma. *Biochim. Biophys. Acta* **263**, 734–46.

Quadros, E.V., Rothenberg, S.P., Pan, Y.-C.E. & Stein, S. (1986) Purification and molecular characterization of human transcobalamin II. *J. Biol. Chem.* **261**, 15455–60.

Rachmilewitz, B., Rachmilewitz, M. Chaouat, M. & Schlesinger, M. (1977) The synthesis of transcobalamin II, a vitamin B12 transport protein by stimulated mouse peritoneal macrophages. *Biomedicine* **27**, 213–14.

Ramanujam, K.S., Seetharam, S. & Seetharam, B. (1991) Synthesis and secretion of cobalamin binding proteins by opossum kidney cells. *Biochem. Biophys. Res. Commun.* **179**, 543–50.

Ramanujam, K.S., Seetharam, S., Dahms, N. & Seetharam, B. (1991*a*) Functional expression of intrinsic factor-cobalamin receptor by renal proximal tubular epithelial cells. *J. Biol. Chem.* **266**, 13135–40.

Ramanujam, K.S., Seetharam, S., Ramasamy, M. & Seetharam, B. (1991*b*) Expression of cobalamin transport proteins and cobalamin transcytosis by colon adenocarcinoma cells. *Am. J. Physiol.* **260** (*Gastrointest. and Liver Physiol.* **23**), G416–22.

Ramanujam, K.S., Seetharam, S., Ramasamy, M. & Seetharam, B. (1990) Renal brush border membrane bound intrinsic factor. *Biochim. Biophys. Acta* **1030**, 157–64.

Ramasamy, M., Alpers, D.H., Saxena, V. & Seetharam, B. (1990) Effect of lectins on the cobalamin-protein binding reactions: Implications for the tissue uptake of cobalamin. *J. Nutr. Biochem.* **1**, 213–19.

Ramasamy, M., Alpers, D.H., Tiruppathi, C. & Seetharam, B. (1989) Cobalamin release from intrinsic factor and transfer to transcobalamin II within the rat enterocyte. *Am. J. Physiol.* **257**, G791–7.

Robertson, J.A. & Gallagher, N.D. (1979) Effect of placental lactogen on the number of intrinsic factor receptor in the pregnant mouse. *Gastroenterology* **77**, 511–17.

Robertson, J.A., & Gallagher, N.D. (1985*a*) In vitro evidence that cobalamin is absorbed by receptor mediated endocytosis in the mouse. *Gastroenterology* **88**, 908–12.

Robertson, J.A. & Gallagher, N.D. (1983) Increased intestinal uptake of cobalamin in pregnancy does not require synthesis of new receptors. *Biochim. Biophys. Acta* **757**, 145–50.

Robertson, J.A. & Gallagher, N.D. (1985*b*) Intrinsic factor-cobalamin accumulates in the ilea of mice treated with chloroquine. *Gastroenterology* **89**, 1353–9.

Rosenblatt, D.S., Hosak, A., Matiazuk, N.V., Cooper, B.A. & Lafnanboise, R. (1985) Defect in vitamin B12 release from lysosomes: Newly described inborn error of vitamin B12 metabolism. *Science* **228**, 1319–21.

Rothenberg, S.P., Weiss, J.P. & Cotter, R. (1978) Formation of transcobalamin II-vitamin B12 complex by guinea pig ileal mucosa in organ culture after *in vivo* incubation with intrinsic factor-vitamin B12. *Br. J. Haematol.* **40**, 401–14.

Savage, C.R. & Green, P.D. (1976) Biosynthesis of transcobalamin II by adult rat liver parenchymal cells in culture. *Arch. Biochem. Biophys.* **173**, 691–702.

Schepp, W., Rouff, H.J. & Miederer, S.E. (1983) Cellular origin and release of intrinsic factor from isolated rat gastric mucosal cells. *Biochim. Biophys. Acta* **763**, 426–33.

Scott, J.M., Bloomfield, F.J., Stebbins, R. & Herbert, V. (1974) Studies on derivation of transcobalamin III from granulocytes: enhancement of lithium and elimination by fluoride of in vitro increments in vitamin B12-binding capacity. *J. Clin. Invest.* **53**, 228–39.

Seetharam, B. & Alpers, D.H. (1982) Absorption and transport of cobalamin (Vitamin B_{12}). *Ann. Rev. Nutr.* **2**, 343–9.

Seetharam, B., Alpers, D.H. & Allen, R.H. (1981a) Isolation and characterization of the ileal receptor for intrinsic factor-cobalamin. *J. Biol. Chem.* **256**, 3785.

Seetharam, B., Bagur, S.S. & Alpers, D.H. (1981b) Interaction of receptor for intrinsic factor–cobalamin complex with synthetic and brush border lipids. *J. Biol. Chem.* **256**, 9813–15.

Seetharam, B., Bagur, S.S. & Alpers, D.H. (1982) Isolation and characterization of proteolytically derived ileal receptor for intrinsic factor-cobalamin. *J. Biol. Chem.* **257**, 183–9.

Seetharam, B., Jimenez, M. & Alpers, D.H. (1983) Effect of bile and bile acids on the binding of intrinsic factor to cobalamin and intrinsic factor-cobalamin complex to ileal receptor. *Am. J. Physiol.* **245** (*Gastrointest. Liver Physiol.* 8), G72–7.

Seetharam, B., Levine, J.S., Ramasamy, M. & Alpers, D.H. (1988) Purification, properties, and immunochemical localization of a receptor for intrinsic factor complex in the rat kidney. *J. Biol. Chem.* **263**, 4443–9.

Seetharam, B., Presti, M., Frank, B., Tiruppathi, C. & Alpers, D.H. (1985) Intestinal uptake and release of cobalamin complexed with rat intrinsic factor. *Am. J. Physiol.* **248** (*Gastrointest. Liver Physiol.* 11), G306–31.

Seetharam, S., Dahms, M., Li, N., Ramanujam, K.S. & Seetharam, B. (1991a) In vitro translation and expression of renal intrinsic factor-cobalamin receptor. *Biochem. Biophys. Res. Commun.* **177**, 751–6.

Seetharam, S., Ramanujam, K.S. & Seetharam, B. (1991b) Intrinsic factor receptor activity and cobalamin transport in bile duct-ligated rats. *Am. J. Physiol.* **262** (*Gastrointest. Liver Physiol.* 25), G210–15.

Seligman, P.A. & Allen, R.H. (1978) Characterization of receptor for transcobalamin II isolated from human placenta. *J. Biol. Chem.* **253**, 1766–72.

Sennett, C., Rosenberg, L.E. & Mellman, I.S. (1981) Transmembrane transport of cobalamin in prokaryotic and eukaryotic cells. *Ann. Rev. Biochem.* **50**, 1053–86.

Serfilippi, D. & Donaldson, R.M. (1986) Production and secretion of intrinsic factor by isolated rabbit gastric mucosa. *Am. J. Physiol.* **14**, 287–92.

Shaw, S., Jayatilleke, E., Meyers, S., Coleman, N., Herzlich, B. & Herbert, V. (1989) The ileum is the major site of absorption of vitamin B12 analogues. *Am. J. Gastrol.* **84**, 22–6.

Simon, J.D., Houck, W.E. & Albala, M.M. (1976) Release of unsaturated vitamin B12 binding capacity from human granulocytes by the calcium ionophore A23187. *Biochem. Biophys. Res. Commun.* **73**, 444–50.

Simons, K. & Weber, T. (1966) The vitamin B12 binding protein in human leukocytes. *Biochim. Biophys. Acta* **117**, 201–8.

Smolka, A. & Donaldson, R.M., Jr. (1990) Monoclonal antibodies to human intrinsic factor. *Gastroenterology* **98**, 607–14.

Steinberg, W.M., King, C.E. & Toskes, P.P. (1980) Malabsorption of protein-bound cobalamin but not unbound cobalamin during cimetidine administration. *Dig. Dis. Sci.* **25**, 188–91.

Stenman, U.H. (1974) Amniotic fluid vitamin B12 binding protein. Purification and characterization with isoelectric focusing and other techniques. *Biochim. Biophys. Acta* **342**, 173–84.

Stenman, U.H. (1975*a*) Vitamin B12-binding proteins of R-type, cobalophilin: characterization and comparison of cobalophilin from different sources. *Scand. J. Haematol.* **14**, 91–107.

Stenman, U.-H. (1975*b*) Characterization of R-type Vitamin B12-binding proteins by isoelectric focusing II. Comparison of cobalophilin from different sources. *Lab. Clin. Invest.* **35**, 147–55.

Tan, C.H. & Hansen, H.J. (1968) Studies on the site of synthesis of transcobalamin II. *Proc. Soc. Exp. Biol. Med.* **127**, 740–4.

Tang, L.H., Chokshi, H., Hu, C., Gordon, M. & Alpers, D.H. (1992) The intrinsic factor (IF)-cobalamin (Cbl) receptor binding site is located in the amino-terminal portion of IF. *J. Biol. Chem.* **267**, 22982–6.

Teo, N.H., Scott, J.M., Neale, G. & Weir, D.G. (1980) Effect of bile on Vitamin B12 absorption. *Br. Med. J.* **281**, 831–3.

Trugo, N.M. & Newport, M.J. (1985) Vitamin B12 absorption in the neonatal piglet. 2. Resistance of the vitamin B12-binding proteins in sow's milk to proteolysis *in vivo*. *Brit. J. Nutr.* **54**, 257–67.

Trugo, N.M., Ford, J.E. & Salter, D.N. (1985) Vitamin B12 absorption in the neonatal piglet. 3. Influences of vitamin B12-binding protein in sows' milk on uptake of vitamin B12 by microvillus membrane vesicles prepared from small intestine of the piglet. *Brit. J. Nutr.* **54**, 269–83.

Vatn, M.H., Berstad, A. & Myren, J. (1974) The effect of exogenous secretin and cholecystokinin on pentagastrin-stimulated intrinsic factor in man. *Scand. J. Gastroenterol.* **9**, 313–17.

Vatn, M.H., Schrumpf, E. & Myren, J. (1975*a*) The effect of carbachol and pentagastrin on the gastric secretion of acid, pepsin, and intrinsic factor in man. *Scand. J. Gastroenterol.* **10**, 55–8.

Vatn, M.H., Semb, L.L.S. & Schrumpf, E. (1975*b*) The effect of atropine and vagotomy on the secretion of gastric intrinsic factor (IF) in man. *Scand. J. Gastroenterol.* **10**, 59–64.

Wahlstedt, V. & Grasbeck, R. (1985) Cobalamin binding proteins in human urine. Identification and quantitation. *J. Lab. Clin. Med.* **106**, 439–46.

Yang, Y., Ducos, R., Rosenberg, A.J., Catrou, P.G., Levine, J.S., Podell, E.R. & Allen, R.H. (1985) Cobalamin malabsorption in three siblings are due to abnormal intrinsic factor that is markedly susceptible to acid and proteolysis. *J. Clin. Invest.* **76**, 2057–65.

Youngdhal-Turner, P., Mellman, I.S., Allen, R.H. & Rosenberg, L.E. (1979) Absorptive endocytosis of transcobalamin II complex by cultured human fibroblasts. *Exp. Cell Res.* **118**, 127–34.

Zucker, K.A., Adrian, T.E., Ballantyne, G.H. & Modlin, I.M. (1988) Prostaglandin E analogue inhibition of intrinsic factor release. *Scand. J. Gastroenterol.* **23**, 650–4.

5

Folate Binding Proteins

S. Weitman, R.G.W. Anderson & B.A. Kamen

Prologue

In the decade since the first recognition of the existence of folic acid, a voluminous literature on this subject has arisen, of such extent and complexity that only the specialist in biochemistry can keep abreast of all its developments. As a member of the vitamin B complex, folic acid equals thiamine, riboflavin and niacin in importance, although its function as a constituent of this essential group has hitherto been to a large extent obscured by its extreme importance in the megaloblastic anemias. Other aspects of folic acid metabolism are now assuming prominence: its relation to other pteroylglutamic acids, to other vitamins and to liver extract; and the effect of the administration of the so-called "folic acid antagonists".

This introduction to *The Nutritional and Clinical Significance of Folic Acid* was published in 1950 (Lederle Laboratories, 1950). It is a monograph of approximately 100 pages, with a bibliography of 367 items.

Indeed, there has been much interest in the folic acid antagonists: a literature search from 1966 to 1990 found over 12 000 citations for methotrexate alone. The purpose of this chapter, however, will not be to review folate biochemistry, folate-mediated reactions or the biochemistry of antifolates; there are a number of recent multivolume texts that accomplish this feat (Blakley & Benkovic, 1984; Sirotnak *et al.*, 1984), but rather to update and discuss folate homeostasis with specific reference to a protein(s) which has a high affinity for folic acid and related compounds[1] but no known enzyme activity.

---

[1] The word folate or folates in this chapter will refer to folic acid and its reduced derivatives. When discussing the specific folates, they will be cited as such, e.g. folic acid, 5-methyltetrahydrofolic acid, 5-formyltetrahydrofolic acid, etc.

Introduction

A folate binding protein(s) (FBP) was initially described over 20 years ago (Ford *et al.*, 1969). Clinical annotations of this early work are available (Waxman, 1975; Colman & Herbert, 1980). In the past several years, three reviews of this literature have also been published (Kane & Waxman, 1989; Henderson, 1990; Antony, 1990, 1992). Reviews of folate and antifolate transport and folate metabolism (Blakley & Benkovic, 1984; Sirotnak, 1985; Dembo & Sirotnak, 1984; Huennekens *et al.*, 1978) and overviews of the clinical pharmacology and use of antifolates are also available (Kamen, 1987; Fleming & Schilsky, 1992). Therefore the specific aims of this chapter will be to update and comment upon recent work and provide some speculations for future studies. However, a historical review of FBP, especially in the first decade of discovery, and the recognition of the FBP as a membrane-bound protein (and subsequently defining it as a receptor) will also be presented, such that the transition from FBP to 'receptor' can be appreciated.

Why a high-affinity folate uptake system?

The external environment can be hostile. Both eucaryotic and procaryotic organisms must have an effective means of separating the environment from the intracellular milieu. Toxins or unwanted compounds must be kept out (or eliminated) and certain molecules must be effectively accumulated. An example of this latter class of compounds is the vitamins ('Vital amines'), compounds absolutely essential for life, but which cannot be synthesized by the cell or organism. For example, some bacteria concentrate cobalamin 100 000-fold from seawater (Sennett & Rosenberg, 1981). Folates also are a part of this group of substances. The folate concentration in human plasma is approximately 5–20 nM; in human cells it is 1–20 μM. Thus it seems as if cells, tissues, organs and organisms should have a means to accumulate and protect folates (and other vitamins) from degradation or loss. The need for vitamins such as folate is assumed, but less is known about the molecular regulation at the cellular level. For example, in the most recent edition of the book *Molecular Biology of the Cell* (Alberts *et al.*, 1989), the word folate is not in the index, and the means by which cells maintain folate homeostasis is not presented.

In organizing this review of folate binders or receptors, defining different requirements or functions for specific types of cells and tissues with regard to folate homeostasis was helpful in classifying a functional role for an FBP. For example, lymphocytes (or other blood cells) are constantly bathed in plasma folate and most likely need to take up folate only as required for their own metabolism (and cell division). Other tissues or organs may have different means and needs for accumulating folate. Intestinal mucosa should be able to absorb folate from the gut lumen and translocate it to the portal system, as the means of entry for 'new' folate into the body. Kidney epithelial cells should trap and 're'-absorb folate that was filtered (indeed, there is little if any folate lost in the urine (Retief *et al.*, 1976)). A specialized means for accumulating and maintaining folate in body compartments also exists; for example, cerebrospinal fluid has a folate concentration 3–4 times that of the plasma (Herbert & Zalusky, 1961; Wells & Casey, 1967). The soluble folate binder, such as found in secretory fluids (milk) may aid in specialized cellular absorption or may perhaps function to block utilization of the cofactor by bacteria (Ford, 1974; Tani *et al.*, 1983; Colman & Herbert, 1980; Hansen *et al.*, 1980; Tani & Iwai, 1984).

Historical review

The first decade

The initial presentation of some properties of a folate binder, first identified in milk, was by Ford *et al.* (1969). Although their work represented the first description of such a binder, earlier investigations (Ghitis, 1966) suggested its existence. The need for such a carrier (binder) was also postulated to explain the rapid plasma clearance of intravenously injected [^{3}H]-PteGlu in rats, accompanied by a high intra-cellular accumulation of radioactivity, especially in liver and kidney (Johns *et al.*, 1961). Further evidence for folate carriers was obtained by Lichtenstein *et al.* (1969) when they demonstrated a specific carrier-mediated uptake system for folic acid and certain naturally occurring derivatives in L1210 cells *in vitro*. Additional evidence for a carrier-mediated transport system for folates had been presented by a number of workers studying both mammalian (Rosenberg *et al.*, 1969; Baugh *et al.*, 1971; Bobzien & Goldman, 1972; Selhub *et al.*, 1973; Spector & Lorenzo, 1975; see older review by Butterworth & Krumdieck, 1975) and bacterial cells (Cooper, 1970; Shane & Stokstad, 1975; Huennekens *et al.*, 1974; McIntyre *et al.*, 1975). A putative 'carrier' has since been identified by photoaffinity and NHS tritium labeling (methotrexate, MTX) but the carrier has not been isolated (Price *et al.*, 1987, 1988; Henderson & Zevely, 1984; Fan *et al.*, 1992) (see below, carrier-mediated folate transport).

Milk folate binding protein

As noted above, the existence of a 'milk folate binder' was originally suggested by Ghitis (1966) and Ghitis & Lora (1967) and later by Metz *et al.* (1968). Ghitis (1966) observed that activated charcoal could not absorb *Lactobacillus casei* growth-promoting activity (i.e. folate) from bovine milk until the sample had been boiled. It was concluded that boiling denatured a binding factor and released the folate. It was further shown that folic acid added directly to fresh milk was not detected by microbiological assay until the milk was heat-denatured. This finding indicated that the milk contained an unsaturated folate binder capable of reacting with exogenously added folic acid. Using this same technique of adding folic acid to a sample allowed Ghitis (1966) to demonstrate that 5-formyltetrahydrofolic acid was not completely available for *L. casei* utilization and also that the binding factor had no significant affinity for the folic acid analog methotrexate (MTX).

Using Sephadex gel filtration to separate the folic acid from the folic acid–binder complex, Metz *et al.* (1968) showed binding of [^{3}H]folic acid to a macromolecular factor in milk. These workers also showed that unlabeled folic acid, but not 5-methyltetrahydrofolic acid, 5-formyltetrahydrofolic acid, *p*-aminobenzoylglutamic acid or methotrexate, could block the binding of [^{3}H]folic acid. Thus the binding of folic acid by the milk factor appeared to be saturable and specific.

Ford *et al.* (1969) partially purified a folate binding factor from bovine milk by ammonium sulfate precipitation followed by DEAE–cellulose and Sephadex gel chromatography. The binder had a molecular mass of 35–40 kDa, based upon Sephadex G75 filtration, but also had a propensity to self-aggregate to produce apparent molecular masses of 78 kDa and over 100 kDa. Contrary to the finding of Metz *et al.* (1968), Ford *et al.* (1969) showed that synthetically prepared 5-methyltetrahydrofolic acid could compete for folic acid binding sites; moreover, they demonstrated that [^{14}C]5-methyltetrahydrofolic acid was bound. However, the binding factor clearly had a preference for oxidized folates. Binding of folates was found to be pH-dependent. Below pH 3.5 the folate–binder complex completely dissociated, but binding activity was regained when the solution was neutralized. Although the binding activity initially chromatographed with β-lactoglobulin on DEAE cellulose, when chromatographed in dilute salt (0.005 M potassium phosphate), the folate binder eluted from the column in the wash volume whereas the β-lactoglobulin was retained.

Waxman & Schreiber (1973) showed that commercial preparations of β-lactoglobulin contained an unsaturated folate binder. This binder and the binder prepared from fresh milk (Waxman *et al.*, 1971; Rosenberg *et al.*, 1969) were used to set up competitive and non-competitive radio-ligand binding assays for folate (see also Archibald *et al.*, 1972; Dunn & Forster, 1973; Mincey *et al.*, 1973; Waxman & Schreiber, 1973).

Folate binders have also been found in the milk of other animals. Ford *et al.* (1969) demonstrated that human colostrum contained a folate binder. The folate in human milk, as determined by localization of *L. casei* growth-promoting activity, appeared in two fractions on Sephadex G75 gel filtration. The smaller factor was similar to that found in bovine milk. The high-molecular-mass factor eluted in the excluded volume, indicating a molecular mass of 75 kDa or greater.

A large concentration of folate binder was found in goat colostrum, immediately postpartum (Ford *et al.*, 1972). The physical–chemical properties of this factor have not yet been reported, but it has been suggested by one of the authors (Ford, 1974) that the factor may aid in the intestinal absorption of folate by the newly born kid. In this regard it was also noted that while the milk contains high levels of folate binder, in excess of the total folate content of the milk, the plasma of the kid does not contain unsaturated binder; however, during the first several days of life the majority of the microbiologically active folate was non-dialyzable and hence considered 'bound' in the plasma.

Following the initial description of the folate binder in cow's milk, Salter *et al.* (1972) described a method for purifying the binder by affinity chromatography. Waxman & Schreiber (1975) reported some of the properties of a folate binder present in commercially available human milk, which they purified 10 000-fold using this technique (folic acid covalently linked to Sepharose; reviewed by Kamen & Caston, 1980). The FBP had an apparent molecular mass of 26.5 kDa based upon Sephadex G200 gel filtration. The specific activity of the preparation was 7.1 μg folic acid bound per milligram of protein.

Assuming the FBP to be univalent with respect to folate, then the ratio of moles folic acid bound to moles protein was 0.42:1.0, i.e., the preparation was 42% 'active binder'. Electrophoretic analysis of the FBP in sodium dodecyl sulfate (SDS) polyacrylamide gel produced three doublets of 80:87 kDa; 36.5:30 kDa; and 19:11.4 kDa. When subjected to 10% polyacrylamide gel electrophoresis at pH 8.6 two protein bands were detected, both of which had folate binding activity. Isoelectric focusing experiments showed the binding activity to be in three discrete bands at pH 6.8, 7.5 and 8.2. The folate–binder complex adhered to concanavalin A – Sepharose and eluted with methyl α-mannoside.

Further, the protein bands on the 10% polyacrylamide gel electrophoresis stained with periodic acid–Schiff (PAS) (more recent studies confirmed the FBP as a glycoprotein by direct carbohydrate and amino acid analysis (see Antony *et al.*, 1981; Selhub & Franklin, 1984; Luhrs *et al.*, 1987)). The specificity and affinity of the FBP for folic acid, derivatives and analogs was not presented. It was reported that folic acid preincubated with the purified folate binder was not available to HeLa cells *in vitro*. This confirmed earlier observations by Waxman & Schreiber (1975). These findings also agreed with the observations of Ford (1974) that the folate binder in milk can prevent bacterial uptake of the vitamin. It should be noted, however, that Ford *et al.* (1969) and Ford (1974) also suggested that the binder may aid in the intestinal absorption of folate in a manner analogous to the action of intrinsic factor of vitamin B_{12}.

In summary, milk from a number of different species of animals was shown to contain a folate binding factor, saturated and unsaturated (with respect to folate). This factor has a high affinity for oxidized folates, but also bound 5-methyltetrahydrofolic acid and other reduced folates. It has been purified to apparent homogeneity by affinity chromatography and has a molecular mass of 28–30 kDa. More recently, both bovine and human species have been sequenced (see below). It has a propensity to aggregate and this appears to affect the affinity for folate (Salter *et al.*, 1981; Pedersen *et al.*, 1980). The function of an FBP in milk has not been ascertained. It is speculated that it may retard the uptake of folate by microflora in the intestinal lumen of suckling animals or perhaps assist in the absorption of folate from the lumen, functioning in a manner analogous to intrinsic factor.

Serum and cellular folate binders

Since the discovery of a folate binder in milk several other tissues containing folate binding proteins have been identified (Table 5.1). The first report of an intracellular and serum folate binder in human tissue was by Rothenberg (1970). Serum and leukocyte lysates from several patients with chronic granulocytic leukemia (CGL) were noted to contain unsaturated folate binder, i.e. the sample bound exogenously added [3H]folic acid. Based upon Sephadex G-75 gel filtration, the FBP in both the cell lysate and serum had a molecular mass of over 50 kDa and was separable from dihydrofolic acid reductase activity that was present in the cell lysate. [3H]folic acid complexed to FBP was also found not to be a substrate for dihydrofolic acid reductase.

In direct competitive binding studies using [3H]folic acid and non-

Table 5.1. *Folate binding proteins*

	Molecular mass (kDa)
MEMBRANE BOUND	
Mammalian cells	
MA104 (monkey kidney)	38
L1210 (mouse leukemia)	36
KB (human epidermoid carcinoma)	50
HeLa	—
Caco-2 (human colon carcinoma)	—
Lymphocytes (human)	—
Reticulocytes/erythrocytes	200; 160[a]
T47D (human mammary tumor)	160
Fibroblast (human)	160
Leukemia (human)	35
Mouse alpha mammary	160
Mammalian tissue	
Placenta	
human	38.5
guinea pig	—
Choroid plexus	
human	45–60
porcine	51
rabbit	360–400
Kidney	
rat	30
porcine	35–40
Intestinal mucosa	
rat	16–24
Liver	
rat	55; 100
Milk	
human	35
SOLUBLE	
Mammalian tissue	
Milk	
human	30
goat	30
bovine	—
Serum	
human	30
umbilical cord	40
Urine	
human	35
Cerebral spinal fluid	
human	25
Kidney	
porcine	35–40
Amniotic fluid	
human	25; 100[a]
Mammalian cells	
Chronic myelogenous leukemia	30–45
Leukocytes	25
KB (human epidermoid carcinoma)	40

[a]Two separate proteins isolated.

radioactive folic acid derivatives, the binder was shown to have determinants for folic acid, dihydrofolic acid, tetrahydrofolic acid and folate polyglutamates. Methotrexate also competed for folic acid binding sites, but not very efficiently. The binding of folic acid was not competed for by 5-methyltetrahydrofolic acid, the main serum folate (Herbert *et al.*, 1962), or by 5-formyltetrahydrofolic acid. The binding of radioactive 5-methyltetrahydrofolic acid was not tested. Binding of folic acid by the cell lysate factor was rapid (2% s^{-1} at 37°C and 10^{-9} M PteGlu) and essentially irreversible at neutral pH but decreased at lower pH. As in the initial studies of the different milk preparations, the presence of a saturated folate binder was not evaluated. Rothenberg & DaCosta (1971) and Fischer *et al.* (1975) elaborated upon the properties of the binder from the cell lysate.

On DEAE–cellulose chromatography starting at 10^{-3} M salt, the folate binder was resolved into two bands. The first (14% of the total activity) eluted at approximately 0.1 M salt, pH 7.4. The former fraction had a molecular mass of 34.5 kDa and demonstrated somewhat reversible binding of folate at nanomolar concentrations. The latter had molecular mass 41 kDa and bound both oxidized and reduced folates apparently irreversibly. Neither fraction had determinants for methotrexate or 5-formyltetrahydrofolic acid, but both did bind 5-methyltetrahydrofolic acid, based upon competitive binding studies using [^{3}H]folic acid. These results are in contradiction to the original report (Rothenberg, 1970) in which the FBP had an apparent molecular mass of over 50 kDa and had a measurable affinity for MTX but not 5-methyltetrahydrofolic acid. Perhaps the differences are due to the fact that prior to the DEAE–cellulose chromatography the lysate was subjected to prolonged dialysis against 0.1 M citric acid. This was noted to increase the total folate binding capacity of the extract (presumably by removing endogenously bound folate). This process could have altered the native binder and/or made different sites available to react with exogenously added folates, binding sites not present in the native state. Furthermore, it was also noted that the factor (both fractions) bound 5-methyl[^{3}H]tetrahydrofolic acid. In addition to showing that there were determinants for a reduced folate, this work also demonstrated that the binder did not discriminate between the R and S forms of tetrahydrofolic acid. The same results were obtained when the milk (Rothenberg *et al.*, 1972) and hog kidney (Kamen & Caston, 1975) folate binders were tested with radiolabeled 5-methyltetrahydrofolic acid. The folate binder in the leukemic cell lysate was unaffected by incubation with DNase and RNase, but trypsin, chymotrypsin and pepsin destroyed folate binding activity.

Other high-affinity binders of folate in human serum were described in the 1970s. Waxman & Schreiber (1973) described properties of folate binding factor(s) in the serum of some patients with a folic acid deficiency. In addition, DaCosta & Rothenberg (1974) reported some properties of folate binding factor (s) in the serum and leukocytes of a group of women who were using oral contraceptives or who were pregnant. Both groups of investigators found the FBP to have properties similar to those already described for milk, serum and cell lysates. In the study by Watabe (1978), the serum folate binding factor(s) had a considerably lower affinity for 5-methyltetrahydrofolic acid than for folic acid. Folate binder from these sources did not bind MTX or 5-formyltetrahydrofolic acid. Sephadex G200 gel filtration separated the binding activity into two macromolecular fractions, one of which was excluded from the gel and hence had a putative molecular mass of more than 200 kDa. The etiology of the folate deficiency or the patient's diagnosis was not given. In the study by DaCosta & Rothenberg (1974) the results with regard to serum are difficult to evaluate because the binding titers were often low. However, 10% (1/10) of the women taking oral contraceptives and 48% (31/64) of the pregnant subjects had unsaturated folate binder in their leukocytes. The binding factor(s) in these cases did not bind MTX, formyl- or methyltetrahydrofolic acid as determined by competitive binding experiments with [³H]folic acid. The binding activity moved as a single band on Sephadex G75 gel filtration with an apparent molecular mass of over 50 kDa. Zettner & Duly (1974) extended these initial findings by extensively studying the binding of folic acid and some derivatives in the serum of approximately 100 normal volunteers and more than 900 patients. It was found that all samples showed some capacity to bind [³H]folic acid. The range was from 0.1 ng ml^{-1} (*ca.* 0.25 pmol ml^{-1}) to greater than 8 ng ml^{-1} (*ca.* 20 pmol ml^{-1}) serum. The binding of folic acid was competed for by 5-methyltetrahydrofolic acid, MTX and 5-formyltetrahydrofolic acid as well as by folic acid. Barbituric acid, salicylate and diphenylhyantoin (Dilantin), which are known to interfere with folate absorption (Bernstein *et al.*, 1970), did not block the binding of folic acid. Despite this apparent specific binding of folate, the binding described here may be different from that previously described (DaCosta & Rothenberg, 1974; Waxman & Schreiber, 1973; Zettner & Duly, 1974), using more than ten times the concentration of [³H]folic acid but the same amount of serum. The serum samples 'bound' only 3–5% of the [³H]folic acid and the binding proved to be reversible: 20% of the pre-bound [³H]folic acid was displaced in 20 min by an equimolar concentration of folic acid.

Additional examples of high-affinity, unsaturated folate binders in serum have been presented (Hines *et al.*, 1973; Mantzos, 1975; Bentsen *et al.*, 1990). These binder(s) were found in the serum from patients with several chronic diseases, specifically uremia, chronic alcoholic liver disease, ulcerative colitis, and several different types of carcinomas and sarcomas.

The presence of unsaturated folate binder(s) in animal plasma was studied by Mantzos *et al.* (1974). Plasmas from sheep, goat, cattle, horse, rabbit, dog, rat, guinea pig, and chicken were examined for specific binding of [^{3}H]folic acid. Only pig plasma and 1/16 of sheep samples (the ewe was pregnant) contained unsaturated FBP. The binding capacity of the pig plasma ranged from 14 to 26 ng ml^{-1} plasma, which is 5–100 times that found in human serum (Waxman & Schreiber, 1973; Zettner & Duly, 1974). It was also noted that the folate binder in pig plasma was only partly saturated with endogenous folate. The unsaturated folate binder in pig plasma had the same apparent specificity for folates as the binder derived from hog kidney (Kamen & Caston, 1975, 1986). Binding activity appeared in the β- and α-globulin fraction of the plasma upon routine serum electrophoresis at pH 8.6. This binding factor has been used to set up a radio-ligand binding assay useful in the determination of serum folate (Mantzos, 1975). The results were comparable to those obtained by Kamen & Caston (1974) using the hog kidney folate binder.

Identifying human or animal serum samples having an FBP is relatively easy when samples contain unsaturated binding sites. Demonstrating endogenously saturated FBP is more difficult. Indirect evidence for saturated folate binders, especially in serum, has been presented by several groups of investigators. Kamen & Caston (1974), with later confirmation by Mantzos (1975), noted that serum folate levels determined by radio-ligand binding assay were consistently higher when a serum extract (prepared by boiling the serum) rather than 'whole' serum was measured. This was interpreted to mean that some folate was complexed to a heat-labile factor in the serum (analogous to the studies of the FBP in milk), thus making it undetectable in a competitive binding assay. This observation led to the identification and partial purification (10 000-fold) of a binder present only in a 'folate' saturated form in human umbilical cord serum (Kamen & Caston, 1975). This FBP had a molecular mass of 38–40 kDa and bound both folic acid and 5-methyltetrahydrofolic acid. FBP in amniotic fluid and umbilical cord blood was also identified (Holm *et al.*, 1990) several years later.

The distribution of endogenous serum folate was studied by several investigators (Markkanen, 1968; Markkanen & Himanen, 1971;

Markkanen *et al.*, 1971, 1972*a*,*b*, 1973*a*,*b*,*c*). Using the microbiological assay, some serum folate, even in normal samples, was non-dialyzable, was precipitated with mild boiling, and on Sephadex G200 gel filtration had an apparent molecular mass of 200, 80 and 60 kDa. The 60 kDa fraction was considered to be albumin, simply because it partitioned with albumin. One or both of the other fractions could be similar to the binders previously described (Waxman & Schreiber, 1973; DaCosta & Rothenberg, 1974). Altered patterns of *L. casei* growth-promoting activity in serum fractionated on Sephadex G200 were noted in samples from some patients who were pregnant (Markkanen *et al.*, 1973*a*), taking oral contraceptives, taking anticonvulsants with liver disease, and with multiple myeloma (Markkanen *et al.*, 1973*b*). Other studies (Markkanen *et al.*, 1973*a*,*b*, 1974) using DEAE–Sephadex A25 showed that serum folate was divided into three main fractions, again based upon the microbiological assay for folate. These were in the α_2-macroglobulin, transferrin and albumin regions.

Summary of serum FBP(s)

By the mid 1970s, the work of many investigators, using several different techniques, had demonstrated that both saturated and unsaturated folate binder existed in serum. Moreover, the concentrations of FBP were elevated in serum from patients with certain pathophysiological conditions, namely those associated with pregnancy, neonates, some malignancies and other chronic diseases. The relationship between the saturated and unsaturated forms and the difference in size and relationship to the membrane receptor (see below) has not yet been fully evaluated.

Other soluble binders

The observation that cerebrospinal fluid (CSF) folate concentrations are 3–4 times higher than concentrations in matched serum samples (Herbert & Zalusky, 1961; Wells & Casey, 1967; Markkanen & Himanen, 1971), coupled with the finding of a saturable, specific transport mechanism for folate in hog and rabbit choroid plexus (Chen & Wagner, 1975; Spector & Lorenzo, 1975) led to the study of spinal fluid folate levels and folate binders in children with acute lymphoblastic leukemia (Kamen *et al.*, 1975). The results showed that 60% of spinal fluid folate in 'normals' was complexed to a heat-labile, non-dialyzable factor. More recently, a FBP of molecular mass *ca.* 26 kDa has been identified in CSF (Hansen *et al.*, 1985).

Other cellular folate binders

Isolated small intestinal cells from rats have been shown to bind folic acid and a number of derivatives (Leslie & Rowe, 1972). The binding constant for folic acid (3.98 × 10^{-5} M) agreed well with the transport constant obtained by Burgen & Goldberg (1962) who studied folate transport in perfused segments of rat small intestine. These folate binders were considered to be embedded in the membrane matrix and to aid in the transport of folate from the gut lumen. The relationship to the high-affinity folate binders such as those described earlier has not yet been established. More recent studies (Reisenauer, 1990) using affinity labeling of the folate binding protein in pig intestine have concluded that the intestinal FBP and transport protein 'are identical and that the function of the FBP is to transport folate into the cell'. The K_m for transport is 0.45 μM.

Evidence for the presence of folate binders in liver has been presented by Zamierowski & Wagner (1974) and Corrocher *et al.* (1974). Corrocher and co-workers studied the binding of [^{3}H]folic acid injected intraperitoneally or intravenously (hepatic vein) in rats. The distribution of radiolabeled material was studied by fractionating cell lysates on Sephadex G150. The radioactivity appeared in fractions known to contain the Y and Z proteins (Levi *et al.*, 1969) which are considered to be 'general' binders of anions, and an additional protein termed 'X'. The binding of the radioactivity was coincident with bound bilirubin and sulfobromophthalein. No competitive binding studies or kinetic analysis of binding were presented.

Zamierowski & Wagner (1974) studied the distribution of radioactivity in rats that were injected with [^{3}H]folic acid. They found that liver, kidney and intestine accumulated most of the radioactivity whereas heart accumulated little. The liver preparation, which was most extensively studied, had radiolabeled material and microbiologically active folate in the 350, 150, 90 and 25 kDa size regions, using Sephadex G150 gel fractions. The 90 kDa peak was of nuclear origin and the 25 kDa fraction had dihydrofolic acid reductase activity. Based upon the marked increase in microbiologically active material after the fractions were treated with hog kidney conjugase, the majority of the endogenous folate was considered to be in the folylpolyglutamate form. The affinity and specificity of these 'binders' was not studied directly. In the past it has been shown that the hepatic folate binders are enzymes (Cook & Wagner, 1984). The affinity for folate polyglutamate is in the micromolar, not low nanomolar, range. Other folate enzymes have also been shown to bind folylpolyglutamates (Matherly *et al.*, 1990). These intracellular 'binders' will not be further discussed here but are reviewed in Henderson (1990) and Kane & Waxman (1989). A high affinity FBP in rat brush border of kidney and porcine kidney extracts has been purified and characterized (Selhub & Franklin, 1984; Kamen & Caston, 1986) and probably has a similar function, to absorb folate from urine.

Carrier-mediated folate transport

Folate and antifolate transport

The 1970s also saw the pharmacokinetic description of folate and antifolate (predominantly methotrexate) transport in murine lymphoma cells as well as hepatocytes (Gewirtz *et al.*, 1980; Galivan, 1981; Henderson *et al.*, 1987; Sirotnak *et al.*, 1987; Hill *et al.*, 1979). The model that has evolved primarily through the work of several laboratories has been reviewed (Sirotnak, 1985; Fan *et al.*, 1992). The essential components are a carrier for influx (facilitated transport) and a separate energy-dependent efflux system. Transport is inhibited by selected anions such as sulfobromophthalein and phthalate, further suggesting the importance of an anion pump in folate transport (Henderson & Tsuji, 1987; Henderson & Zevely, 1985). Current work (see Henderson & Strauss, 1990) has shown that there are at least three pathways for MTX influx–efflux as assessed by sensitivity to various metabolic inhibitors or anion pump blockade. The model was developed almost entirely through the study of cellular transport of pharmacological rather than physiological concentration of folate *in vitro*. Typically, the cells studied are maintained in tissue culture media, which contain 100–1000 times the folate concentration found in plasma. Moreover, this folate is the 'wrong' folate: plasma folate is 5-methyltetrahydrofolate whereas tissue culture medium contains folic acid. Thus cells may not have to concentrate folate from the milieu, as they do *in vivo*. For example, in one study of 5-methyl[^{3}H]tetrahydrofolate accumulation by murine leukemia cells, the endogenous intracellular pool of folate was 'expanded' by less than 0.1% *in vitro* (Nixon *et al.*, 1973). Even those cells grown under more natural conditions (such as in ascites) may not accumulate significant amounts of folate if they are not folate-deficient (see below). Further, many studies of transport and accumulation of folates in eukaryotes have been done using the antifolate methotrexate as the ligand. The advantage of using [^{3}H]methotrexate are that (i) metabolism of this drug is low compared with that of the natural folates and (ii) methotrexate is stable compared with reduced folates. A potential disadvantage, in addition to its being a cytotoxic agent, is that methotrexate can be a poor substrate for folate polyglutamate synthetase (FPGS) compared with reduced folates.

Thus the large number of studies briefly noted here, and definitively reviewed by those responsible for the work, defined the kinetics for folate (and antifolate) transport and documented the presence of a carrier-mediated or a facilitated diffusion process for folate uptake, but some of the experimental conditions precluded a study of overall folate homeostasis and perhaps masked the detection (and further analysis) of a process for folate accumulation and metabolism under physiological conditions of growth. Since the apparent K_d for reduced folates is in the range of 1–10 μM, unless there is a large amount of carrier and or rapid trapping of transported folate (perhaps as a folate polyglutamate), at physiological concentrations of extracellular folate (1–10 nM) net accumulation may be adequate but not very efficient, depending on the cellular requirement for folate.

Receptor-coupled transmembrane transport

The second decade: identification of the FBP as a membrane-bound protein

A critical observation relating the FBP to folate accumulation was made by McHugh & Chen (1979). They showed that the high-affinity, specific binding of folic acid as well as some of the reduced derivatives was primarily confined to a particulate (plasma membrane) fraction. During the 1980s several laboratories more completely characterized the properties of a FBP associated with the particulate fraction in human placenta (Antony *et al.*, 1981; Zwiener *et al.*, 1992), KB cells (Elwood *et al.*, 1986; Kane *et al.*, 1986*a,b*), and rat kidney (Selhub & Franklin, 1984). Some of the properties of these factors are listed in Table 5.1 and described in detail in previous reviews (Kane & Waxman, 1989; Henderson, 1990). In addition, our laboratory presented a detailed analysis of the binding specificity of the porcine kidney factor, purified to homogeneity (Table 5.2).

Knowing that the FBP is a glycoprotein, has bound fatty acids and has been shown to be attached to the membrane via a glycosylphosphatidylinositol (GPI) anchor as discussed below, most likely explains some of the apparent differences in the molecular size reported by different laboratories, especially since the direct and predicted amino acid sequence based upon the cDNA (see below) is approximately 28–29 kDa.

Showing that the FBP is a receptor and important in folate

Table 5.2. *Effect of folic acid, derivatives or analogs on [^{3}H]PteGlu binding*

Purified hog kidney folate binder (0 or 75×10^{-12} mol PteGlu binding capacity) was incubated with 1×10^{-12} mol [^{3}H]PteGlu and competitor as noted, in a total volume of 1 ml. Thus, 10^{-12} mol competitor was equal to the [^{3}H]PteGlu. If there was equimolar competition, then 10^{-12} mol competitor would reduce binding of [^{3}H]PteGlu by 50%. The percentage inhibition was calculated as follows:

$$\% \text{ inhibition} = (\text{dpm control} - \text{dpm sample/dpm bound in control}) \times 100.$$

Control was binding in the absence of competing compound. The reaction was started by adding binder to the solution of [^{3}H]PteGlu and competitor and was stopped after 15 min using charcoal coated with dextran. The results are from duplicate experiments run in duplicate on two different days with freshly prepared compounds.

Compound	Percentage inhibition at amounts shown (pmol)			
	1.0	2.0	10.0	100.0
Folic acid	52	66	85	98
Dihydrofolic acid	55	64	80	99
Tetrahydrofolic acid	50	60	75	95
Diopterin	53	65	85	97
Teropterin	54	67	85	98
Folylheptaglutamate	52	65	84	98
5-Methyltetrahydrofolate	20	40	68	95
5-Formyltetrahydrofolate	0	0	0	5
5-Methyltetrahydrofolyltriglutamate	20	42	65	95
10-Formyltetrahydrofolate	30	45	70	90
Aminopterin	0	0	0	0
Methotrexate	0	0	0	5
3',5'-Dichloromethotrexate	0	0	0	0
Pterin			0	0
Pterin-6-carboxylic acid			0	0
Xanthopterin			0	0
p-Aminobenzoic acid			0	0
Glutamic acid			0	0
p-Aminobenzoylglutamic acid			0	0
Pteroic acid	0	0	5	10
Thymidine			0	0
Adenosine			0	0
NADH			0	0
NADPH			0	0

accumulation (process of folate homeostasis) at least in some cells has been more difficult, although intuitively correct. For example Watkins & Cooper (1983) showed that K562 cells had enough endogenous folate (when growing in typical medium, containing 1–10 μM folic acid) to undergo five more cell divisions (i.e. the cells could have had a 32-fold excess of folate) when placed in a folate-free medium. We (Kamen & Capdevila, 1986) had shown that cells that are folate-replete would not accumulate significant amounts of 5-methyltetrahydrofolic acid, despite the fact that most cells concentrate folate 10–100 times from plasma *in situ*. Therefore, in order to study folate homeostasis, it seems as if cells have to be grown in physiological folate and perhaps even be made somewhat folate-deficient before they can be studied with regard to acquisition of folate.

In addition, recent evidence for a very strict control of the endogenous folate pool *in vivo* was presented by Houghton *et al.* (1990). Studying the potentiation of 5-fluorouracil by 5-formyltetrahydrofolic acid in human xenografts, it was found that in order to increase the (tumor) tissue pool of folate only 3–5-fold, and only temporarily, since tissue kinetics paralleled the decline in plasma folate, required a dose of *ca.* 100 mg kg^{-1}, i.e. 20 000 – 50 000 times the daily requirement for normal folate homeostasis.

Receptor-mediated folate transport in vitro

Studying cells grown in low folate (concentrations less than 1 nM), several laboratories have identified a folate binding protein, here-after referred to as the folate receptor,[1] on the membrane of KB (human nasopharyngeal carcinoma) cells, CACO2 (colon carcinoma) cells, MA104 (monkey kidney epithelial) cells and more recently in both murine and human leukemia cells (Henderson *et al.*, 1988; Jansen *et al.*, 1989*a,b*). The latter took growth for many passages in low (physiological) folate medium before it was detectable.

Using high-specific-activity 5-methyl[^{3}H]tetrahydrofolate we have shown that MA104 cells can concentrate folate over 100-fold from the media and rapidly convert it to higher forms (pentaglutamate) in only a few hours *in vitro* (Kamen *et al.*, 1988, 1989). These studies identified several essential steps in the accumulation of 5-methyltetrahydrofolate by folate-deficient MA104 cells, a monkey kidney cell line (reviewed in

---

[1] Based upon work reviewed by Wiley (1985) and initially proposed by Kaplan (1981). Receptors can be considered to be in one of two classes: class 1 is involved in hormonal binding and signal transduction and class 2 are molecules that facilitate uptake of molecules by cells.

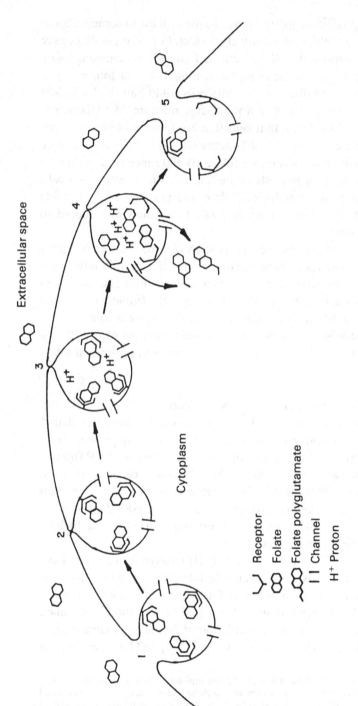

Figure 5.1. A model for receptor-mediated uptake of folate. Caveolae contain receptors that bind folate when the compartment is open (1). The caveola closes and possibly detaches from the membrane (2), and a proton gradient is generated that causes the folate to dissociate from the receptor (3). The high concentration of folate generated in the caveola space creates a gradient that favors movement across the membrane by an anion carrier. The folate is polyglutamated in the cytoplasm to retain the vitamin within the cell (4). The caveola reopens to initiate another round of folate uptake (5). From Rothberg *et al.* (1990) as reviewed by Anderson *et al.* (1992).

Anderson *et al.*, 1992). First, the 5-methyl[^{3}H]tetrahydrofolate is bound at the cell surface by a high-affinity, high-specificity receptor (this folate is removable with an acid–saline wash); second, the receptor–ligand complex is internalized into an acid-resistant, but still membrane-bound, compartment; third, the compartment is acidified and the folate dissociates from the receptor and is transferred to the cytoplasm, and fourth, the receptor returns to the cell surface (acid-labile fraction) to begin another round of folate binding and internalization (Figure 5.1).

Recent studies (Kamen *et al.*, 1991) of the effect of probenecid on 5-methyl[^{3}H]tetrahydrofolic acid accumulation by MA104 cells have provided more evidence for the existence of an anion transporter coupled to the receptor. Probenecid (10 mM) blocked cytoplasmic accumulation of 5-methyl[^{3}H]tetrahydrofolic acid but had no significant effect on cell binding or internalization of folate into a protected compartment (Zwiener *et al.*, 1992). Moreover, probenecid also blocked efflux of 5-methyl[^{3}H]tetrahydrofolic acid when the cells were first loaded with folate using pharmacological concentrations of ligand (2 μM) in the extracellular medium. With regard to the uptake of ligand, 5-methyl[^{3}H]tetrahydrofolic acid would accumulate in MA104 cells in which the receptor was first blocked with folic acid (to eliminate receptor binding of ligand) only when it was present at large extracellular concentrations. The K_t and V_{max} for receptor-independent, probenecid-sensitive uptake of 5-methyl[^{3}H]tetrahydrofolic acid were 1 μM and 20 pmol h^{-1} per 10^6 cells, respectively. Comparing receptor-mediated and -independent folate accumulation by folate-depleted MA104 cells reveals a marked increase in the efficiency of receptor-coupled transport at physiological concentrations of ligand. For example, at an extracellular concentration of 2 nM (0.002 μM), 5-methyl[^{3}H]tetrahydrofolic acid is accumulated at a rate of 0.7–0.8 pmol h^{-1} per 10^6 cells when the receptor is operative, but less than 0.04 pmol h^{-1} per 10^6 cells when the receptor is blocked, a difference of approximately 20-fold.

In addition to increasing the efficiency of folate uptake, receptor-coupled transport is under strict control (Kamen *et al.*, 1989). After 4–6 h of folate repletion, the rate of accumulation decreases significantly and the cells already have 75–80% of the total folate measured at 24 h of uptake. The rate of accumulation seems to be parallel to the synthesis of folylpentaglutamates. Thus folate accumulation by these cells is a self-limiting process, the acute regulation of which may be at the level of the transport or metabolism stage. It is more difficult to assess the efficiency of receptor-coupled transport in KB cells, which have a surface binding capacity of 50–100 pmol folate per 10^6 cells (compared with 1–2 in MA104 cells). This amount greatly exceeds the intracellular folate requirements as assessed by replication time and morphology; thus the efficiency of receptor-coupled folate accumulation process could easily be underestimated.

Receptor distribution in vivo

Recently, Colnaghi and her associates (1987, 1990) identified a glycosylphosphatidylinositol-anchored, 38 kDa glycoprotein (GP38) overexpressed in human ovarian carcinoma. GP38 is recognized by two monoclonal antibodies, MOv18 and MOv19. Purification of the protein and cloning of the cDNA revealed this glycoprotein to be the folate receptor (Campbell *et al.*, 1991; Coney *et al.*, 1991). Using Western blot and immunohistochemistry with MOv19, we have been able to detect as little as 40 pg (*ca.* 1 fmol) of receptor protein. In a survey of over 20 normal tissues we found immunoreactivity limited to the epithelium of the choroid plexus, fallopian tube, uterus and epididymis. The acinar cells of the breast, submandibular salivary and bronchial glands also showed intense staining, as did the trophoblastic cells of the placenta (Weitman *et al.*, 1992a,b). The distribution of this receptor has been confirmed by Northern blot and radioimmunoassay (Willis *et al.*, 1992; Weitman *et al.*, 1992a). The apparent distribution of the receptor in these tissues would again suggest a transport or storage role.

As noted earlier, receptor expression may be enhanced on malignant tissues (Rothenberg, 1970). Using Western and Northern blot analysis and [^{3}H]folic acid binding, we and others have found overexpression of the folate receptor on malignant cell lines and tissues (Vegglan *et al.*, 1989; Willis *et al.*, 1992; Weitman *et al.*, 1992a). Specifically, malignant cells bound 20 pmol or more of [^{3}H]folate per 10^6 cells, while normal epithelial cells and fibroblasts bound under 1 pmol of radioligand per 10^6 cells (Weitman *et al.*, 1992a). We have also surveyed more than 50 fresh human tumor samples (Table 5.3) for receptor by immunoblotting with MOv19. These data suggest that folate receptor expression occurs in a variety of tumor types, primarily epithelial carcinomas. The wide distribution of the folate receptor on human malignancies has recently been confirmed by Garin-Chesa *et al.* (1993).

Molecular biology and biochemistry of the folate receptor

In 1989 four laboratories reported isolation of a cDNA for the folate receptor (Sadasivan & Rothenberg, 1989; Lacey *et al.*, 1989;

Table 5.3. *Folate receptor Western blot reactivity*

Tumor type	Total number	Positive	Negative
Brain	17		
astrocytoma	7	2	5
ependymoma	5	4	1
other	5	2	3
Breast	6		
carcinoma	5	3	2
fibrocystic	1	0	1
Lung	3		
carcinoma		3	0
Kidney	4		
renal cell	4	2	0
Wilm's	2	0	2
Ovarian	9		
carcinoma	7	5	2
adenoma	2	1	1
Cervical	6		
carcinoma		6	0
Lymphoma	1		
Hodgkin's			1
Uterus	4		
carcinoma	3	0	3
other	1	1	0
Bone	1		
osteosarcoma		0	1

Elwood, 1989; Ratnam *et al.*, 1989). The cDNA sequence is approximately 1100 base pairs and codes for a polypeptide that contains 250–275 amino acids with three potential sites for glycosylation. The gene(s) or pseudogenes have been mapped to chromosome 11 q 13.3–q 14.1 (Campbell *et al.*, 1991; Rogoussis *et al.*, 1993). The cDNAs from CACO2, KB, and human placenta libraries are identical and the predicted amino acid sequence of the membrane form is more than 99% identical to the actual analysis of the FBP from milk (Svendsen *et al.*, 1982, 1984). This correlation at the nucleic acid level of analysis, and [^{35}S]methionine labeling data showing a product–precursor relationship of the soluble and membrane FBP *in vitro* (Kane *et al.*, 1986a), supports the hypothesis that the soluble form of the receptor is a processed product of the membrane form; that is, the soluble folate binding protein is folate receptor released from the cell membrane. This has been confirmed using immunoprecipitation and SDS–PAGE electrophoresis and more directly by RNA blot analysis which detected only one poly(A)$^+$ mRNA (Sadasivan & Rothenberg, 1989). Of interest is that Ratnam *et al.* (1989) and Wang *et al.* (1992) noted that two homologous forms of folate binding protein (MFR-1 and MFR-2) are expressed in placental tissue, one of which is identical to the KB cell receptor and the other speculated to represent an embryonic or fetal type used during cellular proliferation. The characteristics of this second sequence have not yet been presented. Elwood (1989) reported only one message in the placental library. However, Page *et al.* (1993) recently confirmed the presence of an additional cDNA isolated from the human placenta. The function(s) and significance of these multiple cDNAs is unclear.

The cDNA sequence also predicted that the receptor would be coupled to the membrane via a glycosylphosphatidylinositol (GPI) linkage (Lacey *et al.*, 1989). This was confirmed by releasing the receptor with a glycosylphosphatidylinositol-specific phospholipase C (Lacey *et al.*, 1989; Luhrs & Stomiany, 1989). The latter researchers also determined that the glycerol backbone of the GPI anchor consisted of a fatty acyl group (docosanoic acid 22:0) and an alkyl component consisting of a mixture of fatty alcohols attached to the C1 carbon. The presence of covalently bound fatty acids on the inositol ring was also suggested by lipid analysis of the products of nitrous acid deamination.

Epilogue

Folate binding protein: carrier, receptor, growth factor?

The folate binding protein was discovered in milk over two decades ago. Our understanding of its role in folate homeostasis is still evolving. Indeed, another recent review questions whether it should be called a receptor or just a binding protein (Henderson, 1990). Since, at least in one type of epithelial cell, it is important in the accumulation of folate under physiological conditions of growth *in vitro*, and since it is inducible in lymphoblasts grown in low folate concentrations (Henderson *et al.*, 1988; Jansen *et al.*, 1989*a*,*b*) and found in fetal tissue, we believe the designation of receptor is appropriate, as this connotes a function, more than just ligand binding, with respect to foliate homeostasis. Further, since Antony *et al.* (1987) have shown that anti-placental folate receptor is mitogenic for normal human bone marrow in soft agar and the receptor is GPI-linked at least in KB and MA104 cells, as well as in transfected

COS (Lacey *et al.*, 1989) cells, it is not unreasonable that some sort of signaling or message for cell division is related to the accumulation of a vitamin essential for purine and thymidine synthesis. In addition, the fact that the receptor can become a plasma (soluble) folate binding protein through the action of a specific phosphatidylinositol phospholipase C or D (the latter found in plasma), and the possibility of the receptor becoming a carrier (such as transcobalamin I for vitamin B_{12}) and having paracrine- or autocrine-like effects, needs to be explored (see reviews of GPI-linked proteins by Ferguson & Williams (1988), Lisanti *et al.* (1990), Low (1989) and Thomas *et al.* (1990) for a comprehensive review of the versatility of GPI-anchored proteins). A minimal amount of investigation has suggested that folate–binder complex will be cleared by the liver (Rubinoff *et al.*, 1981), but uptake into other tissues is lacking.

The next decade should, with the availability of modern molecular techniques to study the receptor at the DNA, RNA and protein level, allow confirmation and clarification of the earlier, descriptive studies of folate binding, the relationship of soluble and membrane-anchored 'binders' and the role of folate receptor in cellular and/or organism homeostasis.

Pharmacological and medical implications

Folate is an essential molecule for normal cell growth and division. Understanding its regulation (homeostasis) should lead to a more complete understanding of antifolate pharmacodynamics and perhaps more specific therapy for diseases marked by inappropriate proliferation of cells (e.g. cancer, psoriasis, autoimmune diseases) in which antifolates are already known to be effective pharmacologic agents. Studies by several investigators (Antony *et al.*, 1987; Dixon *et al.*, 1992; Luhrs *et al.*, 1992) also raise the possibility that folate accumulation is intricately involved in the process of cell replication (and maybe transformation). In this regard, several laboratories (Dixon *et al.*, 1992; Luhrs *et al.*, 1992; Matsue *et al.*, 1992) have shown that transfection of the cDNA for the receptor results in a selective growth advantage for transfected cells grown in low folate. Therefore, modulation of the folate receptor result in signal transduction of as yet undefined events leading to appropriate or inappropriate cell division?

Folate, like other vitamins, is required for cell growth under normal conditions. It seems reasonable to assume that cells have derived efficient and specific means of hoarding these required molecules. Further, it is not unreasonable to ask whether the control mechanism for folate homeostasis is part of the regulatory mechanism(s) in the cell cycle. Every time a cell undergoes division, the folate pool will theoretically be halved; increasing the folate pool may therefore be an important control point in cell proliferation. The effective use of antifolates over the past four decades has proven this empirically. Perhaps a better understanding of folate homeostasis will further lend itself to specific therapies as well as to expanding our basic knowledge of how cells manage small but critical molecules.

References

Alberti, S., Miotti, M., Fornaro, M., Mantovani, S., Canevari, S., Menard, S. & Colnaghi, M.L. (1990) The CA-MOv 18 molecule, a cell-surface marker of human ovarian carcinomas, is anchored to the cell membrane by phosphatidylinositol. *Biophys. Res. Comm.* **17**, 1051–5.

Alberts, B., Bray, D., Lewis, J., Raff, M., Roberts, K. & Watson, J.D. (1989) *Molecular Biology of the Cell.* Garland Publishing, New York.

Anderson, R.G.W., Kamen, B.A., Rothberg, K.G. & Lacey, S.W. (1992) Potocytosis: sequestration and transport of small molecules by caveola. *Science* **255**, 410–11.

Antony, A.C. (1992) The biochemical chemistry of folate receptors. *Blood* **79**, 2807–20.

Antony, A.C., Bruno, E., Briddell, R.A., Brandt, J.E., Verma, R.S. & Hoffman, R. (1987) Effect of perturbation of specific folate receptors during in vitro erythropoiesis. *J. Clin. Invest.* **80**, 1618–23.

Antony, A.C., Utley, C. & Van Horne, K.C. (1981) Isolation and characterization of a folate receptor from human placenta. *J. Biol. Chem.* **256**(18), 9684–92.

Archibald, E.L., Mincey, E.K. & Morrison, R.T. (1972) Estimation of serum folate levels by competitive protein binding assay. *Clin. Biochem.* **6**, 274–84.

Baugh, C.M., Krumdieck, C.L., Baker, H.J. & Butterworth, C.E. (1971) Studies on the absorption and metabolism of folic acid. I. Folate absorption following exposure of isolated intestinal segments of synthetic pteroyl-poly-glutamates of various chain lengths. *J. Clin. Invest.* **50**, 2009–17.

Bentsen, K.D., Hansen, S.I., Holm, J. & Lyngbye, J. (1990) Abnormalities in folate binding pattern of serum from a patient with megaloblastic anemia and folate deficiency. *Clin. Chim. Acta* **109**, 225–8.

Bernstein, L.H., Gutstein, S., Weiner, S. & Efron, G. (1970) The absorption and malabsorption of folic acid and its polyglutamates. *Am. J. Med.* **48**, 570–9.

Blakley, R.L. & Benkovic, S.J. (1984) *Folates and Pterins.* John Wiley & Sons, New York.

Bobzien, W.F. & Goldman, I.D. (1972) The mechanism of folate transport in rabbit reticulocytes. *J. Clin. Invest.* **51**, 1688–96.

Burgen, A.S.V. & Goldberg, N.J. (1962) Absorption of folic acid from the small intestine of the rat. *Brit. J. Pharm. Chem.* **19**, 313–20.

Butterworth, C.E. & Krumdieck, C.L. (1975) Intestinal absorption of folic acid monoglutamates and polyglutamates: a brief review of some recent developments. *Brit. J. Haemat.* **31**(suppl), 111–18.

Campbell, I.G., Jones, T., Foulkes, W.D. & Trowsdale, J. (1991) High-affinity folate binding protein is a marker for ovarian cancer. *Cancer Res.* **51**, 5329–38.

Chen, C. & Wagner, C. (1975) Folate transport in the choroid plexus. *Life Sci.* **16**, 1571–82.

Colman, N. & Herbert, V. (1980) Folate-binding proteins. *Ann. Rev. Med.* **31**, 433–9.

Coney, L.R., Tomassetti, A., Carayannopoulos, L., Frasca, V., Kamen, B.A., Colnaghi, M. & Zurawski, V.R. (1991) Cloning of a tumor-associated antigen: MOv18 and MOv19 antibodies recognize a folate binding protein. *Cancer Res.* **52**, 3396–401.

Cook, R.J. & Wagner, C. (1984) Glycine N-methyltransferase is a folate binding protein of rat liver cytosol. *Proc. Natl Acad. Sci. U.S.A.* **81**, 3631–4.

Cooper, B.A. (1970) Studies of ^{3}H-folic acid uptake by *Lactobacillus casei*. *Biochim. Biophys. Acta* **208**, 99–109.

Corrocher, R., De Sandre, G. & Pacor, M.L. (1974) Hepatic protein binding of folate. *Clin. Sci. Molec. Med.* **46**, 551–4.

DaCosta, M. & Rothenberg, S.P. (1974) Appearance of a folate binder in leukocytes and serum of women who are pregnant or taking oral contraceptives. *J. Lab. Clin. Med.* **83**, 207–14.

Dembo, M. & Sirotnak, F.M. (1984) Membrane transport of folate compounds in mammalian cells. In *Folate Antagonists as Therapeutic Agents* (ed. F.M. Sirotnak, J.J. Burchall, W.B. Ensminger & J.A. Montgomery), pp. 173–217. Academic Press, Orlando, Florida.

Dixon, K.H., Mulligan, T., Chung, K.N., Elwood, P.C. & Cowan, K.H. (1992) Effects of receptor expression following stable transfection into wild type and methotrexate transport-deficient ZR-75-1 human breast cancer cells. *J. Biol. Chem.* **267**, 24140–7.

Dunn, R.T. & Foster, L.B. (1973) Radioassay of serum folate. *Clin. Chem.* **19**, 1101–5.

Elwood, P.C. (1989) Molecular cloning and characterization of the human folate-binding protein cDNA from placenta and malignant tissue culture (KB) cells. *J. Biol. Chem.* **264**, 14893–901.

Elwood, P.C., Kane, M.A., Portillo, R.M. *et al.*, (1986) The isolation, characterization, and comparison of the membrane-associated and soluble folate-binding proteins from human KB cells. *J. Biol. Chem.* **261**(33), 15416–23.

Fan, J., Vitols, K.S. & Huennekens, F.M. (1992) Multiple folate transport systems in L1210 cells. *Adv. Enzyme Reg.* **32**, 3–15.

Ferguson, M.A.J. & Williams, A.F. (1988) Cell-surface anchoring of proteins via glycosyl-phosphatidylinositol structures. *Ann. Rev. Biochem.* **57**, 285–320.

Fischer, C.D., da Costa, M. & Rothenberg, S.P. (1975) The heterogeneity and properties of folate binding proteins from chronic myelogenous leukemia cells. *Blood* **46**(6), 855–67.

Fleming, G.F. & Schilsky, R.L. (1992) Antifolates: the next generation. *Sem. Oncol.* **19**, 707–19.

Ford, J.E. (1974) Some observations on the possible nutritional significance of vitamin B_{12} and folate-binding proteins in milk. *Brit. J. Nutr.* **31**, 243–7.

Ford, J.E., Knaggs, G.S., Salter, D.N. & Scott, K.J. (1972) Folate nutrition in the kid. *Brit. J. Nutr.* **27**, 571–83.

Ford, J.E., Salter, D.N. & Scott, K.J. (1969) The folate-binding protein in milk. *J. Dairy Res.* **36**, 435–48.

Galivan, J. (1981) Transport of methotrexate by primary cultures of rat hepatocytes: Stimulation of uptake in vitro by the presence of hormones in the medium. *Arch. Biochem. Biophys.* **206**(1), 113–21.

Garin-Chesa, P., Campbell, I., Saigo, P.E., Lewis, J.L., Old, L.J. & Rettig, W.J. (1993) Trophoblast and ovarian cancer antigen lk26. *Am. J. Path.* **142**, 557–67.

Gewirtz, D.A., White, J.C., Randolph, J.K. & Goldman, I.D. (1980) Transport, binding, and polyglutamation of methotrexate in freshly isolated rat hepatocytes. *Canc. Res.* **40**, 573–8.

Chitis, J. (1966) The labile folate of milk. *Amer. J. Clin. Nutr.* **18**, 452–7.

Ghitis, J. & Lora, C. (1967) The folate binding in milk. *Amer. J. Clin. Nutr.* **20**(1), 1–4.

Hansen, S.I., Holm, J. & Lyngbye, J. (1980) High-affinity protein binding of folate in urine. *IRCS Med. Sci.* **8**, 846.

Hansen, S.I., Holm, J. & Lyngbye, J. (1985) A high-affinity folate binding protein in human cerebrospinal fluid. *Acta. Neurol. Scand.* **71**, 133–5.

Henderson, G.B. (1990) Folate-binding proteins. *Ann. Rev. Nutr.* **10**, 319–35.

Henderson, G.B. & Strauss, B.P. (1990) Characteristics of a novel transport system for folate compounds in wild-type and methotrexate-resistant L1210 cells. *Canc. Res.* **50**, 1709–14.

Henderson, G.B. & Tsuji, J.M. (1987) Methotrexate efflux in L1210 cells. *J. Biol. Chem.* **262**(28), 13571–8.

Henderson, G.B., Tsuji, J.M. & Kumar, H.P. (1987) Transport of folate compounds by leukemic cells. *Biochem. Pharm.* **36**(18), 3007–14.

Henderson, G.B., Tsuji, J.M. & Kumar, H.P. (1988) Mediated uptake of folate by a high-affinity binding protein in sublines of L1210 cells adapted to nanomolar concentrations of folate. *J. Memb. Biol.* **101**, 247–58.

Henderson, G.B. & Zevely, E.M. (1984) Affinity labeling of the 5-methyltetrahydrofolate/methotrexate transport protein of L1210 cells by treatment with an N-hydroxysuccinimide ester of [^{3}]Methotrexate. *J. Biol. Chem.* **259**(7), 4558–62.

Henderson, G.B. & Zevely, E.M. (1985) Characterization of the multiple transport routes for methotrexate in L1210 cells using phthalate as a model anion substrate. *J. Memb. Biol.* **85**, 263–8.

Herbert, V., Larrabee, A.R. & Buchanan, J.N. (1962) Studies on the identification of folate compound of human serum. *J. Clin. Invest.* **41**, 1134–8.

Herbert, V. & Zalusky, R. (1961) Selective concentration of folic acid activity in cerebrospinal fluid. *Fed. Proc. Fedn Am. Socs Exp. Biol.* **20**, 453.

Hill, B.T., Bailey, B.D., White, J.C. & Goldman, I.D. (1979) Characteristics of transport of 4-amino antifolates and folate compounds by two lines of L5178Y lymphoblasts, one with impaired transport of methotrexate. *Canc. Res.* **39**, 2440–6.

Hines, J.D., Kamen, B.A. & Caston, J.D. (1973) Abnormal folate binding proteins in azotemic patients. *Blood* **42**, 997.

Holm, J., Hansen, S.I. & Hoier-Madsen, M. (1990) A high-affinity folate binding protein in human amniotic fluid. Radioligand binding characteristics, immunological properties and molecular size. *Biosci. Rpts* **10**, 79–85.

Houghton, J.A., Williams, L.G., de Graaf, S.S.N. *et al.* (1990) Relationship between dose rate of [6RS] Leucovorin administration, plasma concentrations of reduced folates, and pools of 5,10-

methylenetetrahydrofolates and tetrahydrofolates in human colon adenocarcinoma xenografts. *Canc. Res.* **50**, 3493–502.

Huennekens, F.M., Vitols, K.S. & Henderson, G.B. (1978) Transport of folate compounds in bacterial and mammalian cells. In *Advances in Enzymology*, vol. 47 (ed. A. Meisler), pp. 313–45. John Wiley and Sons, New York.

Huennekens, F.M., DiGirolamo, P.M. & Fujii, K. (1974) Folic acid and vitamin B_{12}: Transport and conversion to coenzyme forms. In *Advances in Enzyme Regulation*, vol. 12 (ed. G. Weber), pp. 131–53. Pergamon Press, Oxford.

Jacob, E. & Herbert, V. (1974) Evidence against transferrin as a binder of either vitamin B_{12} or folic acid. *Blood* **43**, 767–8.

Jansen, G., Kathmann, I., Rademaker, B.C., Braakuis, B.J.M., Westerhof, G.R., Rijksen, G. & Schornagel, J.H. (1989a) Expression of a folate binding protein in L1210 cells grown in low folate medium. *Canc. Res.* **49**, 1959–63.

Jansen, G., Westerhof, G.R. & Kathmann, I. (1989b) Identification of a membrane-associated folate-binding protein in human leukemic CCRF-CEM cells with transport-related methotrexate resistance. *Canc. Res.* **49**, 2455–9.

Johns, D.G., Sperti, S. & Burgen, A.S.V. (1961) The metabolism of tritiated folic acid in man. *J. Clin. Invest.* **40**, 1684–96.

Kamen, B.A. (1987) Folic acid antagonists. In *Metabolism and Action of Anti-cancer Drugs* (ed. G. Powis & R.A. Prough), pp. 141–55. Taylor & Francis, London.

Kamen, B.A. & Capdevila, A. (1986) Receptor-mediated folate accumulation is regulated by the cellular folate content. *Proc. Natl Acad. Sci. U.S.A.* **83**, 5983–7.

Kamen, B.A. & Caston, J.D. (1974) Direct radiochemical assay for serum folate: competition between ^{3}H-folic acid and 5-methyltetrahydrofolic acid for a folate binder. *J. Lab. Clin. Med.* **83**, 164–74.

Kamen, B.A. & Caston, J.D. (1975) Identification of a folate binder in hog kidney. *J. Biol. Chem.* **250**(6), 2203–5.

Kamen, B.A. & Caston, J.D. (1980) Purification of folate binding factors. In *Methods in Enzymology* (ed. D.B. McCormick & L.D. Wright), vol. 66, pp. 678–94. Academic Press, New York.

Kamen, B.A. & Caston, J.D. (1986) Properties of a folate binding protein (FBP) isolated from porcine kidney. *Biochem. Pharm.* **35**(14), 2323–9.

Kamen, B.A., Gross, S. & Caston, J.D. (1975) Distribution of folate binding proteins in serum and spinal fluid of patients with acute lymphoblastic leukemia. *Am. Soc. Ped. Res.* **9**, 389 Abs.

Kamen, B.A., Johnson, C.A., Wang, M.-T. & Anderson, R.G.W. (1989) Regulation of the cytoplasmic accumulation of 5-methyltetrahydrofolate in MA104 cells is independent of folate receptor regulation. *J. Clin. Invest.* **84**, 1379–86.

Kamen, B.A., Smith, A.R. & Anderson, R.G.W. (1991) The folate receptor works in tandem with a probenecid-sensitive carrier in MA104 cells in vitro. *J. Clin. Invest.* **87**, 1442–9.

Kamen, B.A., Wang, M.T. & Streckfuss, A.J. (1989) Delivery of folates to the cytoplasm of MA104 cells is mediated by a surface membrane receptor that recycles. *J. Biol. Chem.* **263**(27), 13602–9.

Kane, M.A., Elwood, P.C., Portillo, R.M., Antony, A.C. & Kolhouse, J.F. (1986) The interrelationship of the soluble and membrane-associated folate-binding proteins in human KB cells. *J. Biol. Chem.* **261**(33), 15625–31.

Kane, M.A., Portillo, R.M., Elwood, P.C., Antony, A.C. & Kolhouse, J.F. (1986*b*) The influence of extracellular folate concentration on methotrexate uptake by human KB cells. *J. Biol. Chem.* **261**(1), 44–9.

Kane, M.A. & Waxman, S. (1989) Role of folate binding proteins in folate metabolism. *Lab. Invest.* **60**, 737–46.

Kaplan, J. (1981) Polypeptide-binding membrane receptors: Analysis and classification. *Science* **212**, 14–20.

Lacey, S.W., Sanders, J.M., Rothberg, K.G., Anderson, R.G.W. & Kamen, B.A. (1989) Complementary DNA for the folate binding protein correctly predicts anchoring to the membrane by glycosyl-phosphatidylinositol. *J. Clin. Invest.* **74**, 715–20.

Lederle Laboratories (1950) *The Nutritional and Clinical Significance of Folic Acid.*

Leslie, G.I. & Rowe, P.B. (1972) Folate binding by the brush border membrane proteins of small intestinal epithelial cells. *Biochemistry* **11**(9), 1696–703.

Levi, A.J., Gatmaitan, Z. & Arias, I.M. (1969) Two hepatic cytoplasmic protein fractions, Y and Z, and their possible role in the hepatic uptake of bilirubin, sulfobromophthalein and other anions. *J. Clin. Invest.* **48**, 2156–67.

Lichtenstein, N.S., Oliverio, V.T. & Goldman, I.D. (1969) Characteristics of folic acid transport i the L1210 leukemia cell. *Biochim. Biophys. Acta* **193**, 456–67.

Lisanti, M.P., Rodriguez-Boulan, E. *et al.* (1990) Emerging functional roles for the glycosyl-phosphatidylinositol membrane protein anchor. *J. Memb. Biol.* **117**, 1–10.

Low, M.G. (1989) Glycosyl-phosphatidylinositol: a versatile anchor for cell surface proteins. *FASEB J.* **3**, 1600–8.

Luhrs, C.A., Pitiranggon, P., DaCosta, M., Rothenberg, S.P., Slomiany, B.L., Brink, L., Tous, G.I. & Stein, S. (1978) Purified membrane and soluble folate binding proteins from cultured KB cells have similar amino acid compositions and molecular weights but differ in fatty acid acylation. *Proc. Natl Acad. Sci. U.S.A.* **84**, 6546–9.

Luhrs, C.A. & Slomiany, B.L. (1989) A human membrane associated folate binding protein is anchored by a glycosylphosphatidylinositol tail. *J. Biol. Chem.* **264**, 21446–9.

Luhrs, C.A., Raskin, C.A., Durbin, R., Wu, B., Sadasivan, E., McAllister, W. & Rothenberg, S.P. (1992) Transfection of a glycosylated phosphatidylinositol-anchored folate-binding protein complementary DNA provides cells with the ability to survive in low folate medium. *J. Clin. Invest.* **90**, 840–7.

Mantzos, J. (1975) Radioassay of serum folate with use of pig plasma folate binders. *Acta Haemat.* **54**, 289–96.

Mantzos, J.D., Alevizou-Terzaki, V. & Gyftake, E. (1974) Folate binding in animal plasma. *Acta Haemat.* **51**, 204–10.

Markkanen, T. (1968) Pteroylglutamic acid (PGA) activity of serum in gel filtration. *Life Sci.* **7**(2), 887–95.

Markkanen, T. & Himanen, P. (1971) Pteroylglutamic acid in cerebrospinal fluid. *Int. J. Vit. Nutr. Res.* **41**, 79–85.

Markkanen, T., Himanen, P. & Pajula, R.-L. (1974) Binding of folic acid to serum proteins III. The effect of pernicious anaemia. *Acta Haemat.* **51**, 193–203.

Markkanen, T., Himanen, P., Pajula, R.L. Ruponen, S. and Castren, O. (1973*a*) Binding of folic acid to serum proteins I. The effect of pregnancy. *Acta Haemat.* **50**, 85–91.

Markkanen, T., Himanen, P., Pajula, R.-L. & Molnar, G. *et al.* (1973*b*) Binding of folic acid to serum proteins II. The effect of diphehylhydantoin treatment of various diseases. *Acta Haemat.* **50**, 284–92.

Markkanen, T., Pajula, R.-L., Himanen, P. & Virtanen, S. *et al.* (1973*c*) Serum folic acid activity (*L. casei*) in sephadex gel chromatography. *J. Clin. Path.* **26**, 486–93.

Markkanen, T., Pajula, R.-L., Virtanen, S. & Himanen, P. (1972*a*) New carrier protein(s) of folic acid in human serum. *Acta Haemat.* **48**, 145–50.

Markkanen, T., Peltola, O. & Himanen, P. (1971) Metabolic aspects of serum PGA in megaloblastic states. *Int. J. Vit. Nutr. Res.* **41**, 457–63.

Markkanen, T., Virtanen, S., Himanen, P. & Pajula, R.L. (1972*b*) Transferrin, the third carrier protein of folic acid activity in human serum. *Acta Haemat.* **48**, 213–17.

Matherly, L.H., Czajkowski, C.A., Muench, S.P. & Psiakis, J.T. (1990) Role for cytosolic folate-binding proteins in the compartmentation of endogenous tetrahydrofolates and the 5-formyl tetrahydrofolate-mediated enhancement of 5-fluoro-2'-deoxyuridine antitumor activity in vitro. *Canc. Res.* **50**, 3262–9.

Matsue, H., Rothberg, K.G., Takashima, A., Kamen, B.A., Anderson, R.G.W. & Lacey, S.W. (1992) Folate receptor allows cells to grow in low concentrations of 5-methyltetrahydrofolate. *Proc. Natl. Acad. Sci. USA* **89**, 6006—9.

McHugh, M. & Cheng, Y.-C. (1979) Demonstration of a high affinity folate binder in human binder in human cell membranes and its characterization in cultured human KB cells. *J. Biol. Chem.* **254**, 11312–18.

McIntyre, L.J., Dow, J.W. *et al.*, (1975) Cellular acquisition of folate: a compartmental model. *J. Mol. Med.* **1**, 3–10.

Metz, J., Zalusky, R. & Herbert, V. (1968) Folic acid binding by serum and milk. *Amer. J. Clin. Nutr.* **21**, 289–97.

Mincey, E.K., Wilcox, E. & Morrison, R.T. (1973) Estimation of serum and red cell folate by a simple radiometric technique. *Clin. Biochem.* **6**, 274–84.

Miotti, S., Canevari, S., Menard, S., Mezzanzanica, D., Porro, G., Pupa, S.M., Regazzoni, M., Tagliabue, E. & Colnaghi, M.I. (1987) Characterization of human ovarian carcinoma-associated antigens defined by novel monoclonal antibodies with tumor restricted specificity. *Int. J. Cancer.* **39**, 297–303.

Nixon, P.F., Slutsky, G., Nahas, A. & Bertino, J.R. (1973) The turnover of folate coenzymes in murine lymphoma cells. *J. Biol. Chem.* **248**, 5932–6.

Page, S.T., Owen, W.C., Price, K. & Elwood, P.C. (1993) Expression of the human placental folate receptor transcript is regulated in human tissues. *J. Molec. Biol.* **229**, 1175–83.

Pedersen, T.G., Svendsen, I., Hansen, S.I., Holm, J. & Lyngbye, J. (1980) Aggregation of a folate-binding protein from cow's milk. *Carlsberg Res. Commun.* **45**, 161–6.

Price, E.M., Ratnam, M., Rodeman, K.M. & Freisheim, J.H. (1988) Characterization of the methotrexate transport pathway in murine L1210

134   *S. Weitman* et al.

leukemia cells: Involvement of a membrane receptor and a cytosolic protein. *Biochemistry* **27**, 7853–8.

Price, E.M., Smith, P.L., Klein, T.E. & Freisheim, J.H. (1987) Photoaffinity analogues of methotrexate as folate antagonist binding probes. 1. Photoaffinity labeling of murine L1210 dihydrofolate reductase and amino acid sequence of the binding region. *Biochemistry* **26**, 4751–6.

Ragoussis, J., Senger, G., Trowsdale, J. & Campbell, I.G. (1992) Genomic organization of the human folate receptor genes on chromosome 11q13. *Genomics* **14**, 423–30.

Ratnam, M., Marquardt, H., Duhring, L. & Freisheim, J.H. (1989) Homologous membrane folate binding proteins in human placenta: Cloning and sequence of a cDNA. *Biochemistry* **28**, 8249–54.

Reisenauer, A.M. (1990) Affinity labeling of the folate-binding protein in pig intestine. *Biochem.J.* **267**, 249–52.

Retief, F.P., Heyns, A.Du P., Oosthuizen, M., van Reenen, O.R. & Badenhorst, C.J. (1976) In vitro binding of folates by body fluids. *Brit. J. Haemat.* **32**, 113–21.

Rosenberg, I.H., Streiff, R.R., Godwin, H.A. & Castle, W.B. (1969) Absorption of polyglutamic folate: participation of deconjugating enzymes of the intestinal mucosa. *New Engl. J. Med.* **280**, 985–8.

Rothberg, K.G., Ying, K., Kolhouse, J.F., Kamen, B.A. & Anderson, R.G.W. (1990) The glycophospholipid-linked folate receptor internalizes folate without entering the clathrin-coated pit endocytic pathway. *J. Cell Biol.* **110**, 637–49.

Rothenberg, S.P. (1970) A macromolecular factor in some leukemic cells which binds folic acid. *Proc. Soc. Exp. Biol. Med.* **133**(2), 428–32.

Rothenberg, S.P. & DaCosta, M. (1971) Further observations on the folate-binding factor in some leukemic cells. *J. Clin. Invest.* **50**, 719–26.

Rothenberg, S.P., DaCosta, M. & Rosenberg, Z. (1972) A radioassay for serum folate: use of a two-phase sequential incubation, ligand-binding system. *New Engl. J. Med.* **286**, 1335–9.

Rubinoff, M., Abramson, R., Schreiber, C. & Waxman, S. (1981) Effect of a folate-binding protein on the plasma transport and tissue distribution of folic acid. *Acta Haemat.* **65**, 145–52.

Sadasivan, E. & Rothenberg, S.P. (1989) The complete amino acid sequence of a human folate binding protein from KB cells determined from the cDNA. *J. Biol. Chem.* **264**, 5806–11.

Salter, D.N., Ford, J.E., Scott, K.J. & Andrews, P. (1972) Isolation of the folate-binding protein from cow's milk by the use of affinity chromatography. *FEBS Lett.* **20**, 302–6.

Salter, D.N., Scott, K.J., Slade, H. & Andrews, P. (1981) The preparation and properties of folate-binding protein from cow's milk. *Biochem. J.* **193**, 469–76.

Selhub, J., Brin, H. & Grossowicz, N. (1973) Uptake and reduction of radioactive folate by everted sacs of rat small intestine. *Eur. J. Biochem.* **33**, 433–8.

Selhub, J. & Franklin, W.A. (1984) The folate-binding protein of rat kidney. *J. Biol. Chem.* **259**, 6601–6.

Sennett, C. & Rosenberg, L.E. (1981) Transmembrane transport of cobalamin in prokaryotic and eukaryotic cells. *Ann. Rev. Biochem.* **50**, 1053–86.

Shane, B. & Stokstad, E.L.R. (1975) Transport of folates by bacteria. *J. Biol. Chem.* **250**, 2243–53.

Sirotnak, F.M. (1985) Obligate genetic expression in tumor cells of a fetal membrane property mediating 'folate' transport: Biological significance and implications for improved therapy of human cancer. *Canc. Res.* **45**, 3992–4000.

Sirotnak, F.M., Burchall, J.J., Ensminger, W.B. & Montgomery, J.A. (1984) *Folate Antagonists as Therapeutic Agents*. Academic Press, Orlando, Florida.

Sirotnak, F.M., Goutas, L.J., Jacobsen, D.M. *et al.* (1987) Carrier-mediated transport of folate compounds in L1210 cells. *Biochem. Pharm.* **36**(10), 1659–67.

Spector, R. & Lorenzo, A.V. (1975) Folate transport by the choroid plexus *in vitro*. *Science* **187**, 540–2.

Svendsen, I. (1984) The complete amino acid sequence of the folate-binding protein from cow's milk. *Carlsberg. Res. Commun.* **49**, 123–31.

Svendsen, I., Hansen, S.I., Holm, J. & Lyngbye, J. (1982) Amino acid sequence homology between human and bovine low molecular weight folate binding protein isolated from milk. *Carlsberg Res. Commun.* **47**, 371–6.

Svendsen, I., Hansen, S.I., Holm, J. & Lyngbye, J. (1984) The complete amino acid sequence of the folate-binding protein from cow's milk. *Carlsberg Res. Commun.* **49**, 123–31.

Tani, M., Fushiki, T. & Iwai, K. (1983) Influence of folate-binding protein from bovine milk on the absorption of folate in gastrointestinal tract of rat. *Biochim. Biophys. Acta* **757**, 274–81.

Tani, M. & Iwai, K. (1984) Some nutritional effects of folate-binding protein in bovine milk on the bioavailability of folate to rats. *J. Nutr.* **114**, 778–85.

Thomas, J.R., Dwek, R.A. & Rademacher, T.W. (1990) Structure biosynthesis, and function of glycosylphosphatidylinositols. *Biochemistry* **29**, 5413–22.

Vegglan, R.S., Fasolato, S., Menard, S., Minucci, D., Pizzetti, P., Regazzoni, E., Tagliabue, E. & Colnaghi, M.I. (1989) Immunohistochemical reactivity of a monoclonal antibody prepared against human ovarian carcinoma on normal and pathological female genital tissues. *Tumori* **75**, 510–13.

Wang, X., Shen, F., Freisheim, J.H., Gentry, L.E. & Ratnam, M. (1992) Differential stereospecificities and affinities of folate receptor isoforms for folate compounds and antifolates. *Biochem. Pharmacol.* **44**, 1898–901.

Watabe, S. (1978) Purification and characterization of tetrahydrofolate protein complex in bovine liver. *J. Biol. Chem.* **253**(19), 6673–9.

Watkins, D. & Cooper, B.A. (1983) A critical intracellular concentration of fully reduced non-methylated folate polyglutamates prevents macrocytosis and diminished growth rate of human cell line K562 in culture. *Biochem. J.* **214**, 465–70.

Waxman, S. (1975) Folate binding proteins. *Brit. J. Haemat.* **29**, 23–9.

Waxman, S. & Schreiber, C. (1973) Characteristics of folic acid-binding protein in folate-deficient serum. *Blood* **42**(2), 291–301.

Waxman, S. & Schreiber, C. (1975) The purification and characterization of the low molecular weight human folate binding protein using affinity chromatography. *Biochemistry* **14**(25), 5422–8.

Waxman, S., Schreiber, C. & Herbert, V. (1971) Radioisotopic assay for measurement of serum folate levels. *Blood* **38**, 219–28.

Weitman, S.D., Lark, R.H., Coney, L.R., Fort, D.W., Frasca, V., Zurawski, V.R. & Kamen, B.A. (1992a) Distribution of the folate receptor (GP38) in normal and malignant cell lines and tissues. *Cancer Res.* **52**, 3396–401.

Weitman, S.D., Weinberg, A.G., Coney, L.R., Zurawski, V.R., Jennings, D.S. & Kamen, B.A. (1992*b*) Cellular localization of the folate receptor: Potential role in drug toxicity and folate homeostasis. *Cancer Res.* **52**, 6708–11.

Wells, D.G. & Casey, H.J. (1967) *Lactobacillus casei* CSF folate activity. *Brit. Med. J.* **3**, 834–6.

Wiley, H.S. (1985) Receptors as models for the mechanisms of membrane protein turnover and dynamics. In *Current Topics in Membranes and Transport* (ed. F. Bronner), pp. 369–412. Academic Press, New York.

Willis, S.A., Lacey, S.W., Weitman, S.D., Kamen, B.A. & Nisen, P.D. (1992) Folate receptor gene expression is tissue-specific and temporally-regulated. *Cancer Res. Ther. Control* **2**, 223–30.

Zamierowski, M. & Wagner, C. (1974) High molecular weight complexes of folic acid in mammalian tissues. *Biochem. Biophys. Res. Commun.* **60**(1), 81–7.

Zettner, A. & Duly, P.E. (1974) New evidence for a binding principle specific for folates as a normal constituent of human serum. *Clin. Chem.* **20**, 1313–19.

Zwiener, R.J., Johnson, C.A., Anderson, R.G.W. & Kamen, B.A. (1992) Purified folate receptor-5-methyltetrahydrofolic acid interaction at neutral and acid pH. *Cancer Res. Ther. Control* **3**, 37–42.

6

Riboflavin Carrier Protein in Reproduction

P. Radhakantha Adiga

Introduction

One of the major advances in our current understanding of the processes evolved to ensure optimal bioavailability of the fat- and water-soluble vitamins for growth, metabolism and reproduction of the vertebrates is the discovery of a group of soluble proteins in the blood and other body fluids which stoichiometrically and reversibly bind the respective vitamins with high affinity and receptor-like specificity, and often carry them to distant sites of utilization. A great deal of information is now available regarding these proteins' basic structural features, ligand-binding characteristics and physiological functions; the latter may include dietary absorption, storage, circulatory transport and prevention of rapid loss due to excretion or catabolism. Vitamins are known to remain biologically inert during the tight, yet non-covalent, association with their respective carriers, thereby permitting a reversible dissociation of the unmodified vitamins for utilization in a form and at sites most appropriate physiologically. Additionally, such an interaction of the fat-soluble vitamins with their respective soluble carrier proteins facilitates their transport in circulation in a water-soluble form.

Some of these vitamin carriers, notably those specific to fat-soluble vitamins, folic acid and cobalamine, are constitutively present ubiquitously throughout the animal kingdom and their elaboration may be significantly enhanced by appropriate stimuli to meet the accelerated demand during growth and reproduction. In contrast, others such as those specific to riboflavin and thiamin are apparently induced *de novo*, mostly, if not solely, as a reproductive stratagem to facilitate vitamin deposition in the developing oocytes in oviparous species and to transport the micronutrients through the physiological barriers (such as that offered by the placenta) in the mammals. This chapter attempts to summarize the current status of the molecular characteristics and the biological role of one such inducible vitamin carrier, riboflavin carrier protein, with emphasis on its evolutionary conservation in mammals and its roles in reproductive biology.

A few remarks on the current nomenclature of the protein which was originally discovered in the chicken and is synonymously referred to as riboflavin-binding protein (RBP or RfBP) (White & Merrill, 1988) and riboflavin-carrier protein (RCP) (Adiga & Murty, 1983; Adiga *et al.*, 1988*a,b*) may be appropriate. The relative merits of these names, particularly in terms of the usage of their respective abbreviated forms, has been discussed (White & Merrill, 1988). We still prefer the name 'riboflavin carrier protein' because 'carrier protein' implies dual functions, namely initial 'binding' and subsequent 'transport' to specific sites inaccessible to other proteins which may bind the vitamin rather non-specifically. The term riboflavin receptor may be preferable, but this nomenclature does not distinguish between the soluble carrier proteins from the membrane-bound vitamin binders, which catalytically mediate vitamin uptake. In this article, therefore, the term 'riboflavin carrier protein' (RCP) is uniformly used to denote the inducible protein with a special function in reproduction.

Riboflavin carrier protein in the chicken

Historical aspects

It has been now established that the low concentration of free riboflavin normally encountered in the blood of vertebrates is mostly protein-bound (hence conserved from excretory loss) since serum albumin (Jusko & Levy, 1975) and a fraction of IgG (McCormick *et al.*, 1987) are known to interact with the vitamin with reasonable affinity, albeit non-specifically. However, such a plasma transport machinery *per se* may not be sufficient to meet the vitamin requirements of the growing vertebrate embryo within the constraints imposed by the reproductive processes. That a special delivery mechanism might become operative to carry or 'target' the vitamin in a protein-bound form to subserve embryonic nutrition (presumably circumventing the competing demands from the maternal organism) is suggested by the discovery of a unique flavoprotein in the chicken egg white (Rhodes *et al.*, 1959) and the yolk (Ostrowski *et al.*, 1962) and in the plasma of egg-laying hens (Blum, 1967; Murthy & Adiga, 1978*a*). As expected from its special function in embryonic nutrition, the vitamin carrier is undetectable in the blood of the adult male or immature birds of either sex (Murthy & Adiga, 1978*a*). Based on these and other observations, a model was proposed (Blum, 1967) to explain its function in the chicken system, essential features of which include: (i) *de novo* hepatic induction and secretion of RCP under estrogenic stimulation during egg laying; (ii) tight complex formation with the vitamin in circulation and deposition in the yolk of the developing oocyte with the firmly bound riboflavin for subsequent use by the prospective embryo. As the mature ovulated oocyte transverses the oviduct, additional RCP (partly saturated with riboflavin) secreted along with other egg white proteins by the oviductal magnum is incorporated around the oocyte in the form of albumin (Blum, 1967; White, 1987; White & Merrill, 1988).

Genetic and immunological studies (Clagett, 1971) and recent elucidation of their primary structures (Norioka *et al.*, 1985) have shown that RCPs from the two egg compartments as well as in the blood of the laying hen are products of a single gene, as is the case with the egg yolk transferrin and the conalbumin of the egg white (Williams, 1962). The indispensibility of this carrier protein for embryonic vitamin nutrition was realized (Maw, 1954; Cowan *et al.*, 1964, 1966; Winter *et al.*, 1967), when the fertilized eggs of a mutant strain of the chicken with an inherited disease 'avian riboflavinuria' were found to lack a functional RCP. The eggs therefore failed to hatch, since acute flavin deficiency, brought about by the maternal failure (due to the genetic defect), to deposit adequate vitamin in the egg, results in embryonic mortality. The only way to rescue the affected progeny is to inject either riboflavin or FMN (flavin mononucleotide) to the fertilized eggs (Kozik, 1985; Buss, 1969). It would therefore appear that the requirement for embryonic development is for the vitamin only, yet in the absence of the RCP-mediated yolk deposition of the vitamin, the very survival of the embryo is jeopardized. In these mutant birds, an autosomal recessive gene (rd) is responsible for the inherited inability to transport the vitamin in a carrier-mediated form from the maternal system to the developing oocyte. These observations imply that (i) the avian oocyte plasma membrane exhibits a barrier to the free vitamin as well as to other protein-bound forms and that RCP is obligatory for yolk deposition of the vitamin; and (ii) RCP plays no significant function in the normal physiology of the birds since the growth rate, attainment of sexual maturity and egg productivity of these mutant birds are unaffected. This premise is further supported by the findings that the male or rapidly growing immature pullets of either sex do not harbour measurable RCP in their circulation (Murthy & Adiga, 1977*a*).

Among the various vitamin carriers hitherto examined in the avian eggs, RCP is the most studied entity apparently because of its relative abundance (0.8% of total protein) and the simple purification protocols involved in its isolation to homogeneity in good yields. Other attractive characteristics associated with this flavin carrier are: (i) its occurrence, unlike most of the major egg proteins, in both the yolk and albumin components (this implies two independent biosynthetic loci, namely the liver and the oviduct, respectively, of the egg-laying birds); (ii) its inducibility *de novo* by sex steroids and hence its potential as a model system for steroid-hormone-modulated gene expression; (iii) its revers- ible binding at a hydrophobic pocket with riboflavin in 1 : 1 molar ratio ($K_a = 10^8 \text{ M}^{-1}$) in preference to its coenzyme forms (FMN and FAD) with complete quenching of flavin fluorescence renders it an unique model system to understand structural features involved in the protein–flavin interaction (Murthy et al., 1976; Adiga & Murty, 1983).

Biochemical aspects

The chicken RCP, a phosphoglycoprotein (molecular mass 30– 37 kDa) has been purified to homogeneity from the egg yolk, and white and the serum of egg-laying or estrogenized chickens (Becvar & Palmer, 1982; Farrell et al., 1969; Miller & White, 1986; Murthy et al., 1979). Among these, the egg white RCP is most studied because of its relative abundance (1% of total protein) and because it occurs less than half saturated with flavin, unlike its counterparts in yolk or serum (Rhodes et al., 1959; White et al., 1986).

Determination of the amino acid composition of the egg white RCP has showed the presence of all common amino acids (Norioka et al., 1985); the protein is particularly rich in glutamic acid, serine and aromatic amino acids. The egg white RCP has been sequenced (Figure 6.1) and shown to contain 219 amino acids with several post-translational modifications (Hamazume et al., 1984, 1987). Recent elucidation of the complete nucleotide sequence of the full length cDNA clone corresponding to egg white RCP and prediction therefrom of the amino acid sequence is in conformity with the structure proposed earlier (Zheng et al., 1988). Additionally, the nucleotide sequence also predicts an N-terminal 17 amino acid hydrophobic signal peptide and a dipeptide sequence of Arg– Arg at the C-terminus during translation, but missing in the secreted protein. Amino acid architectures of the RCPs from the egg white and the serum are identical, confirming the earlier premise based on genetic and immunochemical analysis (Clagett, 1971; Murthy & Adiga, 1978a). As expected from its precursor–product relationship with the serum RCP, yolk RCP also has an identical sequence except that it lacks 11–13 amino acids at the C-terminus. This is presumably due to specific but limited proteolytic cleavage during oocyte uptake or yolk deposition in a manner similar to that of vitellogenin (Christmann *et al.*, 1977) and the yolk apoprotein B (Evans & Burley, 1987). The vitamin carriers from all three sources otherwise exhibit identical characteristics, which include N-terminal pyroglutamic acid, polymorphism in the amino acid sequence (Lys/Asn at the fourteenth residue from the N-terminus) and N-linked carbohydrate chains at Asn-36 and Asn-147 residues. Phosphate groups are also bound to the same serine residues, which occur in a cluster between positions 187 and 197. It is intriguing that all the phosphoserine residues are localized in a restricted, highly anionic region (Fenselan *et al.*, 1985; Hamazume *et al.*, 1984); thus within this segment are found 8 phosphoserine and 5 glutamic acid residues. This phosphopeptide has been purified from the tryptic digest of the reduced-carboxymethylated RCP (Miller *et al.*, 1984) and has lysine at its C-terminus and histidine at

Figure 6.1. The sequence of amino acids (single letter code) of the chicken egg white riboflavin carrier protein. <Q, N-terminal pyroglutamic acid; CHO, asparagine-linked oligosaccharide; P, phosphate residue. (From Hamazume *et al.* (1984). Reproduced with the permission of the authors.)

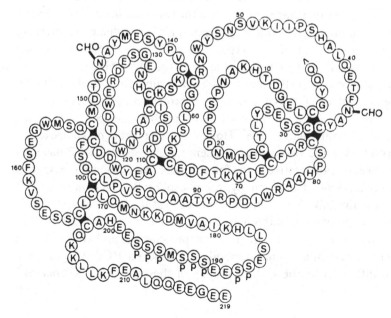

its N-terminus. Sandwiched between these two basic amino acids is a sequence of 21 amino acids among which 14 carry 1 or 2 negative charges at physiological pH. In view of the expected high degree of charge repulsion, it is assumed that this phosphopeptidyl sequence is rigid, with no ordered secondary structure. Another conspicuous feature of this phosphopeptide is the palindromic sequence around Met-194, which is flanked by 6 phosphoserine and 4 glutamic acid residues in a defined sequence. The biological significance of such a palindromic sequence is unknown at present.

The compositions of the oligosaccharide moieties located on Asn_{36} and Asn_{147} of the plasma and yolk RCP are similar, as expected from their precursor–product relationship, but differ from those on the egg white RCP, showing that the post-translational glycosylation is tissue-specific. The major difference is that the oligosaccharides on the plasma- and the yolk-RCP are more complex, with higher contents of sialic acid, galactose and several residues of fucose compared with the egg white RCP, which itself differs from other egg white proteins hitherto known. On the other hand, the location of phosphoserine residues on these RCP proteins is identical, implying that the specificity of the protein kinase involved is independent of the tissue of origin.

Another significant observation concerning chicken egg white RCP (Zheng *et al.*, 1988) is that it shares significant sequence homology (*ca.* 30%) with the bovine milk folate binding protein (Svendsen *et al.*, 1984) suggesting a common ancestral gene for the two vitamin binders. Accordingly, locations of eight out of the nine pairs of cysteines involved in disulfide bridges, all of the tryptophan residues and one of the two glycosylation sites on RCP are common to the folate binder, leading to the hypothesis that the two may belong to a family of high-affinity vitamin transport proteins, just as retinol-binding protein and β-lactoglobulin are members of the superfamily of hydrophobic transporter proteins (Godovac-Zimmermann, 1988). However, alignment to maximize identical residues in the two protein sequences suggests that RCP has a C-terminal extension relative to the folate binder, which in turn is extended at the N-terminus. Contrastingly, the phosphorylated region of RCP is missing in the milk vitamin binder (Svendsen *et al.*, 1984). However, it is relevant to point out at this juncture that there is no detectable immunological cross-reactivity between these proteins as probed with both polyclonal and monoclonal antibodies to chicken RCP, suggesting a major difference in the surface folding of these proteins (Kuzhandhai Velu, 1990).

Riboflavin binding characteristics

A large body of information is available concerning the flavin binding characteristics of RCP, which has been extensively used as a model system to understand the sites and mechanics of flavin–protein interactions. RCP is unique among flavoproteins in that it preferentially binds to riboflavin in 1:1 stoichiometry with complete quenching of flavin fluorescence and 80% quenching of protein fluorescence (Nishikimi & Kyogoku, 1973; Murthy *et al.*, 1976). Some flavin analoges and coenzymes also bind but with relatively lower affinities; nuclear magnetic resonance (NMR) data reveal little differences in the binding sites for the isoalloxazine ring in the yolk and white RCP (Nishikimi & Kyogoku, 1973). Thermodynamic analysis indicates that the flavin-binding hydrophobic cavity may be of relatively smaller size in the yolk RCP (Matsui *et al.*, 1982*a*). Investigations using model compounds with well-defined modifications on the flavin molecule suggest that the dimethyl benzenoid moiety of the isoalloxazine ring is the major site of hydrophobic interaction (Choi & McCormick, 1980). Additionally, N-10 and the ribityl side chain of the flavin seem to participate strongly in this binding through hydrogen bonds (Matsui *et al.*, 1982*b*; Moonen *et al.*, 1984) whereas N-3 of the flavin remains solvent-assessible (Figure 6.2).

While the detailed nature and structural features of the hydrophobic cavity of RCP involved in flavin binding have to await the delineation of its three-dimensional structure by X-ray crystallography (Zanette *et al.*, 1984), the data available from chemical modification of the protein reveal that tryptophan residues are essential for ligand-binding (Murthy *et al.*, 1976; Blankenhorn, 1978). Modification of five tryptophan residues completely abolishes flavin binding to yolk protein; among these residues 1–2 tryptophans are essential for vitamin interaction with the egg white RCP (Murthy *et al.*, 1976). More recent studies (Blankenhorn, 1978) reveal that one of these residues, critically involved in binding, is not protected by bound flavin against chemical modification. Earlier, tyrosine involvement in flavin binding was considered unlikely because extensive iodination or nitration fails to alter the flavin interaction with the protein (Farrell *et al.*, 1969). Further investigations, however, reveal that one tyrosine is protected against chemical modification due to bound flavin, hence its location at the active site (Blankenhorn, 1978). Involvement of carboxyl functions in flavin binding was presumed since flavin binding was drastically reduced at pH 4.0, i.e. near the pK of the carboxyl group on the protein (Murthy *et al.*, 1976). More direct evidence for this premise stems from the observation (Kozik, 1982*a*) that bound flavin protects a critical carboxyl group, located at the active site, against inactivation by carbodiimide.

The integrity of all of the nine disulfide bridges for the properly folded rigid structure of the egg white RCP, including that involved at the hydrophobic pocket where the flavin binds, has been emphasized (Murthy *et al.*, 1977; Kozik, 1982*b*) since reduction of even a single disulfide bond leads to flavin release. Complete reduction of all the disulfide bonds results in protein aggregation, presumably owing to hydrophobic collapse (Kozik, 1982*b*; Kuzhandhai Velu, 1990). Concurrently, significant alterations in the surface contour occur, as revealed by the inability of all the monoclonal antibodies (MAbs) directed against highly conformation-dependent discontinuous epitopes of the native RCP to bind to the reduced unfolded protein. However, it could be clearly shown that reoxidation of disulfide bonds at alkaline pH under appropriate conditions results in complete restoration of the native folded structure with concomitant regain of flavin-binding characteristics and of the ability to interact the MAbs (Kuzhandhai Velu, 1990). These results emphasize that the protein folding, including regain of flavin-

Figure 6.2. Schematic representation of riboflavin carrier protein, interaction at the hydrophobic pocket. Dotted lines, hydrophobic interaction; dashed lines, hydrogen bonding.

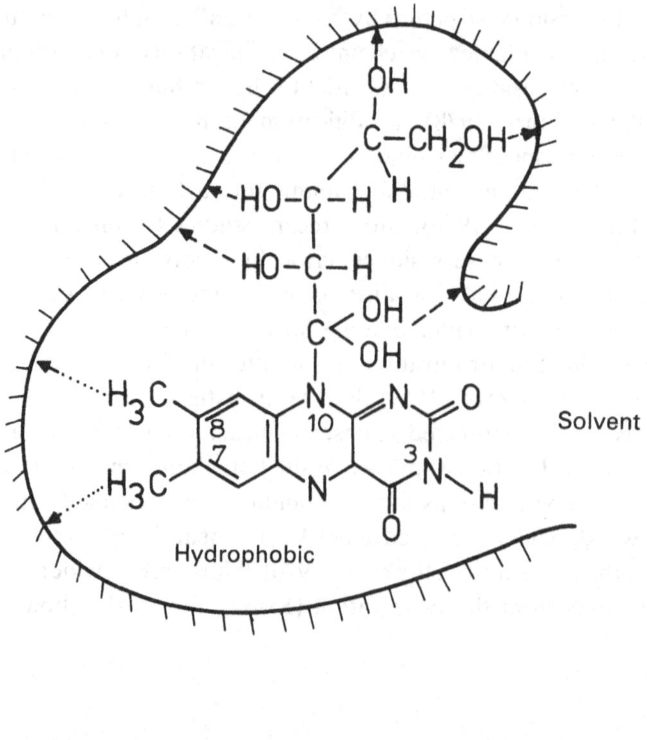

binding characteristics at the hydrophobic interior, and the overall surface topology is primarily dictated by the primary amino acid sequence of the protein and that the post-translational modifications introduced on the mature RCP have no perceptible influence on re-folding (Kuzhand-hai Velu, 1990).

Recognition sites for oocyte uptake

It is generally believed that, like conalbumin, RCP, mostly present as apo-protein in the egg white, may have a bacteriostatic role as a scavenger for flavin that may be released during embryonic growth (Tranter & Board, 1982). On the other hand, the vitamin carrier that occurs in the yolk as holoprotein at six times higher concentration than in the blood is presumably involved in meeting the vitamin requirement of the developing embryo. Direct evidence for yolk deposition of RCP from the blood with the vitamin firmly bound stems from experiments using mutant hens afflicted with the hereditary syndrome 'avian riboflavinuria' (Hammer *et al.*, 1973). The general phenomenon of yolk deposition of the proteins is now well characterized (Perry *et al.*, 1978). The vitelline membrane encompassing the oocyte is highly invaginated and covered with coated pits which are characteristics of cell membranes active in receptor-mediated endocytosis (Cuatrecasas & Roth, 1983). While it is conceivable that the oocyte uptake and yolk deposition of RCP involves endocytosis through the oocyte plasma membrane in a manner analogous to that of vitellogenin, immunoglobulin and very low density lipoprotein (Roth *et al.*, 1976; Yusko *et al.*, 1981; Perry *et al.*, 1984), attempts to locate and characterize chicken RCP-specific receptors have been inconclusive. An alternative mode of yolk deposition of RCP has been envisaged (White & Merrill, 1988). The essential features of this mode of concentrative uptake involves the formation of Ca^{2+}-mediated ionic bridges between phosphoserine residues of RCP and those of vitellogenin, as in casein micelles (Schmidt, 1982) and endocytosis of the aggregate through the well-known vitellogenin receptor located on the oocyte membrane (Woods & Roth, 1984). Such an alternative mechanism is attractive and has been, in fact, earlier implicated in the yolk deposition of the vitamin D binding protein in the chicken egg (Fraser & Emtage, 1976). The requirement for the integrity of the phosphorylated state of RCP for oocyte uptake (Miller *et al.*, 1982*b*) is in line with this premise. However, the operation of a similar mechanism appears unlikely in higher vertebrates like the mammals wherein the evolutionarily conserved RCP mediates transplacental riboflavin transport during gestation

(Adiga & Murty, 1983) in the absence of vitellogenin, which is restricted to the eggs of oviparous species (Tata *et al.*, 1983).

Among the structural features of RCP required to target the vitamin through the vitelline membrane of the oocyte en route to yolk deposition, the oligosaccharide moiety has received significant attention. Recent investigations into the sequence of sugar residues of the N-linked carbohydrate moieties of the egg yolk RCP reveal that most of them (*ca.* 68%) belong to the biantennary complex type of oligosaccharide and the remainder to one triantennary type, both having terminal sialic acid residues in 2–6 linkages with penultimate galactose residues (M. Tarutani, N. Norioka, T. Mega, S. Hase & T. Ikenaka, personal communication). These structural features of the oligosaccharide moieties are characteristic of secretory glycoproteins (Kornfeld & Kornfeld, 1985). Since the egg white RCP with a carbohydrate composition different from that of the plasma RCP is less efficiently incorporated into the yolk, it can be assumed that subtle structural differences in bound oligosaccharide moieties modify the glycoprotein targeting to specific sites on the oocyte. Additionally, the removal of sialic acid, N-acetylglucosamine and galactose or chemical oxidation of galactose leads to diminished transmembrane transport without concomitant alterations in flavin binding capacity. This lends credence to the importance of carbohydrate in the recognition phenomenon (Miller *et al.*, 1982*a*, 1981*a*).

Another important structural feature of RCP implicated in oocyte plasma membrane recognition is the phosphopeptide region. Enzymatic dephosphorylation of the egg white or yolk RCP has no effect on its ligand binding, but oocyte uptake is greatly reduced (Miller *et al.*, 1982*b*). Removal of a single phosphate residue from the yolk RCP decreases its oocyte deposition by 60% and this cannot be restored by the compensatory addition of anionic groups such as by succinylation. Decreased uptake is proportional to the number of phosphate residues hydrolyzed, thereby implying a specific role for the clustered phosphoserine residues. These appear to function autonomously, independently of the rest of the protein, and could be involved in the recognition of the putative receptor on the oocyte membrane either through direct interaction or by directing the protein in such a way as to facilitate subsequent interaction with the receptor (Miller *et al.*, 1984).

It is noteworthy that chemical modification of lysine residues of RCP by succinylation also drastically reduced yolk deposition (Miller *et al.*, 1981*b*) indicating that the uptake also involves other segments of the polypeptide chain. It would appear therefore the RCP uptake by the oocytes may be a complex sequence of multiple interactions involving bound phosphate, sialic acid and lysine residues; the elucidation of underlying mechanisms awaits further research.

Hormonal modulation of biosynthesis

There is now a large body of evidence to show that, in oviparous vertebrates, *de novo* biosynthesis of most of the major yolk-specific proteins in the liver is in response to enhanced circulatory levels of estrogen during the reproductive phase and this can be mimicked by the administration of the steroid either to the adult males or to immature animals of either sex (Tata & Smith, 1979). It is also known that egg white proteins are synthesized exclusively in the oviduct under instruction from the same steroid, although other steroids can substitute for estrogen, once the tissue is differentiated by exposure to the latter (Palmiter & Gutman, 1972). Since, unlike other major egg proteins, RCP is encountered in both the yolk and the white compartments, dual sites of elaboration are anticipated. Earlier studies on hepatic RCP synthesis in roosters and immature chickens revealed its estrogen dependence (Clagett *et al.*, 1970). The induced egg proteins are secreted by the respective biosynthetic loci even in the absence of a developing oocyte (Cecchini *et al.*, 1979) and hence the plasma and oviductal RCP concentrations directly reflect the synthetic capabilities of liver and oviduct, respectively, since no storage capacity exists in these tissues for the proteins.

Detailed studies on the kinetics and hormonal specificity of hepatic induction of RCP following estrogen administration to immature male chickens is rendered possible due to the development of a specific and sensitive radioimmunoassay (RIA) for RCP in the blood (Murthy & Adiga, 1977*a*). Following a single injection of estradiol 17-β, the plasma RCP is enhanced several-fold after 6 h, reaching peak levels around 48 h to decline thereafter to basal levels. A 2-fold amplification of the response manifests on secondary stimulation with the same dose of the steroid hormone, which is qualitatively similar to that observed during vitellogenin induction in the avian and amphibian liver (Gruber *et al.*, 1976; Tata & Smith, 1979). While the rate of RCP induction is hormonal dose-dependent, the time-course of response remains unaltered with no appreciable change in half-life of the protein, which, however, is modulated by the thyroid status of the birds. Thus in hyperthyroid chicks, higher concentrations of estrogen are required to attain the same inductive response as in normal birds; this result has been explained as due to accelerated catabolism of the inducer as well as of the induced protein

(Murthy & Adiga, 1977b). Progesterone cannot substitute for the estrogen for inductive response in the liver, nor is it able to affect the kinetics of estrogen-induced RCP production. Antiestrogens such as clomiphene citrate are, however, capable of blocking RCP induction in the liver.

A comparison of RCP induction in the liver and oviduct reveals subtle qualitative differences in the hormonal regulation of the RCP gene in the two estrogen-dependent avian tissues, which may reflect differential modulation of tissue-dependent regulatory elements governing the gene expression in the two biosynthetic loci (Durgakumari & Adiga, 1986a). The major difference is that progesterone can substitute for estrogen during secondary stimulation in the estrogen-primed oviduct in analogy with ovalbumin synthesis, unlike the hepatic induction of RCP, which shows absolute estrogen specificity. Similar patterns in the kinetics and hormonal specificity of hepatic induction have been observed in two other chicken egg vitamin carriers, namely thiamin- and biotin-carrier proteins (Muniyappa & Adiga, 1980c; Murty & Adiga, 1985). A comparison of the kinetics of *de novo* estrogen induction of RCP with that of estrogen modulation of retinol binding protein in the immature chicken liver has revealed several contrasting features (Durgakumari & Adiga, 1986c). Major differences are: (i) unlike RCP induction, enhanced retinol binding protein accumulation is not strictly hormonal dose-dependent, although a minimal threshold level of the steroid is required to elicit a measurable response; (ii) the memory effect during secondary stimulation of RCP induction with attendant amplification of inductive response is not discernible with retinol binding protein; and (iii) administration of α-amanitin and cycloheximide completely curtail RCP induction for prolonged periods whereas inhibition of RBP accumulation is partial and relatively short-lived. The only common factors are: (i) that neither process is influenced by progesterone in the liver; and (ii) that antiestrogens are inhibitory, although to different extents. Of relevance in this context are the observations that riboflavin deficiency has little effect on RCP production and its yolk deposition (Benore-Parsons *et al.*, 1985) whereas both retinol (Smith *et al.*, 1975) and biotin (White, 1987) seem to significantly affect the secretion of their respective carrier proteins by the liver.

Cell-free translation of poly (A^+)RNA from both the liver and the oviduct from estrogen-treated birds show that enhanced RCP mRNA levels correlate with increased RCP synthesis in the two tissues (Durgakumari & Adiga, 1986b). Recent studies in our laboratory (Murti, 1990) using ^{32}P-labeled cDNA probes for chicken RCP mRNA by dot–blot hybridization and Northern analysis have confirmed that increased RCP production during estrogenic stimulations reflects enhanced mRNA activities (Murti, 1990) due to stimulated transcription and possibly to cytoplasmic stabilization as shown for other major egg proteins (Shapiro *et al.*, 1988; O'Malley *et al.*, 1969).

Immunological characteristics

Information hitherto available on the antigenicity and antigenic domains of chicken RCP is rather restricted; more detailed analysis is awaited. The protein is highly immunogenic and antibodies can be raised in a variety of mammalian species (Ramanathan *et al.*, 1979) including subhuman primates (Adiga *et al.*, 1991; Natraj, 1991). Chemical modifications of RCP show that the major contribution to immunogenicity and antigenicity is from the conformational characteristics dictated by the protein folding rather than from the post-translational modifications such as glycosylation and phosphorylation (Ramanathan *et al.*, 1980a; Kuzhandhai Velu, 1990). Protein unfolding by total disulfide reduction, oxidation of tryptophan residues and blocking by succinylation or dinitrophenylation of peptidyl lysine residues lead to a loss of both flavin binding and antigenicity of the apo-RCP. Curiously enough, a dinitrophenyl (DNP) derivative of the holoprotein exhibits some degree of antigenic similarity with the native protein while DNP-apo-RCP fails to give a similar reaction. Amidation of 88% of the lysines in the apo-protein is accompanied by an 80% decrease in potential antigenicity, with complete retention of flavin binding activity. However, a comparison of the slopes of the inhibition curves obtained in RIA with the amidated and unmodified RCP reveals weakening of the binding characteristics of most of the antigenic determinants. Modifications of tryptophan and tyrosine residues do not greatly alter the antigenic properties, but lead to a complete loss of flavin binding properties. Furthermore, the holo- and apo-RCP react similarly in RIA and immunodiffusion analysis; the apo-protein bound to its antibody still interacts with the flavin at 97% of the predicted amount (Murthy *et al.*, 1976). These results suggest that antigenic sites are mostly localized in areas different from the ligand-binding sites and are in line with several instances reported in the literature where the biologically active site is independent of the antigenic site, in analogy with the much studied phenomenon of the conservation of enzyme active sites through various stages of phylogenetic development (Wilson *et al.*, 1977). It is therefore attractive to speculate at this stage that RCP has also largely retained through evolution the conformational characteristics of the hydrophobic environment harboring the flavin binding site.

We developed a battery of MAbs to chicken egg white RCP as probes (Visweswariah et al., 1987; Kuzhandhai Velu, 1990) in order to study the immunotopological characteristics of chicken RCP in greater detail. Our main objectives have been: (i) to assess whether the immunological characteristics are retained on a determinant-to-determinant basis during evolution, and (ii) to delineate, if possible, the location and the nature of some of the biologically important regions such as those involved in recognition of putative receptor sites implicated in transport across the oocyte–placental plasma membranes. Among the seven categories of these MAbs, each recognizing a distinct, non-overlapping epitopic conformation, six recognize conformation-dependent (or assembled) epitopes, while the remaining one (6B2C12) is directed against a segmental (or sequential) epitope. The MAb 6B2C12 recognizes the native or deglycosylated egg white RCP in enzyme-linked immunosorbent assay (ELISA) or Western blot analysis but not the yolk RCP. This MAb also binds a synthetic oligopeptide corresponding to the C-terminal 17 amino acid residues (Gln_{202}–Glu_{219}) (Figure 6.1) of the egg white and the plasma RCP which is partly missing in the yolk RCP (Norioka et al., 1985). This epitope, specified by the MAb 6B2C12, like those corresponding to all other MAbs, is conserved through evolution in the mammals, including humans (Karande et al., 1991). The functional importance of this epitopic conformation located at the C-terminus of RCP was revealed when the ascites fluid of the MAb 6B2C12 was administered to pregnant mice to immunoneutralize their endogenous RCP. Since early fetal resorption occurs as a consequence of passive immunoneutralization of RCP, it would appear that this C-terminal sequence of RCP is important for transplacental vitamin transport, presumably by functioning as a recognition site for the putative placental receptor for the vitamin carrier. Alternatively, it is conceivable that by forming a stable immune complex at the appropriate site on RCP, the MAb 6B2C12 may be sterically blocking the surface recognition sites spatially proximal to its location (Karande et al., 1991). A similar approach using other MAbs may reveal additional important surface-exposed structural features of RCP involved in the carrier-mediated riboflavin transport phenomenon.

Details concerning the location and conformational characteristics of the epitopes recognized by other MAbs are at present unknown and await further study. Available information suggests that at least three other determinants may be contributed mainly by the C-terminal half of RCP. In a recent study (Kuzhandhai Velu, 1990) a 133 amino acid C-terminal fragment (Pro_{87}–Glu_{219}) (M_r = 25 000) has been purified by reduction and resolution on SDS–PAGE from the chicken egg white RCP after selective cleavage of the acid-labile Asp_{86}–Pro_{87} bond with 75% formic acid in the presence of 7 M guanidine–HCl. This C-terminal fragment gave positive reactions in ELISA with three of the MAbs (5B1D3, 5A2E6 and 6B2C12). These observations imply that: (i) two conformation-dependent epitopes corresponding to MAbs like 5B1D3 and 5A2E6 are located contiguous to the sequential epitope specified by MAb 6B2C12 and located at the C-terminal end of the protein; and (ii) these have greater propensity to renature to assume native-like structures. Preliminary data also reveal that the phosphopeptide (His_{182}–Lys_{202}) purified from tryptic digests of disulphide-reduced egg white RCP (Miller *et al.*, 1984) also harbors a major part of the antigenic determinant recognized by the MAb 5B1D3 in solid-phase RIA; the binding of the phosphopeptide to this MAb is dose-dependent. However, it is unlikely that phosphate residues contribute to its antigenicity, since dephosphorylation of RCP with acid phosphatase has little effect on the binding of this MAb to the native protein (N. Kuzhandhai Velu, A.A. Karande & P.R. Adiga, unpublished observations).

Riboflavin carrier protein in other egg-laying species

From the above discussion, it is clear that RCP satisfies several criteria expected of a carrier protein specifically designed to transport a vital nutrient like riboflavin for oocyte deposition during the reproductive phase to ensure uninterrupted embryonic growth. Its high affinity and specificity for free riboflavin should enable it to selectively bind and scavenge low concentrations of the vitamin in the maternal blood and carry it through the circulation to be recognized for selective uptake at the oocyte plasma membrane as an essential maternal investment to subserve embryonic development. Similarly, estrogen regulation of its gene expression should ensure its adequate availability at the appropriate time during the oocyte maturation process. Since such an exquisite combination of functional properties are not generally acquired all at once during evolution of important proteins, it has been surmised (White, 1991) that the vitamin carrier proteins like RCP must have been recruited as a result of gene duplication and modification of pre-existing entities such as the membrane-associated vitamin binders, which function catalytically to capture uncomplexed vitamins from the environment for cellular utilization, although there is no direct evidence for this hypothesis so far.

Since the embryonic development process within the shelled eggs of oviparous species is independent of maternal influence, it is anticipated that the mechanism involving RCP-mediated vitamin deposition in the developing chicken oocyte also operates in other egg-laying species. This has been amply substantiated by the detection, isolation and characterization of the vitamin carrier in the eggs of several species of bird (Feeney & Allison, 1969) including domestic ducks (Muniyappa & Adiga, 1980*a*; Abrams *et al.*, 1988). Both the egg compartments contain RCP, but the concentration and the proportion of apo-RCP in egg white varies widely among avian species. The vitamin carrier has been characterized also from the cleidoic eggs of the American alligator, painted turtle and Indian python (Abrams *et al.*, 1989). All these RCPs are glycoproteins with affinities for riboflavin comparable to that of their chicken counterpart, but of variable carbohydrate contents. RCP is confined to the yolk in alligator eggs. Earlier evidence for RCP in fish eggs was inconclusive (White & Merrill, 1988), but recent studies on the carp (*Cyprinus carpio*) show that an RCP exhibiting immunological homology and physicochemical properties similar to those of the chicken protein can be purified from eggs and the serum from this freshwater fish during the breeding season. All the MAbs raised against the chicken RCP bind with the piscine protein; evidence has been adduced for the presence, in polysomal RNA of the gravid fish liver, of RCP mRNA as detected by filter hybridization with ^{32}P-labeled RCP cDNA (Malhotra *et al.*, 1991). These results imply that RCP occurs in the eggs of the primitive vertebrates lower on the evolutionary scale than the birds and reptiles. It is conceivable that RCP, like vitellogenin, is conserved among all the egg-laying species, including insects, to serve as a reproductive stratagem to ensure embryonic nutrition, although no direct evidence has been adduced for the latter hypothesis.

Mammalian riboflavin carrier proteins and their roles in reproduction

Rationale for involvement in fetal nutrition
In higher mammals, including humans, the rapidly developing fetoplacental unit demands a continuous and unremitting supply of nutrients from the mother for its well-programmed development and uninterrupted growth, unperturbed to a large extent by the competing demands from the maternal system. Consequently, it accumulates, often against concentration gradients, several nutrients including water-soluble vitamins (Dancis & Schneider, 1975; Miller *et al.*, 1976). Riboflavin is among these micronutrients, which exhibits a high fetal : maternal ratio (4 : 1); the molecular mechanisms underlying this facilitated transport process have remained enigmatic particularly since the placental

membranes are known to prevent the transit of free riboflavin or its coenzyme forms. Furthermore, it is well established that increasing amounts of estrogens originating from the fetoplacental unit are secreted into the maternal circulation to regulate the maternal physiology during gestation, although the molecular role of this steroid hormone remains to be established.

In view of these lacunae, it is attractive to visualize that this placental barrier to free riboflavin in mammals may be similar to that offered by the plasma membrane of the avian oocyte and hence that a carrier-mediated riboflavin delivery mechanism involving RCP may be operative during mammalian reproduction. In support of such a hypothesis is the retention, during evolutionary transition from oviparity to viviparity, of several aspects of reproductive strategies, such as the involvement of gonadotropins and gonadal hormones in sexual maturation, the process of ovulation, spermatogenesis and fertilization, etc. It appears logical therefore to hypothesize that the gene coding for RCP may be conserved during the evolutionary transition and may be expressed under instruction from the same hormonal signal, namely estrogen. If this hypothesis is true, it is to be anticipated that the major structural features, particularly those relevant to the biological functions, would be retained in the putative mammalian RCP (in analogy with several other important conserved proteins with vital functions such as cytochrome c, histones and serum albumin), opening the possibility that such a conserved protein may be detected and quantified by immunochemical analysis.

Detection, isolation, and physicochemical and immunological characterization

The above working hypothesis is borne out by the immunological evidence that in pregnant (but not in male or immature) rats, measurable amounts of an RCP-like entity can be identified by a specific RIA developed for the chicken egg vitamin carrier (Adiga & Muniyappa, 1978; Nutrition Reviews, 1979). In conformity with this is the finding that the rodent pregnancy sera harbor a protein which is immunoprecipitable, along with its tightly bound [^{14}C]riboflavin, by the specific antiserum to chicken RCP. Incidentally, this represents the first demonstration of RCP in a female mammal, contrary to the claim made elsewhere (Merrill *et al.*, 1979). Biochemical evidence for the rodent RCP is provided by its purification by bioaffinity chromatography on immobilized lumiflavin, although its M_r was ambiguous at that time (Muniyappa & Adiga, 1980*a*). More recent data from our laboratory, however, reveal that the rodent RCP further purified by fast protein liquid chromatography has a molecular mass (36 kDa) comparable to that of the chicken RCP (A.A. Karande & P.R. Adiga, unpublished results). Its high-affinity interaction specifically with free riboflavin in preference to FMN and FAD and its existence in the rodent circulation during gestation or following estrogen treatment clearly favor the view that its function is that of transplacental vitamin transport to subserve embryonic growth, as in the oviparous vertebrates (Adiga & Murty, 1983).

If RCP is evolutionarily conserved from the oviparous vertebrates to viviparous mammals like the rodents, despite the marked difference in their patterns of embryonic growth, such an essential transport mechanism can be expected to be retained with or without modification during further evolution of the species. This is amply substantiated by the finding that the sera from pregnant mice, guinea pigs and cows exhibited immunological cross-reactivity in RIA using chicken RCP and its antiserum. The observed parallelism in displacement of ^{125}I-labeled chicken RCP from its antibody in a dose-dependent manner by different pregnancy sera is indicative of conservation of the majority of the immunochemical characteristics expected from the homologous proteins during evolution (Wilson *et al.*, 1977). Recently, RCPs from pregnant bonnet monkeys (*Macaca radiata*) and human (maternal and umbilical cord) sera have been purified to apparent homogeneity and characterized (Visweswariah & Adiga, 1987*a,b*). The primate RCPs display marked similarities to chicken RCP with regard to several characteristics including the molecular size and isoelectric point, preferential binding to riboflavin, and immunological cross-reactivity. Sequence similarities among the RCPs are supported by the overall distribution patterns of several iodopeptides on two-dimensional maps of ^{125}I-labeled tryptic peptides (Adiga *et al.*, 1988*a*; Visweswariah, 1987). Some differences could, however, be observed in the distribution patterns of peptides during fingerprinting, which may indicate subtle differences in the primary structures of RCP during evolution. This is anticipated since the rates of protein evolution depend on the probability that such substitution will be compatible with the molecular functions of the protein and on whether it is essential for the survival of the organisms.

The recent availability of MAbs recognizing seven different epitopic conformations on the surface of the chicken egg white RCP (Visweswariah *et al.*, 1987; Kuzhandhai Velu, 1990) afforded an opportunity to compare the immunotopological characteristics of various mammalian RCPs with those of the avian protein. It is remarkable to find that all the MAbs recognize the various mammalian RCPs in either liquid- or solid-phase RIA, with displacement curves being parallel to that elicited with the chicken protein (Figure 6.3). However, the amounts of the various mammalian RCPs required for 50% displacement of radiolabeled chicken RCP from the different MAbs vary depending upon the MAb and the source of RCP. These observations lead to the inevitable conclusion that by and large the surface contour of RCP, represented by the various epitopic conformations recognized by the MAbs, is mostly conserved during evolution, with substitutions if any being conservative and not greatly altering the overall immunotopological characteristics of the vitamin carrier.

Figure 6.3. Displacement of ^{125}I-labeled chicken egg white RCP from different monoclonal antibodies by various RCP preparations. The volume of culture supernatants binding 25% of input radioactivity was incubated with unlabeled chicken RCP (open circles), or RCP purified from rat (triangles), bovine (squares) or human (filled circles) pregnancy sera, and ^{125}I-labeled chicken RCP (*ca.* 2 ng). Immunoprecipitation of the bound radioactivity was carried out by the double antibody–PEG precipitation method (Visweswariah *et al.*, 1987).

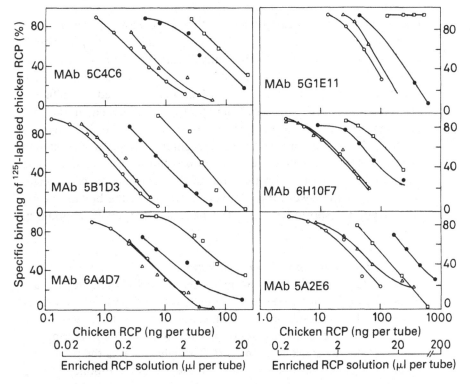

Other riboflavin binding proteins

Another interesting report on the occurrence of riboflavin binding proteins in mammalian circulation is that of Merrill *et al.* (1979) who purified multiple protein species from bovine plasma by affinity chromatography using N-3-carboxymethylriboflavin coupled to AH–Sepharose. At least three major protein bands could be observed migrating in the regions of β and γ globulins, following resolution by cellulose acetate electrophoresis. The M_r of one of these protein species was 150 000, but an interesting observation was made on to a small amount of another species (M_r 37 000). All the three proteins bound [^{14}C]riboflavin avidly with high affinity ($K_d = 10^{-6}$ M l^{-1}). The presumed pregnancy-specific low-M_r (37 000) protein from bovine serum was purified to homogeneity and shown to have even higher affinity. No further analysis of this protein, particularly its relationship with the chicken egg RCP, was reported. It is surmised that the protein binding riboflavin with high-affinity is associated with pregnancy in higher mammals with a function analogous to that of the serum RCP in laying hens.

Also of relevance in this context is the observation with human sera (Merrill *et al.*, 1981; McCormick *et al.*, 1987) that besides albumin, which is known to associate with riboflavin with relatively low affinity (Jusko & Levy, 1975), a certain fraction of immunoglobulin G (IgG) also binds riboflavin with a reasonably high affinity (4 μM). This IgG fraction is non-specific in the sense that it is present in the sera of male and female non-pregnant and pregnant individuals. It has been suggested that albumin and a fraction of IgG may be sufficient for transplacental flavin transport in humans. While the operation of such a mechanism in humans is conceivable, the IgG-mediated transplacental transport of the vitamin cannot be the evolutionarily conserved mechanism since no detectable flavin binding activity is encountered in the IgG fraction of pregnant monkey serum (Visweswariah & Adiga, 1987a); in the bovine species IgG is not transported through the placenta (Farber & Thornburg, 1983).

Hormonal regulation of hepatic synthesis

When the plasma concentrations of RCP in female rats are assessed using the heterologous RIA during different periods of the 4 day estrous cycle and gestation, it is clear that estrogen is the physiological modulator of the protein (Muniyappa & Adiga, 1980c). During the estrous cycle, the concentration of RCP is highest during the proestrous phase, coincident with high levels of the steroid hormone. During pregnancy, significant concentrations of RCP are observed in the plasma at day 4, i.e. before implantation, suggesting the maternal origin of the protein. Between days 10 and 16 of pregnancy, when intense embryonic growth and organogenesis ensues, RCP is manifested at high concentrations (Muniyappa & Adiga, 1980c). Ovariectomized females also respond to estrogen administration and elicit progressively increasing concentrations of the plasma RCP to reach a maximum around day 4 after stimulation. Unequivocal proof for estrogen as the primary signal for hepatic induction stems from experiments wherein male rats are administered estradiol-17β at a pharmacological dose (10 mg kg^{-1} body mass). The kinetic analysis of RCP induction shows several overall similarities in inductive response between the rodent and the avian systems, which include ineffectiveness of progesterone and inhibition by antiestrogens and cycloheximide of estrogen modulation of RCP gene expression (Murty & Adiga, 1982a).

In conformity with the above are the recent findings in the author's laboratory (Murti, 1990) that the changing plasma concentrations of RCP in the female rat during estrous cycle and pregnancy bear a reasonable correlation with corresponding hepatic concentrations of RCP mRNA as assessed by dot–blot filter hybridization of total RNA using ^{32}P-labeled chicken RCP cDNA as the probe. However, what is most intriguing, and rather unexpected, is the observation that the fetal liver itself harbors measurable amounts of hybridizable RCP mRNA (particularly from day 17 of gestation), which progressively increase till term. Moreover, the fetal liver RCP mRNA appears translationally active since [^{35}S]methionine-labeled immunoprecipitable RCP can be recovered from the cytosol of 20 day fetal liver cultured *in vitro* with the labeled amino acid (G.L. Prasad & P.R. Adiga, unpublished observations). The physiological functions of the fetoplacental RCP and its contributions, if any, to the maternal circulatory RCP concentrations during pregnancy are at present conjectural. It is conceivable that it performs a housekeeping function in the developing fetus as a vitamin conservation mechanism. Of relevance in this context is the observation (Krishnamurthy *et al.*, 1984) in this laboratory that free riboflavin added to the fetal liver cytosol is very unstable and is rapidly catabolized into as yet unidentified flavin derivatives. Furthermore, riboflavin bound to RCP appears to be a better substrate for fetal flavokinase since a twofold higher enzymatic activity is observed compared with free flavin (Table 6.1). In accordance with these findings is the recent report (Slomezynska & Zak, 1987) that the flavokinase forms a complex with chicken egg white RCP and can be purified using immobilized RCP as the affinity matrix. It is therefore conceivable that a RCP–flavokinase complex may be physiologically relevant to

Table 6.1 *Flavokinase activity in the rat embryonic liver tissue; the relative efficacies of riboflavin–RCP complex and free riboflavin as substrates.*

Enzyme activity was assayed by the method of Merrill & McCormick (1980). Substrate concentration is expressed in terms of riboflavin either bound to RCP in 1:1 molar ratio or free FMN formed is separated from other flavins by paper chromatography and quantitated (Krishnamurthy, 1984).

Substrate (0.2 mM)	Enzyme activity (nmol FMN mg protein^{-1} h^{-1})
RCP–riboflavin complex	0.39
Free riboflavin	0.18

facilitate flavin coenzyme biosynthesis *in vivo* in the fetal liver. The endocrine signal for the fetal RCP gene expression *in utero* may also involve estrogen since a major portion of the steroid hormone produced in the placenta is shunted to the fetal circulation, particularly during later stages of pregnancy (Lax *et al.*, 1983; Buster & Marshall, 1979). Additionally, the hepatic concentration of estrogen receptors increases in the fetus during later stages of gestation, in parallel with maternal estrogens (Lax *et al.*, 1983). Another most intriguing finding is that after parturition, both the neonatal and maternal livers continue to exhibit high concentrations of RCP mRNA up to day 8, which slowly decline thereafter to reach low concentrations by day 15. These exciting findings accord with the earlier observations (Lax *et al.*, 1983; Buster & Marshall, 1979) that high concentrations of estrogen and its high-affinity nuclear receptors in both the maternal and the fetal liver during late pregnancy persist after parturition.

The plasma concentrations of RCP in non-human primates are also modulated in concert with physiological changes in estrogen that occur during the menstrual cycle and pregnancy. Thus in bonnet monkeys (*M. radiata*), maximum RCP levels are encountered during days 16–19 of the menstrual cycle, i.e. 3–4 days after the pre-ovulatory surge of estrogen. The administration of pharmacological doses of estradiol-17β to immature male or female monkeys is also able to enhance the plasma concentrations of RCP in concert with the pharmacokinetics of the hormone, the male monkeys exhibiting a relatively sluggish response as the consequence of more rapid decline of circulatory estrogen (Adiga *et al.*, 1986; Visweswariah & Adiga, 1988). Preliminary results (Visweswariah, 1987) with human volunteers reveal that the physiological modulations of RCP during the menstrual cycle are estrogen-dependent; two peaks of RCP concentration are encountered in the plasma, one major peak 4–5 days following the expected period of ovulation and a second shallow peak early during the follicular phase. These peaks are in phase with the similar patterns of assayable plasma estrogen during ovulation and the late luteal phase (Diczfalusy & Landgren, 1977).

Functional importance in early fetal development

From the foregoing, it is amply clear that the evolutionary conservation of RCP encompasses the estrogen regulation of its gene expression at the biosynthetic locus, the hepatocytes. Since the evolution of a protein depends on whether it is essential for the survival of the organism (Wilson *et al.*, 1977), it may not be presumptuous to assume that RCP is involved in the evolutionarily conserved primary mechanism of transplacental delivery of the vitamin in mammals. Supportive evidence for this premise stems from experiments (Murty & Adiga, 1982*b*; Muniyappa & Adiga, 1980*d*) in which endogenous RCP in pregnant rats is passively immunoneutralized by administering adequate amounts of potent antibodies to either homologous or heterologous vitamin carrier, with profoundly deleterious effects on pregnancy progression. Within 24–48 h after antibody administration, there occurs a precipitous fall in the plasma progesterone concentration, indicative of abrupt pregnancy termination. Autopsy reveals complete fetal resorption or abortion. Similarly, active immunization of proven fertile female rats with the chicken egg white RCP induces a good titer of anti-RCP antibodies; when such animals are mated with fertile males, their conceptuses are resorbed during days 8–9 of gestation as reflected by an abrupt decrease in the plasma progesterone, presumably due to fetal death and hence shutting off of the luteotropic stimulus (Murty & Adiga, 1982*b*). The specificity and selectivity of the adverse effects of RCP immunoneutralization on the rodent embryo is revealed by unimpaired maternal well-being in terms of mass gain and vitamin status as determined by erythrocyte flavin content and glutathione reductase activity coefficient. Similar results are observed in mice (Natraj *et al.*, 1987*b*).

Extension of these studies to non-human primates provides unequivocal evidence for the operation of similar phenomena (Seshagiri & Adiga, 1987; Adiga *et al.*, 1991; Natraj, 1991). Active immunization of proven fertile female bonnet monkeys (*M. radiata*), exhibiting normal menstrual cyclicity, with the purified chicken egg white RCP elicits high titers of antibodies, showing thereby that despite overall similarities with

the primate RCP, the chicken vitamin carrier possesses certain immunologically distinguishable domains capable of eliciting antibodies.
Although the immunopotencies of the different antisera are variable, all
the animals tested respond to the heteroantigen with a definite fraction of
the antibodies interacting with ^{125}I-labeled monkey RCP. Furthermore,
when the monkey anti-chicken RCP antibodies are administered intravenously to an unimmunized monkey during its early pregnancy, preferential and progressive neutralization of these antibodies by the
endogenous RCP manifests in its circulation. These observations lend
credence to the premise that the circulating anti-RCP antibodies immunoneutralize the endogenous RCP in the monkeys actively immunized.
As predicted from the data obtained in the rodent model (Adiga &
Murty, 1983) immunization *per se* has no adverse effect on maternal well-
being, menstrual cyclicity, circulating gonadal hormone levels or vitamin
status (as assessed by erythrocyte flavin content and glutathione reductase activity coefficient). These findings favor the view that, as in the
chickens and the rodents, RCP is not essential for maternal well-being of
the macaque and its role is confined to reproductive processes, a view
further strengthened by the plasma appearance of RCP in high concentrations during only the luteal phase of the menstrual cycle and pregnancy
(Visweswariah & Adiga, 1988).

An examination of the reproductive performance of these actively
immunized monkeys reveals that pregnancy termination occurs generally
when their circulatory anti-RCP antibodies are high, although the onset
of abortion, when experienced, varies from animal to animal. Despite
this lack of predictable correlation between total antibody repertoire and
outcome of pregnancy, more than 50% of recognized pregnancies were
terminated in these animals, clearly testifying that the primate RCP plays
a vital role as an obligatory vitamin carrier through the placental barrier
(Seshagiri & Adiga, 1987). More recent experiments using variously
denatured chicken RCPs as the immunogen suggest that one immunogenic efficacy of chicken RCP in terms of elicitation of bioneutralizing
antibodies can be substantially improved and that early pregnancy
suppression due to active immunization can approach nearly 100%.
These observations suggest that the native conformation of the chicken
RCP eliciting heterogenous antibodies directed against the assembled
epitopes is largely redundant for bioneutralization, which can be accomplished by a restricted population of antibodies recognizing only sequential
epitopes (Adiga *et al.*, 1991). Confirmatory evidence for these observations stems from experiments with another primate species from the
New World, the marmoset (*Callithrix jacchus*) wherein early pregnancy

termination can be demonstrated by both passive and active immuniz-ation with antibodies to the chicken RCP (Natraj, 1991).

Mechanism of fetal wastage in rodents

The acute fetal wastage culminating in abrupt pregnancy termin-ation following passive immunoneutralization of RCP in rats 11 days pregnant could be traced to drastic curtailment (>90%) of [^{14}C]riboflavin influx from the maternal supply line to the fetoplacental unit over a period of several hours (Murty & Adiga, 1981). The severe fatal flavin deficiency so precipitated seems to set in motion a series of intra-embryonic events characterized by major disturbance in relative contents and concentrations of flavin coenzymes, the depletion of FAD levels being most dramatic (Krishnamurthy *et al.*, 1984) (Figure 6.4). In attempts to decipher the enzymic basis of these changes, it is intriguing to

Figure 6.4. Influence of immunoneutralization of maternal RCP in the pregnant rat on the relative contents of various flavin coenzymes in the embryonic tissue after 24 h. Day 11 pregnant rats were administered either rabbit non-immune serum (control) or rabbit RCP antiserum; 24 h later, the animals were killed and embryonic flavins separated and quantitated by fluorimetry. Results (mean ± SEM, *n* = 5) are expressed as a percentage of the total flavin in the control embryo (Krishnamurthy, 1984).

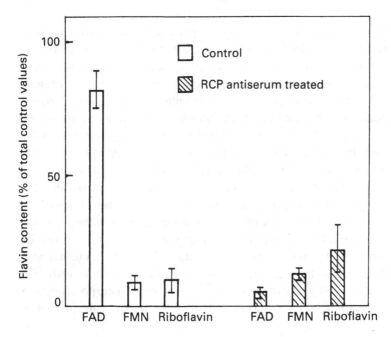

find that induced severe flavin deficiency depresses the biosynthetic FAD pyrophosphorylase in the affected fetuses, which also harbor relatively higher levels of FAD-degrading phosphatase. The combined consequence is the depletion of FAD to below the critical content required to maintain development. Thus, in the absence of sustained carrier-mediated flavin delivery, fetal survival is jeopardized (Surolia et al., 1985). Histological examination of the affected fetoplacental unit reveals (i) detachment of the placental membrane from the decidua; (ii) drastic mitotic arrest in the neural tube; (iii) leukocyte infiltration into both the maternal blood vessels and fetal liver, characteristics of degenerative tissues; and (iv) significant trophoblast degeneration (Adiga et al., 1988a). Cytological examination of the affected fetal liver in mice treated with RCP-antiserum shows extensive degenerative changes with the arrest of differentiation of hemopoietic cells, i.e. the process of erythrocyte aplasia (Natraj et al., 1987b). Additionally, elevated activities of several catabolic enzymes, such as cathepsin D and aryl phosphatase, are encountered in the fetoplacental unit as early as 6 h after antiserum administration, suggesting lysosomal mobilization characteristic of degenerative tissues (Natraj et al., 1987a).

RCP in human pregnancy

As mentioned earlier, human RCP isolated from pregnancy sera and umbilical cord sera has a similar molecular mass to, and shares other physicochemical and immunological properties including flavin binding characteristics with, its avian and rodent counterparts. Its presence in umbilical cord sera in higher concentrations than in the maternal circulation accords with similar relative concentrations of the vitamin measured in the fetal circulation. The detailed mechanisms underlying RCP-mediated transplacental flavin transport as well as the source of umbilical cord RCP present in higher concentrations are yet to be delineated. Preliminary data obtained in our laboratory do indicate the existence of a specific, high-affinity ($K_d = 1.6 \times 10^{-8}$M) membrane receptor capable of binding ^{125}I-labeled RCP on human first-trimester and term placentae, implying an endocytotic mechanism of transport (Prasad et al., 1990a). On the other hand, the propensity of the placental villus tissue to synthesize its own RCP is suggested by: (i) the detection of RCP mRNA in the polysomal RNA fraction of the first-trimester and term placentae by filter hybridization and Northern analysis with ^{32}P-labeled chicken RCP cDNA; and (ii) ^{35}S]methionine incorporation into immunoprecipitable RCP during in vitro short-term culture. What is most interesting is the finding that estradiol-17β added to the culture medium significantly stimulates RCP-mRNA levels as well as the translational machinery of the vitamin carrier. This stimulation can be curtailed by simultaneous addition of tamoxifen, a potent estrogen antagonist. Since large amounts of estrogens are produced by the placenta itself, it is reasonable to assume that RCP synthesis and secretion by the placenta is autoregulated by the steroid hormone through an autocrine–paracrine pathway (Prasad *et al.*, 1990*a*).

Limited clinical data concerning the relative RCP concentrations in the maternal and umbilical vein blood (representing the maternal supply line) in women at normal term delivery reveal that the umbilical vein RCP levels are 4–5-fold higher (Figure 6.5). In contrast, under conditions of clinically diagnosed complicated pregnancies such as pre-eclamptic toxemia, intrauterine growth retardation (IUGR) and fetal distress syndrome (FDS), the umbilical vein RCP concentrations markedly

Figure 6.5. Concentrations of RCP in the maternal and cord blood during vaginal delivery at term in pregnant women. Toxemia, pre-eclamptic toxemia without complication; IUGR, intrauterine growth retardation (mean baby mass at birth = 1.8 kg); FDS, fetal distress syndrome. RCP concentrations in the plasma were measured by radioimmunoassay and expressed in terms of chicken RCP equivalent. Blood samples were collected at the Vani Vilas Maternity Hospital, Bangalore, India, under expert clinical supervision.

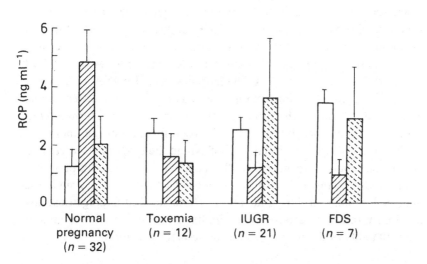

decrease with concomitant increase in the maternal blood. It is possible that these alterations represent the placental insufficiency as the cause or the consequence of these pathological conditions (D. Arul, Prakash Rao & P.R. Adiga, unpublished observations).

RCP function in other reproductive systems

RCP in the mammary gland

From the above discussion, it is clear that one of the major reproductive functions of the estrogen-inducible RCP is to transport and target the essential vitamin through the physiological barrier offered by the placenta to ensure uninterrupted rapid proliferation of embryonic tissues. Now the question arises as to whether RCP exists and becomes functional in other reproductive systems in the female where established physiological barriers are functional. For example, in the mammary gland (which is also estrogen-dependent for its growth and differentiated function) (Lippman & Dickson, 1989) there exist circumferential tight junctions between epithelial cells, which offer a major barrier for the passage of blood constituents other than small ions (Pitilka *et al.*, 1973; Linzell & Peaker, 1971). Since riboflavin concentration in the mammary secretions such as the milk is several-fold higher than in the maternal serum (Jenness, 1974; Wolfrit *et al.*, 1987) it is attractive to visualize that RCP occurs and functions as the vitamin carrier for vitamin secretion into milk. This working hypothesis is supported by (i) the detection by radioimmunoassay of the vitamin carrier in the milk of the rat, guinea pig, cow, monkey and human; (ii) its purification from bovine milk and elucidation of physicochemical and immunological properties which are similar to those of the chicken egg RCP, and (iii) the ability of the rat mammary gland to synthesize the vitamin carrier during pregnancy and lactation as evidenced by the quantitation of RCP mRNA by dot–blot and Northern analysis using ^{32}P-labeled chicken RCP cDNA as the probe and by [^{35}S]methionine incorporation into a protein immunoprecipitable with the antibodies to chicken egg vitamin carrier in short-term *in vitro* culture (Prasad *et al.*, 1990*b*). The proposed function of the mammary RCP is to sequester the available riboflavin in the tissue and secrete it in the protein-bound form into the milk. It appears likely that this process is in turn aided by the higher concentrations of RCP-bound vitamin in the maternal circulation during lactation. These seminal findings raise the possibility that the carrier-bound riboflavin in milk facilitates greater bioavailability of the micronutrient for neonatal nutrition.

RCP in spermatogenesis

Recent investigations carried out in the author's laboratory on testicular tissue have focused on the possibility that RCP plays an important, though restricted, role in male reproduction. This idea is based on the following rationale. In adult male mammals, spermatogenesis involves the most active and continuous cellular proliferation and differentiation; however, an effective blood–testis barrier in the form of tight junctions between adjacent Sertoli cells in the seminiferous tubules isolates the germ cells from the extratubular environment; hence all nutrients and growth factors must be supplied by or transported across the Sertoli cells (Bardin *et al.*, 1988). Furthermore, there is now evidence for: (i) the operation of protein-mediated nutrient transport (e.g. by transferrin) to support germinal cell development (Bardin *et al.*, 1988; Skinner & Griswold, 1980); and (ii) the production of estrogen by both the Sertoli cells and the Leydig cells and its involvement in the regulation of testicular function (Dufau, 1985). It is therefore intriguing to find significant amounts of RCP mRNA in total testicular polysomal RNA as quantitated by hybridization with a chicken RCP cDNA probe. During *in vitro* culture of rat testis incubated with ^{3}H-labeled amino acids, synthesis of labeled RCP immunoprecipitable with specific antibodies to the chicken protein (M_r 37 000) can be unequivocally demonstrated. Further, RCP could be immunocytochemically localized in both the Leydig and the Sertoli cells as well as the pachytene and the round spermatids in rat testis and in the acrosomal region of the mature spermatozoa from rats, cattle, monkeys and humans. This could be confirmed by analysis of solubilized sperm extracts by Western blotting (Prasadan *et al.*, 1990). These exciting findings strongly suggest the involvement of RCP in the transport to, and sequestration of the vitamin by, the germinal cells during cellular proliferation and differentiation, along the lines suggested for iron transport involving testicular transferrin (Griswald, 1988).

Conclusions and future prospects

It is clear that RCP utilized by the avian system for yolk deposition of riboflavin for use by the prospective embryo is remarkably conserved not only down the phylogenic scale to fishes, but also during evolutionary transition to viviparity as well as further evolution all the way to humans. The fact that the mammals have developed special mechanisms for providing nutrition continuously to their embryos *in utero* and yet have not eliminated the carrier-mediated riboflavin delivery mechanism to support fetal growth emphasizes the indispensability of RCP in the survival of the species. It is noteworthy that many aspects of basic structural features, immunochemical characteristics and function, including such details as the biosynthetic loci, hormonal specificity of induction and modulation, are scrupulously retained throughout. The absolute estrogen specificity of RCP gene modulation confers on this steroid hormone an important molecular function: the continued production during gestation of the vitamin carrier at its biosynthetic loci, and transport and delivery of riboflavin in adequate amounts to ensure optimal fetal nutrition, uninfluenced by the competing demands of the mother. This strongly suggests that the progressive increase in estrogen concentrations in the maternal circulation during gestation represents one of the endocrine communications originating from the fetoplacental unit to command the mother to provide adequate amounts of the micronutrient for sustained growth of the fetus. Since immunointerference with the carrier function terminates pregnancy, the possibility exists that genetic or endocrine defects in the production, functionality or transport of RCP may explain some of the ill-defined causes of habitual or spontaneous abortions and of intrauterine growth retardation. Finally, the finding that immunoneutralization of RCP produces acute yet selective fetal riboflavin deficiency opens new vistas of research on fetal vitamin metabolism in relation to its development. It is conceivable that, during fetal development, riboflavin has other special roles in addition to its function as a part of the flavin coenzymes. These roles may be analogous to those suggested for biotin in modulating the cell cycle and cellular differentiation through growth factor synthesis (Dakshinamurti & Chauhan, 1989).

Based on the available data, a working model is presented (Figure 6.6) to account for the role of RCP in fetal development. It is proposed that, during gestation, sustained production of RCP in the maternal liver is dictated by estrogen produced either directly by the fetoplacental unit or through luteotropic stimulation of the ovary. The vitamin carrier secreted into circulation tightly binds riboflavin and carries it to the placental villous membrane where, the carrier–riboflavin complex is endocytosed. The free vitamin thus released is then complexed with the placental RCP to be transported for fetal utilization. Estrogen produced by the placenta and shunted to the fetal circulation induces RCP in the fetal liver, which carries and conserves the micronutrient in fetal circulation for effective utilization during its rapid growth.

This vitamin carrier has served well as a model system to understand some of the structural features involved in flavin–protein interaction.

Although several of the structural characteristics of both the protein and its ligand involved in their mutual interaction are known, they warrant further elucidation, particularly with regard to the amino acid sequences and protein folding characteristics at the hydrophobic pocket involved in flavin binding and the energetics involved therein. RCP has been crystallized (Zanette *et al.*, 1984); the three-dimensional protein folding pattern as revealed by X-ray diffraction is expected to reveal some of these parameters. The availability of several MAbs to RCP, recognizing different epitopic conformations on its surface, may lead to further delineation of some of the conformational characteristics recognized by the placental receptor; one such region located at the C-terminus has been identified (Karande *et al.*, 1991). Further progress in understanding the post-receptor intracellular events and the fate of endocytosed vitamin–carrier complex, and the role and rationale of endogenous RCP production at the post-barrier sites of delivery, should be rewarding. Application of recombinant DNA technology to decipher the genomic organization of RCP, with particular reference to the location and

Figure 6.6. Schematic representation of mechanism of induction and involvement in transplacental flavin transport of RCP in pregnant mammals. E, estrogen; CG, chorionic gonadotropin.

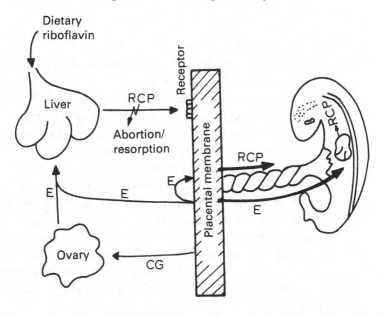

sequence of estrogen-response elements and promoter sequences responsible for hormonal modulation of its gene expression, may unravel some of the intricacies involved in RCP biosynthesis. Approaches such as site-directed mutagenesis may be useful in achieving greater understanding of the structure–function relations of this remarkable vitamin carrier.

Evidence has been adduced in this chapter for the occurrence and functionality of RCP in tissues and processes other than those directly involved in fetal or embryonic development, which, however, fall within the gambit of the general area of reproduction. Thus the earlier working hypothesis that RCP's role in reproduction is confined to flavin transport through the oocyte or placental barrier to subserve embryonic or fetal development in the female needs to be broadened to include other reproductive functions involving physiological barriers; the recent discoveries of RCP synthesis and secretion in the mammary gland and the testicular tissue reveal additional biological roles such as in lactation and spermatogenesis. The only characteristics shared by these systems are: (i) that they all have well-developed physiological barriers; and (ii) that they all involve estrogen-influenced target tissue undergoing cellular proliferation and differentiation. It is conceivable that there may be other physiological systems satisfying all the above parameters in the body where RCP is produced and functionally required; these await fruitful future research.

Original research from the author's laboratory, referred to in this chapter, was carried out with the financial support of the Department of Biotechnology, the Government of India, the Council of Scientific and Industrial Research, the Indian Council of Medical Research, and the Family Planning Foundation, New Delhi. The author wishes to place on record research contributions from several of his former and current associates; in particular he thanks Drs Anjali A. Karande, P. Devi Prasad and K.G. Bhat for their consent to include unpublished observations and useful discussion. Thanks are also due to Professor Harold B. White III, University of Delaware, Newark, Delaware, USA, for the kind gift of the RCP-cDNA clone used in various investigations cited above; to Dr T. Mega and co-workers for permission to reproduce their figure; and to Mrs Vani R. Iyer for the preparation of the manuscript.

References

Abrams, V.A.M., Bush, L., Kennedy, T., Sherwood, T.A. & White, H.B. (1989) Vitamin transport proteins in alligator eggs. *Comp. Biochem. Physiol.* **93B**, 241–7.

Abrams, V.A.M., McGahan, T.J., Rohrer, J.S., Bero, A.S. & White, H.B. (1988) Riboflavin binding proteins from reptiles: A comparison with avian riboflavin binding protein. *Comp. Biochem. Physiol.* **90B**, 243–7.

Adiga, P.R., Karande, A.A., Visweswariah, S.S. & Seshagiri, P.B. (1988*a*) Estrogen-induced riboflavin carrier protein and its role in fetal development. In: *Progress in Endocrinology,* vol. 11 (ed. H. Imura, K. Shizume & S. Yoshida), pp. 343–8. Excerpta Medica, Amsterdam.

Adiga, P.R., Karande, A.A., Visweswariah, S.S. & Seshagiri, P.B. (1991) Carrier protein mediated transplacental riboflavin transport in the primates. In: *Perspectives in Primate Reproductive Biology* (ed. N.R. Moudgal, K. Youshinaga, A.J. Rao & P.R. Adiga), pp. 129–40. Wiley Eastern, New Delhi.

Adiga, P.R. & Muniyappa, K. (1978) Estrogen induction and functional importance of carrier proteins for riboflavin and thiamin in the rat during gestation. *J. Steroid Biochem.* **9**, 829.

Adiga, P.R. & Murty, C.V.R. (1983) Vitamin carrier protein during embryonic development in birds and mammals. In *Molecular Biology of Egg Maturation* (ed. R. Porter & J. Whelan) (Ciba Foundation Symposium 98), pp. 111–36. Pitman, London.

Adiga, P.R., Seshagiri, P.B., Malathy, P.V. & Sandhya, S.V. (1986) Reproduction-specific vitamin carrier protein involved in transplacental vitamin transport in mammals including primates. In: *Pregnancy Proteins in Animals.* (ed. J. Hau), pp. 317–29. de Gruyter, Berlin.

Adiga, P.R., Visweswariah, S.S., Karande, A.A. & Kuzhandhai Velu, N. (1988*b*) Biochemical and immunological aspects of riboflavin carrier protein. *J. Biosci.* **13**, 87–104.

Bardin, C.W., Cheng, C.Z., Musto, N.A. & Gunsalus, G. (1988) The Sertoli cell. In: *Physiology of Reproduction* (ed. E. Knobil & J. Neil), pp. 933–73. Raven Press, New York.

Becvar, J. & Palmer, G. (1982) The binding of flavin derivatives to the riboflavin binding protein of egg white. *J. Biol. Chem.* **257**, 5607–17.

Benore-Parsons, M., Yonno, L., Mulholland, L., Saylor, W.W. & White, H.B. (1985) Transport of riboflavin binding protein in the hen oocyte. Bound vitamin is not required for protein deposition. *Nutrition Res.* **8**, 789–800.

Blankenhorn, G. (1978) Riboflavin binding in egg white flavorprotein: The role of tryptophan and tyrosine. *Eur. J. Biochem.* **82**, 155–60.

Blum, J.C. (1967) *Le metabolisme de la riboflavine chez la poule ponduse.* Hoffmann-La Roche, Paris, 161 pp.

Buss, E.G. (1969) Genetic interference in egg transfer, utilization and requirement of riboflavin by the avian embryo. In: *The Fertility and Hatchability of the Hen's Egg* (ed. T.C. Carter & B.M. Freeman), pp. 109–16. Oliver & Boyd, Edinburgh.

Buster, J.E. & Marshall, J.R. (1979) Conception, gamete and ovum transport, implantation, fetal placental hormones, hormonal preparation for parturition, control. In *Endocrinology,* vol. 3 (ed. L.J. Degroot, G.F. Cahill Jr, W.D. Odell, L. Martini, J.T. Potts, U. Steinberger & A.T. Winegrad), chapter 130, pp. 1601–2. Greene & Stratton.

Cecchini, G., Perl, M., Lipstick, J., Singer, T.P. & Kearney, E.B. (1979) Transport and binding of riboflavin by *Bacillus subtilis. J. Biol. Chem.* **254**, 7295–301.

Choi, J.D. & McCormick, D.B. (1980) Interaction of flavins with egg white riboflavin-binding protein. *Arch. Biochem. Biophys.* **204**, 41–51.

Christmann, J.L., Grayson, M.J. & Huang, R.C.C. (1977) Comparative study of hen's yolk phosvitin and plasma vitellogenin. *Biochemistry* **16**, 3250–6.

Clagett, C.O. (1971) Genetic control of the riboflavin binding protein. *Fed. Proc. Fedn. Ass. Socs exp. Biol.* **30**, 127–9.

Clagett, C.O., Buss, R.G., Saylor, E. & Girsh, S.J. (1970) The nature of the biochemical lesion in avian renal riboflavinuria. Hormonal induction of riboflavin binding protein in roosters and young chickens. *Poultry Sci.* **49**, 1468–72.

Cowan, J.W., Boucher, R.V. & Buss, E.G. (1964) Riboflavin utilization by a mutant strain of a single comb White Leghorn chickens. 2. Absorption of radioactive riboflavin from the digestive tract. *Poultry Sci.* **43**, 172–4.

Cowan, J.W., Boucher, R.V. & Buss, E.G. (1966) Riboflavin utilization by a mutant strain of a single comb White Leghorn chickens. 4. Excretion and resorption of riboflavin by the kidney. *Poultry Sci.* **45**, 538–41.

Cuatrecasas, P. & Roth, T.F. (eds) (1983) *Receptor Mediated Endocytosis,* Chapman & Hall, London. 304 pp.

Dakshinamurti, K. & Chauhan, J. (1989) Biotin. *Vitamins and Hormones* **45**, 337–84.

Dancis, J. & Schneider, H. (1975) Physiology: Transfer and barrier function. In: *Placenta and Maternal Supply Line* (ed. P. Grunwald), pp. 98–124. MTP Press, Lancaster.

Diczfalusy, E. & Landgren, B.H. (1977) Hormonal changes in the menstrual cycle. In: *Regulation of Human Fertility* (ed. E. Diczfalusy), pp. 21–71. Scriptor, Copenhagen.

Dufau, M.L. (1985) Endocrine regulation and communicating functions of the Leydig cell. *Ann. Rev. Physiol.* **50**, 483–508.

Durgakumari, B. & Adiga, P.R. (1986a) Hormonal induction of riboflavin carrier protein in the chicken oviduct and liver: A comparison of kinetics and modulation. *Molec. Cell. Endocrinol.* **44**, 285–92.

Durgakumari, B. & Adiga, P.R. (1986b) Correlation between riboflavin carrier protein and its mRNA activity in estrogen stimulated chicken liver and oviduct. *J. Biosci.* **10**, 193–202.

Durgakumari, B. & Adiga, P.R. (1986c) Estrogen modulation of retinol binding protein in immature chicks: Comparison with riboflavin carrier protein. *Molec. Cell. Endocrinol.* **46**, 121–30.

Evans, A.J. & Burley, R.W. (1987) Proteolysis of apo-protein B during transfer of very low density lipoprotein from hens blood to egg yolk. *J. Biol. Chem.* **262**, 501–4.

Farber, J.J. & Thornburg, K.L. (1983) *Placental Physiology: Structure and Function of Fetomaternal Exchange.* Raven Press, New York, 192 pp.

Farrell, H.M. Jr, Mallette, M.F., Buss, E.G. & Clagett, C.O. (1969) The nature of the biochemical lesion in avian renal riboflavinuria III. Isolation and characterization of the riboflavin binding protein from egg white. *Biochim. Biophys. Acta* **194**, 433–42.

Feeney, R.E. & Allison, R.G. (1969) *Evolutionary Biochemistry of Proteins.* 290 pp. Wiley-Intersciences, New York.

Fenselan, C., Heller, D.N., Miller, M.S. & White, H.B. (1985) Phosphorylation sites in riboflavin binding protein characterized by fast atom bombardment mass spectrometry. *Analyt. Biochem.* **150**, 309–14.

Fraser, D.R. & Emtage, J.S. (1976) Vitamin D in the avian egg – its molecular identity and mechanisms of incorporation into yolk. *Biochem. J.* **160**, 671–82.

Godovac-Zimmermann, J. (1988) The structural motif of β-lactoglobulin and retinol binding protein: A basic frame-work for binding and transport of small hydrophobic molecules. *Trends Biochem. Sci.* **13**, 64–6.

Griswald, M.D. (1988) Protein secretions of Sertoli cells. *Int. Rev. Cytol.* **110**, 133–55.

Gruber, M., Bos, E.A. & Ab, G. (1976) Hormonal control of vitellogenin synthesis in the avian liver. *Molec. Cell. Endocrinol.* **5**, 41–50.

Hamazume, Y., Mega, T. & Ikenaka, T. (1984) Characterization of hen's egg white- and yolk-riboflavin binding proteins and amino acid sequence of egg white riboflavin binding protein. *J. Biochem. (Japan)* **95**, 1633–44.

Hamazume, Y., Mega, T. & Ikenaka, T. (1987) Position of disulphide bonds in riboflavin binding protein of hen's egg white. *J. Biochem. (Japan)* **101**, 217–23.

Hammer, C.H., Buss, E.G. and Clagett, C.O. (1973) Avian riboflavinuria. The fate of the riboflavin binding protein riboflavin complex during incubation of the hen's egg. *Poultry Sci.* **53**, 520–30.

Jenness, R. (1974) The composition of milk. In: *Lactation – A Comprehensive Treatise.* (ed. B.L. Larson & V.R. Smith), vol. III, pp. 3–107. Academic Press, New York.

Jusko, W.J. & Levy, G. (1975) Absorption, protein binding and elimination of riboflavin. In: *Riboflavin* (ed. R.S. Rivlin), pp. 99–152. Plenum Press, New York.

Karande, A.A., Velu, N.K. & Adiga, P.R. (1991) A monoclonal antibody recognizing the C-terminal region of chicken egg white riboflavin carrier protein terminates early pregnancy in mice. *Molec. Immunol.* **28**, 471–8.

Kornfeld, R. & Kornfeld, S. (1985) Assembly of asparagine linked oligosaccharides. *Ann. Rev. Biochem.* **54**, 631–64.

Kozik, A. (1982*a*) Carbodiimide modification of carboxyl groups in egg white riboflavin binding proteins. *Biochim. Biophys. Acta* **704**, 542–5.

Kozik, A. (1982*b*) Disulphide bonds in egg white riboflavin binding protein – chemical reduction studies. *Eur. J. Biochem.* **121**, 395–400.

Kozik, A. (1985) Riboflavin binding protein. *Pol. Biochem.* **31**, 263–81.

Krishnamurthy, K. (1984) Studies on vitamin carrier proteins: physicochemical and functional aspects. Ph.D. Thesis, Indian Institute of Science, Bangalore, India.

Krishnamurthy, K., Surolia, N. & Adiga, P.R. (1984) Mechanism of fetal wastage following immunoneutralization riboflavin carrier protein in the pregnant rat: Disturbances in flavin co-enzyme levels. *FEBS Lett.* **178**, 87–91.

Kuzhandhai Velu, N. (1990) Immunochemical studies on riboflavin carrier protein. Ph.D. Thesis, Indian Institute of Science, Bangalore, India.

Lax, R.E., Tamulevicius, P., Muller, A. & Schriefers, H. (1983) Hepatic nuclear estrogen receptor concentrations in the rat – Influence of age, sex, gestation, lactation and estrous cycle. *J. Steroid Biochem.* **19**, 1083–8.

Linzell, J.L. & Peaker, M. (1971) The permeability of mammary gland. *J. Physiol.* **216**, 701–16.

Lippman, M.E. & Dickson, R.B. (1989) Mechanisms of growth control in normal and malignant breast epithelium. *Rec. Prog. Horm. Res.* **45**, 383–440.

Malhotra, P., Karande, A.A., Krishnaprasadan, T.N. & Adiga, P.R. (1991) Riboflavin carrier protein from carp (*C. carpio*) eggs: Comparison with avian riboflavin carrier protein. *Biochem. Int.* **23**, 127–36.

Matsui, K., Sugimoto, K. & Kasai, S. (1982a) Thermodynamics of association of 8-substituted riboflavins with egg white riboflavin binding protein. *J. Biochem.* **91**, 469–75.

Matsui, K., Sugimoto, K. & Kasai, S. (1982b) Thermodynamics of association of egg yolk riboflavin binding protein with 8 substituted riboflavins. Comparison with the egg white protein. *J. Biochem.* **91**, 1357–62.

Maw, A.J.G. (1954) Inherited riboflavin deficiency in chicken eggs. *Poultry Sci.* **33**, 216–17.

McCormick, D.B., Innis, W.S.A., Merrill, A.H. Jr, Bowers-Komro, D., Oka, M. & Chastain, J.L. (1987) An update on flavin metabolism in rats and humans. In: *Flavins and Flavoproteins* (ed. D.E. Edmondson & D.B. McCormick), pp. 459–71. de Gruyter, Berlin.

Merrill, A.H. Jr, Froehlich, J.A. & McCormick, D.B. (1979) Purification of riboflavin binding proteins from bovine plasma and discovery of a pregnancy-specific riboflavin binding protein. *J. Biol. Chem.* **254**, 9362–4.

Merrill, A.H., Froehlich, J.A. & McCormick, D.B. (1981) Isolation and identification of alternative riboflavin binding protein from human plasma. *Biochem. Med.* **25**, 198–206.

Merrill, A.H. Jr & McCormick, D.B. (1980) Affinity chromatographic purification and properties of flavokinase (ATP: Riboflavin 5'-phosphotransferase) from rat liver. *J. Biol. Chem.* **255**, 1335–8.

Miller, M.S., Benore-Parsons, M. & White, H.B. (1982a) Dephosphorylation of chicken riboflavin binding protein and phosvitin decreases their uptake by oocyte. *J. Biol. Chem.* **257**, 688–24.

Miller, M.S., Burch, R.C. & White, H.B. (1982b) Carbohydrate compositional effects on tissue distribution of chicken riboflavin carrier protein. *Biochim. Biophys. Acta* **715**, 126–36.

Miller, M.S., Buss, E.G. & Clagett, C.O. (1981a) The role of oligosaccharide in the transport of egg yolk riboflavin binding protein. *Biochim. Biophys. Acta* **677**, 225–33.

Miller, M.S., Buss, E.G. & Clagett, C.O. (1981b) Effect of carbohydrate modification on transport of chicken egg white riboflavin binding protein. *Comp. Biochem. Physiol.* **69B**, 681–6.

Miller, M.S., Mas, M.T. & White, H.B. (1984) Highly phosphorylated region of chicken riboflavin binding protein: Chemical characterization and ^{31}P-NMR studies. *Biochemistry* **23**, 569–76.

Miller, R.L., Koszalk, T.R. & Brent, R.L. (1976) Transport of molecules across placental membrane. *Cell Surface Rev.* **1**, 145–207.

Miller, M.S. & White, H.B.III (1986) Isolation of avian riboflavin-binding protein. *Meth. Enzymol.* **122**, 227–31.

Moonen, C.T.W., Van den Berg, A.M.W., Boerjan, M. & Muller, F. (1984) Carbon-13 and nitrogen-15 nuclear magnetic resonance study on the interaction between riboflavin and riboflavin binding apoprotein. *Biochemistry* **23**, 4873–8.

Muniyappa, K. & Adiga, P.R. (1980*a*) Purification and properties of a riboflavin binding protein from the egg white of the duck (*Anas platyrhynchos*). *Biochim. Biophys. Acta* **623**, 339–47.

Muniyappa, K. & Adiga, P.R. (1980*b*) Isolation and characterization of riboflavin binding protein from pregnant rat serum. *Biochem. J.* **186**, 537–40.

Muniyappa, K. & Adiga, P.R. (1980*c*) Estrogen induced synthesis of thiamine binding protein in immature chicks. Kinetics of induction, hormonal specificity and modulation. *Biochem. J.* **186**, 201–10.

Muniyappa, K. & Adiga, P.R. (1980*d*) Occurrence and functional importance of a riboflavin carrier protein in the pregnant rat. *FEBS Lett.* **110**, 209–12.

Murthy, U.S. & Adiga, P.R. (1977*a*) Riboflavin binding protein of hen's egg: Purification and radioimmunoassay. *Indian J. Biochem. Biophys.* **14**, 118–24.

Murthy, U.S. & Adiga, P.R. (1977*b*) Oestrogen induction of riboflavin binding protein in the immature chick: Modulation by thyroid status. *Biochem. J.* **166**, 647–50.

Murthy, U.S. & Adiga, P.R. (1978*a*) Estrogen induction of riboflavin binding protein in immature chicks: Nature of the secretory protein. *Biochem. J.* **170**, 331–5.

Murthy, U.S. & Adiga, P.R. (1978*b*) Estrogen induced synthesis of riboflavin binding protein in immature chicks: Kinetics and hormonal specificity. *Biochim. Biophys. Acta* **538**, 364–75.

Murthy, U.S., Podder, S.K. & Adiga, P.R. (1976) The interaction of riboflavin with a protein isolated from hen's egg white: A spectrofluorimetric study. *Biochim. Biophys. Acta* **434**, 69–81.

Murthy, U.S., Sreekrishna, K. & Adiga, P.R. (1979) A simplified method for purification of riboflavin binding protein from hen's egg. *Analyt. Biochem.* **92**, 345–50.

Murthy, U.S., Suresh, M.R., Prasad, M.S.K. & Adiga, P.R. (1977) Oestrogen stimulation of chicken liver ornithine decarboxylase: Relationship between the levels of the enzyme and poly-A-rich RNA. *Ind. J. Biochem. Biophys.* **14**, 319.

Murti, J.R. (1990) Differential gene expression/modulation by estradiol-17*β* of riboflavin carrier protein and cytochrome P-450 in the chicken and rat. Ph.D. Thesis, Indian Institute of Science, Bangalore, India.

Murty, C.V.R. & Adiga, P.R. (1981) Mechanism of fetal wastage following immunoneutralisation of riboflavin carrier protein in the pregnant rat. *FEBS Lett.* **135**, 281–4.

Murty, C.V.R. & Adiga, P.R. (1982*a*) Induction of riboflavin carrier protein in the immature male rat by estrogen: Kinetic and hormonal specificity. *J. Biosci.* **4**, 227–37.

Murty, C.V.R. & Adiga, P.R. (1982*b*) Pregnancy suppression by active immunisation against gestation-specific riboflavin carrier protein. *Science* **216**, 191–3.

Murty, C.V.R. & Adiga, P.R. (1985) Estrogen induction of biotin binding protein in immature chicks : Kinetics, hormonal specificity and modulation. *Molec. Cell. Endocrinol.* **40**, 79–86.

Natraj, U. (1991) Role of riboflavin carrier protein in the maintenance of pregnancy in common marmosets (*C. jacchus*). In: *Perspectives in Primate Reproductive Biology* (ed. N.R. Moudgal, K. Yoshinaga, A.J. Rao & P.R. Adiga), pp. 141–52. Wiley Eastern, New Delhi.

Natraj, U., George, S. & Kadam, P.A. (1987a) Termination of pregnancy with antiserum to chicken riboflavin carrier protein. Alterations in lysosomal enzyme activities. *Ind. J. Exp. Biol.* **25**, 147–50.

Natraj, U., Kuman, R.A. & Kadam, P. (1987b) Termination of pregnancy in mice with antiserum to chicken riboflavin carrier protein. *Biol. Reprod.* **36**, 677–685.

Nishikimi, N. & Kyogoku, Y. (1973) Flavin protein interaction in the egg white flavoprotein. *J. Biochem. (Japan)* **73**, 1233–42.

Norioka, N., Okada, T., Hamazume, Y., Mega, T. & Ikenaka, T. (1985) Comparison of amino acid sequences of hen plasma-, yolk- and white-riboflavin binding proteins. *J. Biochem. (Japan)* **97**, 19–28.

Nutrition Reviews (1979) Riboflavin and thiamin binding proteins, their physiological significance and hormonal specificity. *Nutr. Rev.* **37**, 261–3.

O'Malley, B.W., McGuire, W.L., Kohler, P.O. & Korenman, S. (1969) Studies on the mechanism of steroid hormone regulation of synthesis of specific protein. *Recent Prog. Horm. Res.* **25**, 105.

Ostrowski, W., Skahzynski, B. & Zak, Z. (1962) Isolation and properties of flavoprotein from the egg yolk. *Biochim. Biophys. Acta* **59**, 515–19.

Palmiter, R.G. & Gutman, G.A. (1972) Fluorescent antibody localization of ovalbumin, conalbumin, ovomucoid and lysozyme in chick oviduct magnum. *J. Biol. Chem.* **247**, 6459–61.

Perry, M.M., Gilbert, A.B. & Evans, A.J. (1978) Electron microscopic observations on the ovarian follicle of the domestic fowl during rapid growth phase. *J. Anat.* **125**, 481–97.

Perry, M.M., Griffin, H.D. & Gilbert, A.B. (1984) The binding of very low density and low density lipoproteins to the plasma membrane of the hen's oocyte. *Exptl. Cell Res.* **151**, 433–46.

Pitilka, D.R., Hamamoto, S.T., Duafala, J.G. & Nemanic, M.K. (1973) Cell contacts in the mouse mammary gland, I. Normal gland in post-natal development and secretory cycle. *J. Cell Biol.* **56**, 797–818.

Prasad, P.D., Kumari, U., Rao, A.J. & Adiga, P.R. (1990a) Mode of transplacental transport of riboflavin carrier protein. In: *International Conference on Perspectives in Primate Reproductive Biology* (*Programme and Abstracts*), p. 55. Indian Institute of Science, Bangalore.

Prasad, P.D., Malhotra, P., Karande, A.A. & Adiga, P.R. (1990b) Circulatory riboflavin carrier protein levels in women with breast cancer. In: *International Conference on Perspectives in Primate Reproductive Biology* (*Programme and Abstracts*), p. 50. Indian Institute of Science, Bangalore.

Prasadan, T.N.K., Murti, J.R., Malhotra, P. & Adiga, P.R. (1990) Riboflavin carrier protein during spermatogenesis in the testis. In: *Proceedings of the Second Annual Meeting of the Indian Society for Study of Reproduction and Fertility* (*Programme and Abstracts*), p. 19. Indian Institute of Science, Bangalore.

Ramanathan, L., Guyer, R.B., Buss, E.G. & Clagett, C.O. (1980a) Chemical modifications of riboflavin binding protein: effects on function and antigenicity. *Molec. Immunol.* **17**, 267–74.

Ramanathan, L., Guyer, R.B., Buss, E.G., Clagett, C.O. & Litwak, S. (1980b) Avian riboflavinuria XI. Immunological cross-reaction with cross-reacting protein from normal heterozygous and mutant hens. *Biochem. Genet.* **18**, 1131–48.

Rhodes, M.B., Bunnett, N. & Feeney, R.E. (1959) The flavoprotein-apoprotein system of egg white. *J. Biol. Chem.* **234**, 2054–60.

Roth, T.F., Cutting, J.A. & Atlas, S.B. (1976) Protein transport: A selective membrane mechanism. *J. Supramolec. Structure* **4**, 527–48.

Schmidt, D.G. (1982) Association of caseins and casein micelle structure. In: *Development in Dairy Chemistry* (ed. P.F. Fox), vol. 1, pp. 61–86. Applied Sciences, New York.

Seshagiri, P.B. & Adiga, P.R. (1987) Pregnancy suppression in the bonnet monkey by active immunization with the chicken riboflavin carrier protein. *J. Reprod. Immunol.* **12**, 93–107.

Shapiro, D.J., Barton, M.C., Chang, T.C. & Lew, D. (1988) Estrogen regulation of gene transcription and mRNA stability. In: *Progress in Endocrinology* (ed. H. Imura, K. Shizume & S. Yoshida), vol. 1, pp. 633–8. Excerpta Medica, Amsterdam.

Skinner, M.K. & Griswold, M.D. (1980) Sertoli cells synthesize and secrete transferrin-like protein. *J. Biol. Chem.* **255**, 9523–5.

Slomezynska, M. & Zak, Z. (1987) The effect of riboflavin-binding protein (RBP) on flavokinase catalytic activity. *Comp. Biochem. Physiol.* **87B**, 681–5.

Smith, E.J., Muto, Y. & Goodman, D.S. (1975) Tissue distribution and subcellular localization of retinol binding protein in normal and vitamin deficient rats. *J. Lipid Res.* **116**, 318–23.

Surolia, N., Krishnamurthy, K. & Adiga, P.R. (1985) Enzymic basis of deranged flavin nucleotides consequent on immunoneutralization of the maternal riboflavin carrier protein in the pregnant rat. *Biochem. J.* **230**, 363–7.

Svendsen, I., Hansen, S.I., Holm, J. & Lyngbye, J. (1984) The complete amino acid sequence of the folate binding protein from cow's milk. *Carlsberg Res. Commun.* **49**, 123–31.

Tata, J.R. & Smith, D.F. (1979) Vitellogenesis: a versatile model for hormonal regulation of gene expression. *Rec. Prog. Hor. Res.* **35**, 47–90.

Tata, J.R., Tames, T.C., Watson, C.S., Williams, J.L. & Wolffe, A.P. (1983) Hormonal regulation and expression of vitellogenin multigene family. In: *Molecular Biology of Egg Maturation* (ed. R. Porter & J. Whelan), pp. 96–110. Pitman, London.

Tranter, H.S. & Board, R.G. (1982) The antimicrobial defense in avian egg: Biological perspective and chemical basis. *J. Appl. Biochem.* **4**, 295–338.

Visweswariah, S.S. (1987) Studies on riboflavin carrier proteins: physicochemical, biosynthetic and immunological aspects. Ph.D. Thesis, Indian Institute of Science, Bangalore, India.

Visweswariah, S.S. & Adiga, P.R. (1987*a*) Purification of a circulatory riboflavin carrier protein from pregnant bonnet monkey (*M. radiata*). Comparison with chicken egg vitamin carrier. *Biochim. Biophys. Acta* **915**, 141–8.

Visweswariah, S.S. & Adiga, P.R. (1987*b*) Isolation of riboflavin carrier protein from pregnant human and umbilical cord serum. Similarities with chicken egg riboflavin carrier protein. *Biosci. Rep.* **7**, 563–71.

Visweswariah, S.S. & Adiga, P.R. (1988) Estrogen modulation of riboflavin carrier protein in the bonnet monkey (*M. radiata*). *J. Steroid Biochem.* **31**, 91–6.

Visweswariah, S., Karande, A.A. & Adiga, P.R. (1987) Immunological characterization of riboflavin carrier protein using monoclonal antibodies. *Molec. Immunol.* **24**, 969–74.

White, H.B. (1987) Vitamin binding proteins in the nutrition of the avian embryo. *J. Exp. Zool.* (suppl.) **1**, 53–63.

White, H.B. III (1991) Maternal diet, maternal proteins and egg quality. In: *Egg Incubation, its Effects on Embryonic Development in Birds and Reptiles* (ed. D.C. Deeming & M.W.J. Ferguson), pp. 1.15. Cambridge University Press.

White, H.B., Armstrong, J. & Whitehead, C.C. (1986) Riboflavin binding protein – concentration and fractional saturation in chicken eggs as a function of dietary riboflavin. *Biochem. J.* **238**, 671–5.

White, H.B. & Merrill, A.H. (1988) Riboflavin binding proteins. *Ann. Rev. Nutr.* **8**, 279–99.

Williams, J. (1962) A comparison of conalbumin and transferrin in the domestic fowl. *Biochem. J.* **83**, 355–64.

Wilson, A.C., Carlson, S.S. & White, T.J. (1977) Biochemical evolution. *Ann. Rev. Biochem.* **46**, 573–639.

Winter, W.P., Buss, E.G. & Clagett, C.O. (1967) The nature of the biochemical lesion in avian renal riboflavinuria II. The inherited change of a riboflavin binding protein from blood and eggs. *Comp. Biochem. Physiol.* **22**, 897–906.

Wolfrit, N., Dancis, J., Levetz, M., Jean-Lehanka, B.S. & Young, B.K. (1987) Riboflavin concentration in maternal and cord blood in human pregnancy. *Am. J. Obstet. Gynecol.* **157**, 748–52.

Woods, J.W. & Roth, T.F. (1984) A specific subunit of vitellogenin that mediates receptor binding. *Biochemistry* **23**, 5774–80.

Yusko, S.C., Roth, T.F. & Smith, T. (1981) Receptor mediated vitellogenin binding to chicken oocytes. *Biochem. J.* **200**, 43–50.

Zanette, D., Monaco, H.L., Zanotti, G. & Spadon, P. (1984) Crystallization of hen egg white riboflavin binding protein. *J. Molec. Biol.* **180**, 1185–7.

Zheng, D.B., Lim, H.K., Pine, J.J. & White, H.B. (1988) Chicken riboflavin binding protein cDNA sequence and homology with milk folate binding protein. *J. Biol. Chem.* **263**, 11126–9.

7

Binding Proteins for α-Tocopherol, L-Ascorbic Acid, Thiamine amd Vitamin B_6

Krishnamurti Dakshinamurti and Jasbir Chauhan

α-Tocopherol binding proteins

Vitamin E is the generic term for two groups of naturally occurring fat-soluble compounds, the tocopherols and the tocotrienols (Kasperek, 1980). In the tocopherols, the side chain is saturated; in the tocotrienols, it is unsaturated. The members of each group are designated α, β γ, and δ depending on the number and position of methyl groups attached to the chroman nucleus. The most biologically active of these compounds is α-tocopherol, which inhibits the peroxidation of membrane lipids. This protection is dependent on the incorporation of α-tocopherol into membranes, and the extent of protection is related to the quantity of tocopherol present in membranes. Vitamin E is found primarily in intracellular membranes at concentrations of one to two orders of magnitude higher than that in the soluble fraction of various tissues (Taylor et al., 1976). In tissue homogenates, α-tocopherol is associated with amphiphilic and protein molecules and does not occur in the free form.

The differing biological activities of vitamin E vitamers could be the result of mechanisms operating at various levels: intestinal absorption, lymph and plasma transport, level of uptake and retention by tissues, and finally, their rates of catabolism. Orally administered radioactive tocopherol appears rapidly in the plasma and tissues of the rat and chicken (Krishnamurthy & Bieri, 1963) The vitamin is transported by lymph and plasma proteins (Peake et al., 1972). Despite various studies on the absorption of α-tocopherol in animals and in man, the mechanism of absorption is still not well understood. It is known that only 20–25% of the orally administered dose of α-tocopherol is absorbed, the remainder being excreted in feces (Simon et al., 1956). The intake of γ-tocopherol is 2–4 times that of α-tocopherol because of its predominance in North

American diets (Bieri & Evarts, 1973, 1974). However, several studies have shown that γ-tocopherol in normal adult human plasma averages only about 15% of the concentration of α-tocopherol (Bieri & Prival, 1965; Behrens & Madere, 1985, 1986).

In spite of much investigation the reason for the lower biopotency of γ-tocopherol is not known. Following an oral dose of γ-tocopherol the plasma and tissue levels of this vitamin are lower than found for α-tocopherol in corresponding experiments. Peak *et al.* (1972) found no marked difference in the rates of absorption of α- and γ-tocopherol in rats with a lymphatic cannula and concluded that γ-tocopherol disappeared faster from tissues. It has been suggested that the concentration of γ-tocopherol in plasma and tissues is determined by the relative concentrations of α- and γ-tocopherol (Behrens & Madere, 1983, 1985, 1986). It has also been postulated that the absorption, transport and tissue uptake of vitamin E is specific for α-tocopherol. Experiments in rats indicate that when the concentration of α-tocopherol is relatively low, γ-tocopherol is absorbed, transported and taken up by tissues in increased amounts (Behrens & Madere, 1983). In humans α-tocopherol seems to be the major tocopherol in plasma, accounting for nearly 87% of the total tocopherol concentration (Behrens & Madere, 1985, 1986). Using three groups of rats with low or normal or high concentrations of vitamin E in the diet, Behrens & Madere (1987) observed that a competition for intestinal absorption and plasma transport takes place among the vitamers. γ-Tocopherol seems to be absorbed at a rate equal to that of α-tocopherol provided that the concentration of the α-form in the gut is low. However, in the presence of α-tocopherol, plasma γ-tocopherol levels varied inversely with the levels of α-tocopherol. Total tocopherol (α + γ) in plasma remained the same in the three groups of animals after dosing with γ-tocopherol or with a mixture of α and γ. The total concentration in plasma was of the same order of magnitude as that obtained when α-tocopherol alone was fed to rats in the low vitamin E group, indicating that the absorption and transport are saturable. A similar mechanism seems to apply for the transport of vitamin E in red blood cells. Similar conclusions about the absorption and transport of plasma tocopherols can be drawn from experiments in which the time course of γ-tocopherol appearance in plasma was studied in cannulated rats after an oral dose of 20 mg of γ-tocopherol (Behrens & Madere, 1986).

The adrenal gland accumulates more tocopherol per gram of tissue than other tissues, regardless of the nature of the diet. In plasma, red blood cells, adrenal gland, spleen and lung uptake seems to be saturable because total tissue tocopherol concentration was similar regardless of dose administered.

Involvement of lipoproteins in the absorption and transport of α-tocopherol

There is a similarity between tocopherol and cholesterol uptake. Tocopherol can exchange between the different lipoproteins, and it can also exchange between lipoproteins and erythrocytes. Low-density lipoprotein (LDL) has been shown to enter cells via the specific, high-affinity LDL receptor. Since LDL is a major constituent of serum lipoproteins and functions as a transporter for tocopherol, the LDL receptor mechanism also mediates the delivery of tocopherol to cells.

Bjornson *et al.* (1976) have shown that intravenous administration of chylomicrons on very low-density lipoprotein (VLDL) containing radioactive tocopherol resulted in its rapid distribution among all the plasma lipoproteins, red blood cells (RBC) and tissues. The mechanism of incorporation of tocopherol from lipoproteins into tissues is not known. Tocopherol appears to be exchanged rapidly in a manner similar to the exchange of free cholesterol between plasma lipoproteins and RBC (Silber *et al.*, 1969; Poukka & Bieri, 1970). Tocopherol concentrations in plasma increase in parallel with the total lipid content of human plasma (Rubinstein *et al.*, 1969; Gontzea & Nicolau, 1972). Elevated concentrations have also been reported in disease states associated with elevated plasma lipids. Peak *et al.* (1972) found that the tocopherol content of the plasma lipoprotein of individual rats was more closely correlated with the total lipid content of each lipoprotein than with the distribution of free cholesterol. In normal human subjects the plasma LDL fraction, which contains the largest amount of total lipid, also contains the largest amount of tocopherol. These observations suggest that tocopherol is in dynamic equilibrium between the individual plasma lipoproteins, and between these lipoproteins and the RBC membrane. It is possible that tocopherol may also be in dynamic equilibrium between the plasma lipoproteins and tissues, since concentrations of tocopherol in adipose tissue tend to parallel those in the plasma. In rats, changes in plasma and liver lipid concentrations are paralleled by similar changes in tocopherol concentrations (Davies *et al.*, 1971).

Behrens & Madere (1985) have shown that when the capacities of lipoproteins to carry α- or γ-tocopherol are expressed in micrograms of tocopherol per milligram of protein, the rank order of the major carriers is LDL > VLDL > HDL (high-density lipoprotein). Traber & Kayden (1984) have studied the uptake of α-tocopherol in normal and LDL-receptor-negative fibroblasts. In normal fibroblasts with increasing amount of LDL the tocopherol content increased in a saturable fasion from 19 to 103 ng tocopherol mg^{-1} protein. In LDL-receptor-negative fibroblasts the increase was from 18 to 39 ng tocopherol mg^{-1} protein, suggesting that cells may require LDL as a source of both cholesterol and tocopherol for membrane genesis.

Muscle cells, similar to other cells, have receptors on their surface for LDL and HDL (Goldstein & Brown, 1974). Gurushinghe et al. (1988) studied the binding and uptake of LDL and HDL α-tocopherol and have shown that LDL binds in a saturable manner to specific receptors on skeletal muscle cells. Although HDL also binds to muscle cells, it is not internalized. LDL competes effectively with HDL for binding to HDL binding sites. This is not due to lipoprotein cross contamination, since SDS–PAGE of the LDL and HDL preparations showed that the major protein components were apoprotein B100 and apoprotein A and B, respectively. At low LDL concentrations, about 85% of the α-tocopherol is transferred to the muscle cell, and this value does not change significantly with concentration of HDL or LDL. Little of the α-tocopherol is lost to the surrounding medium or remains associated with the receptor. It appears that the α-tocopherol is transferred to the muscle cell by some mechanism other than direct LDL internalization. The simplest explanation for these results is that there is a lipid-soluble exchange of α-tocopherol between the LDL and the muscle plasma membrane lipids. This explanation is reinforced by results from HDL binding to muscle cells: there was virtually no uptake of HDL by the cells. Despite this, the exchange of α-tocopherol between HDL and muscle cells was similar to that found for LDL, i.e. about 85% of the HDL-incorporated α-tocopherol was transferred to muscle cells. It would seem, therefore, that a direct exchange of α-tocopherol is more important than internalization of lipoproteins for the delivery of α-tocopherol to the cells. In order that this exchange may take place, the lipoprotein would still need to be in close association with the muscle cell plasma membrane; a relationship between receptor binding of LDL and tocopherol uptake into cells has been demonstrated.

Rajaram et al. (1974) observed that the soluble portion of the small-intestinal mucosa of the rat contained appreciable radioactivity 4 h after oral administration of α-[^{3}H]tocopherol. About 40% of the total radioactivity in the mucosa was associated with the cytosol and the remainder with the particulate material. These authors indicate that the absorbed α-tocopherol is associated with the lipoprotein fraction of the intestinal mucosa cytosol separable by Sephadex gel filtration and polyacrylamide-disc electrophoresis. This is accompanied by a simultaneous appearance of the vitamin in the VLDL of serum and a lipoprotein fraction of liver cytosol. The association of the vitamin with these lipoproteins reaches a maximum at 4 h after its administration and declines thereafter. Over 75% of the radioactivity associated with the lipoprotein fraction is due to α-tocopherol alone. The lipoproteins of the intestinal mucosa and liver cytosol are similar to the VLDL of serum in their elution pattern from the Sephadex column and their mobility on polyacrylamide gel, indicating that the molecular mass of the binding protein is very high. Their results incidentally provide a direct proof for the presence of VLDL-like lipoprotein in the mucosal cells of the small intestine, though its presence had been implied in other reports (Windmueller & Levy, 1968; Kessler *et al.* 1970). The existence of specific tocopherol-binding lipoproteins in rat tissues, such as intestinal mucosa, plasma and liver cells, is thus indicated. The mucosal lipoprotein that carries the vitamin probably aids in the intestinal absorption of α-tocopherol. Serum VLDL is involved in the transport of the vitamin to the various tissues in addition to its important role in the transfer of the absorbed vitamin across the intestinal mucosal cells. The transported vitamin is distributed among the intracellular organelles, in association with a lipoprotein of the cytosol which is probably specific for various organs.

An α-tocopherol binding-protein-mediated transfer of vitamin E between membranes

A liver cytosolic protein from the rat has been isolated that specifically binds the most biologically active isomer of vitamin E, α-tocopherol (Catignani, 1975; Catignani & Bieri, 1977). One proposed function of this α-tocopherol binding protein is the transfer of α-tocopherol between membranes in liver cells. The α-tocopherol binding protein binds α-tocopherol *in vivo* (Behrens & Madere, 1982) and mediates the time-dependent transfer of α-tocopherol from liposomes to liver microsomes (Murphy & Mavis, 1981) and to liver mitochondria (Morwi *et al.*, 1982) *in vitro*. The action of this tocopherol binding protein is similar to that of other cytosolic proteins which transfer or exchange phospholipids (Wirtz, 1974), cholesterol (Bloj & Zilversmit, 1977) and fatty acids (Ockner *et al.*, 1982). The α-tocopherol binding protein has a M_r of 31 000 – 32 000; it binds α-tocopherol with high affinity and is not displaced by dl-α-tocopherol acetate, α-tocopherol quinone, or rac-6-hydroxy-2,5,5,8-tetramethyl-chroman-2-carboxylic acid (Catignani & Bieri, 1977).

Role of rabbit heart cytosolic tocopherol binding in the transfer of tocopherol into nuclei

Nair and co-workers (Hauswirth & Nair, 1972; Patnaik & Nair, 1975, 1977) suggested that significant amounts of α-tocopherol were associated with the nuclei of rat liver. This activity was shown to be associated with a high-affinity nuclear tocopherol binding receptor with biochemical characteristics of a non-histone chromosomal protein. A protein similar to the rat liver cytoplasmic α-tocopherol binding protein (Catignani & Bieri, 1977) has been suggested to function in the transfer of α-tocopherol to the nucleus. Guarnieri *et al.* (1980) have isolated a rabbit heart cytosolic tocopherol-binding protein which functions in the transfer of tocopherol into nuclei. Their results suggested that about 70–75% of the original label was recovered in solubilized nuclear acidic protein. They have further shown that the radioactivity of tocopherol was associated with the chromosomal fraction. They postulated that, in order to accumulate α-tocopherol in the nuclei, it is first necessary to form a protein–vitamin complex. This carrier is able to facilitate the translocation of the vitamin from the cytoplasm to the chromatin, where tocopherol is preferentially bound to a non-histone acidic protein.

Bovine milk lipoprotein lipase

Lipoprotein lipase functions in the hydrolysis of triglycerides to fatty acids and the transfer of the fatty acids to tissue (Nelsson-Ehle *et al.*, 1980). This enzyme has been implicated in the transfer of cholesterol esters (Fielding, 1978) or cholesterol linoleyl ether (Chajek-Shaul *et al.*, 1982) from triglyceride-rich lipoproteins to cells. Traber *et al.* (1985) have shown that, *in vitro*, addition of bovine milk lipase (ligase) to chylomicrons in the presence of human erythrocytes or fibroblasts resulted in the hydrolysis of the triglyceride and transfer of both fatty acids and tocopherol to the cells. In the absence of lipase, no increase in cellular tocopherol was detectable. In order to demonstrate that lipoprotein lipase was involved in transferring tocopherol to the cells they used an incubation system that included only fibroblasts, 4% BSA in Dulbecco's minimum essential medium, and intralipid (an artificial emulsion containing 10% soybean oil, which has γ- but not α-tocopherol). The advantage of this system is that it contains no apolipoproteins and that the γ-tocopherol transferred is not present in fibroblasts. The addition of lipase to this system resulted in mass transfer of γ-tocopherol to the fibroblasts. Further addition of apolipoprotein CII to the incubation medium resulted in an increase in the cellular tocopherol content. In corroboration of their findings, Traber *et al.* (1985) further showed that heparin, which prevents lipoprotein lipase from binding to the cell membrane of fibroblasts, prevented the transfer of tocopherol to fibroblasts. Further addition of apolipoprotein CII to the incubation system did not restore the transfer of tocopherol to fibroblasts. Their results show that, for tocopherol to be transferred to cells during the hydrolysis of triglyceride by lipoprotein lipase, the lipase must bind to the cell membrane as has been demonstrated for transfer of the cholesteryl ethers (Chajek-Shaul *et al.*, 1982).

Binding of α-tocopherol to erythrocyte membranes

α-Tocopherol is an antioxidant *in vitro* and has been shown to decrease the hemolytic damage to red blood cells by oxidative stress. Since the disruption of the red blood cell membrane is the likely locus of the oxidative stress, and since α-tocopherol has been proposed to have an important role in the maintenance of membrane structure and function, it is possible that α-tocopherol binding proteins or receptors are present on the red blood cell membrane. Kitabchi & Wimalasena (1982*a,b*) have shown that red blood cell membranes have specific binding sites for α-tocopherol. These binding sites have properties generally expected of receptors: specificity, saturability, moderate affinity, time-, protein- and temperature-dependence, and reversibility of binding. The K_a values for the high-affinity fast dissociation and low-affinity fast dissociation systems were 2.6×10^7 M^{-1} and $1.24 \times 10^6 M^{-1}$, respectively. Binding of α-tocopherol to red blood cell membrane was highly specific. Of a series of tocopherols, only γ-tocopherol competes with α-tocopherol for binding to the membrane. These binding sites may be of physiological importance in the human RBC, possibly in the antihemolytic action of α-tocopherol. It is also possible that the high-affinity, low-capacity site plays a restricted role in red blood cell function, perhaps in specific cellular reactions, and that the low-affinity, high-capacity site has a more generalized structural role in red blood cell function.

The gel chromatography of membrane-bound [³H] α-tocopherol binding site complex, solubilized with Triton X-100, is resolved into two fractions with M_r, values of 65 000 and 125 000 (Wimalasena *et al.*, 1982). The high-molecular-mass species, when re-chromatographed on a gel filtration column, is resolved into a protein of M_r 65 000, suggesting that these complexes may have a monomer–dimer relationship. Since these complexes may have bound detergent, molecular masses may be overestimated. Earlier studies of rat liver cytoplasmic α-tocopherol binding proteins suggested a M_r of 30 000 for this protein (Catignani, 1975; Catignani & Bieri, 1977). The human RBC membrane-bound α-tocopherol binding sites may be oligomers of the cytosolic α-tocopherol binding protein.

L-Ascorbic acid transport

The antioxidant activity of ascorbic acid (vitamin C) in all living organisms is the best-recognized function of this vitamin. Vitamin C and vitamin E eliminate cytotoxic free radicals by redox cycling. The hydrophobic vitamin E reduces free radicals which are formed in membranes. In this process vitamin E is oxidized to the tocopheroxyl radical, which is reduced to tocopherol by ascorbic acid. Unpaired electrons are channeled from reactive free radicals to ascorbate. Njus & Kelley (1991) propose that ascorbate and vitamin E function at physiological pH as donors of single hydrogen atoms, enabling these vitamins to react efficiently with free radicals and not with molecular oxygen.

Ascorbic acid is a cofactor in the synthesis of collagen and proteoglycans, the two major constituents of the extracellular matrix. The synthesis of collagen by fibroblasts involves enzymes that hydroxylate lysine and proline in the procollagen molecule (Peterkofsky & Udenfriend, 1965; Housmann, 1967). Both intra- and extracellular sites of action in the neurological function of ascorbate have been indicated. A role for ascorbate in the dopamine β-hydroxylase reaction [3,4-dihydroxyphenylethylamine, ascorbate : oxygen oxidoreductase (hydroxylating); EC 1.14.17.1; dopamine β-monooxygenase; DBH] which catalyzes the final step in the synthesis of norepinephrine has been established (Friedman & Kaufman, 1965). Ascorbate has also been shown to be a cofactor of the enzyme peptidylglycine α-aminating monooxygenase in the amidation of the neuroendocrine peptides (Mains *et al.*, 1985) The concentration of ascorbate in brain and endocrine glands is high. In addition, the presence of a complex system to regulate its intracellular and intravesicular availability would indicate a critical role for ascorbate in these structures (Diliberto *et al.*, 1991). Ascorbate is released from brain and endocrine tissues and can affect cellular functions by acting on plasma membrane receptors (Gardiner *et al.*, 1985).

Most animals synthesize ascorbate from glucose in the liver; exceptions to this are the primates and the guinea pig. The intestinal transport of ascorbate has special significance in primates and the guinea pig as the source of ascorbic acid. The absorption and transport of ascorbic acid in various organs and tissues as well as in subcellular organelles has been investigated (Rose, 1988). Vitamin C (L-dihydroascorbic acid) is a stereospecific compound. The L stereoisomer is the biologically active compound whereas the D form has only one twentieth its biological activity. Recently it has been shown (Kipp & Schwarz, 1990) that once inside the cell the D epimer (D-isoascorbic acid) is as efficient as the L form, at least in one aspect of its biological action, that of collagen synthesis. Thus, it is the cell membrane which is the primary site of steroselectivity. The carrier proteins in the cell membrane are stereospecific in terms of binding and transporting the vitamin across the cell membrane to its intracellular site of action.

Dietary vitamin C is a mixture of ascorbate and dehydroascorbic acid. Both compounds are antiscorbutic. Dehydroascorbic acid, the oxidized form, is not an acid, does not ionize and is more hydrophobic than ascorbic acid. A net transepithelial absorption and transport of this form was shown in everted and non-everted loops of guinea pig intestine (Choi & Rose, 1989). The vitamer reaching the serosal side was in the reduced state. Such simultaneous absorption and reduction is characteristic of species which do not synthesize ascorbic acid. On the contrary, there was no net absorption of dehydroascorbic acid in the intestine of the rat, a species capable of synthesizing ascorbic acid in the liver.

In contrast to the intestinal transport process, renal handling of ascorbate by various mammalian species seems to be very similar. Ascorbate is filtered in the glomerulus and is reabsorbed in the proximal convoluted tubule. Brush border vesicles from kidney cortex were used to study renal transport (Toggenburger *et al.*, 1981); a 'mobile carrier' was indicated because of the saturation kinetics, competitive inhibition, and cation dependence. Bowers-Komro & McCormick (1990) used isolated kidney cells (predominantly proximal tubular epithelia) to study ascorbate transport and found that the cells accumulated ascorbic acid. The uptake was saturable, inhibited by D-isoascorbate and dehydroascorbic acid, temperature-dependent and susceptible to metabolic inhibitors. Their finding indicates that ascorbate is reabsorbed by the kidney in a sodium-dependent active process not shared by other acid anions.

An earlier study (Mann & Newton, 1975) indicated an inhibition of dehydroascorbic acid uptake into erythrocytes by glucose. However, Bianchi & Rose (1986) suggest that a glucose-independent system that recognizes both ascorbate and dehydroascorbic acid might be more relevant for the transport of dehydroascorbic acid and the release of ascorbate from erythrocytes.

Human neutrophils contain millimolar amounts of ascorbic acid, which is present almost entirely in the cytosol as free (non-protein-bound) ascorbate and not as dehydroascorbic acid. Freshly isolated human neutrophils, when incubated in a medium containing physiological amounts of ascorbic acid, accumulate millimolar amounts using two transporters (Washko *et al.*, 1989). The high-affinity transporter with an apparent K_m of 2–5 μM is saturated at normal extracellular ascorbic acid concentrations. The low-affinity transporter has an apparent K_m of 6–7 mM and should be in the linear phase of uptake at normal extracellular concentrations of ascorbic acid. Thus, the intracellular concentration of ascorbic acid in neutrophils is dependent on its extracellular concentration. The role of ascorbic acid in neutrophils is not yet understood. It may function to preserve neutrophil integrity and also to protect host tissue by inactivating free radicals and oxidants produced by the phagolysosomes.

The human fetal blood has a high concentration of ascorbic acid relative to maternal blood. There seems to be a net transport of ascorbic acid by the syncytiotrophoblast (Anderson *et al.*, 1984). The uptake of dehydroascorbic acid by human placental tissue is much more rapid than that of ascorbic acid. However, most of the material taken up is in the form of ascorbic acid. The cellular uptake transporter for dehydroascorbic acid is not shared by glucose and is not Na^+-dependent but dependent on normal cellular metabolism.

Ascorbic acid is essential for the formation of bone by osteoblasts. *In vivo* studies indicate that administered [^{14}C]ascorbic acid accumulates preferentially in sites of active osteogenesis. *In vitro* studies have shown that ascorbic acid is essential for the formation of mineralized extracellular matrix by rabbit long bone cells (Anderson *et al.*, 1984) and rat calvaria cells (Bellows *et al.*, 1986). Wilson & Dixon (1989) have used two cell systems, a clonal osteoblast-like cell line ROS 17/2.8 and a primary culture of rat calvaria cells, as *in vitro* osteoblast models. Both cell cultures accumulate ascorbate. Consistent with a carrier-mediated mechanism, transport was saturable, temperature-dependent, stereoselective and dependent on external Na^+ concentration. Ascorbate taken up is rapidly compartmentalized in the lumen of the rough endoplasmic reticulum and bound or metabolized to a non-permeant species, in keeping with its putative role as a cofactor in post-translational modification of procollagen (Prockop *et al.*, 1979). In further work, using the clonal osteosarcoma cell line UMR-106, Dixon & Wilson (1992) have shown that transforming growth factor β (TGFB), an important modulator of cell proliferation and differentiation, stimulates the carrier-mediated ascorbate transport. Stimulation of ascorbate uptake was seen at TGFB concentrations similar to that required for stimulating the production of collagenous extracellular matrix. The enhancement by TGFB of ascorbate uptake was associated with an increase in the apparent V_{max} of the rate of transport without any change in the affinity of the transporter for ascorbic acid, indicating that either the number of functional transporters or the rate at which individual transporters translocate ascorbate is increased. The effect of TGFB is completely blocked by cycloheximide. This would suggest that TGFB increases the rate of synthesis of the Na^+ ascorbate cotransporter or of a regulatory protein which interacts with the existing transporters to increase their turnover number.

Ascorbic acid is transported into 3T6 fibroblasts, which synthesize collagen, by a carrier-mediated, saturable, active Na^+-cotransporter process (Padh & Aleo, 1987a), which appears to be the mechanism of transport in most cell types. It was observed that serum from various animals contained a heat-labile factor which, when it interacted with bacterial endotoxin, inhibited ascorbate transport (Padh & Aleo, 1987b), suggesting that the inhibitor may have been generated during endotoxin-induced activation of complement in unheated serum. The inhibitor has been purified from insulin-activated human serum. This heat-stable serum protein factor, which is produced when serum complement is activated, has a molecular mass of 9 kDa (Padh & Aleo, 1989). The reactivity of the purified inhibitor with antiserum to C3a (component 3 of complement) suggests that the inhibitor is C3a or a very similar molecule. The inhibitor of ascorbate transport increased the apparent K_m for ascorbate without affecting the V_{max}. In view of the similarity of ascorbate transport in many cell types, it is possible that this inhibitor could inhibit ascorbate transport in other cell types as well, resulting in depletion of tissue levels of ascorbate during infection or autoimmune processes, when serum complement is activated.

Following intracardiac injection of [14C]ascorbic acid into guinea pigs, the neurohypophysis takes up the label to a high concentration, exceeded only by the adrenals (Thorn et al., 1986). These authors have demonstrated that isolated nerve endings from ox neurohypophysis took up [14C]ascorbic acid by a saturable process and suggest the presence of both a glucose-dependent and a sodium-dependent uptake (Thorn et al., 1991). Cortisol and triiodothyronine inhibited uptake. Post-translational processing of vasopressin and oxytocin precursors involves a trypsin-like enzyme, a carboxypeptidase and an amidating enzyme [peptidylglycyl-α-amino monooxygenase (PAM)]. PAM, which requires copper, molecular oxygen and ascorbic acid, may be limiting in the processing of amidated peptides (Eipper *et al.*, 1987). In a study of pancreatic islet cells, in which several peptides such as pancreastatin, thyrotropin-releasing hormone, pancreatic polypeptide and amylin are C-terminally amidated, the presence of PAM activity has been reported (Scharfmann *et al.*, 1988). Zhou *et al.* (1991) have demonstrated the presence of PAM and a high content of ascorbic acid as well as a carrier-mediated process for its uptake in the islet cells. As in neurohypophyseal nerve terminals, ascorbate uptake was inhibited by triiodothyronine and glucocorticoids.

Ascorbic acid is of particular importance for the brain because it is a cofactor for the synthesis of catecholamines, facilitates neurotransmitter release, and modulates binding of ligands to neural receptors (Wilson, 1989). Astrocytes have a stereospecific, high-affinity and Na^+-dependent system for ascorbate uptake, regulated by external ascorbate. The apparent V_{max} of transport increases rapidly when the cells are cultured in a medium devoid of ascorbic acid. Such regulation of transport may play a role in intracellular ascorbate homeostasis (Wilson *et al.*, 1990).

In the non-stressed animal, ascorbate is concentrated in the adrenal cortex to about three orders of magnitude higher than in plasma. The transport of ascorbate into bovine adrenal cortical cells (Finn & Johns, 1980) and in primary cultures of adrenal cortical cells (Leonard *et al.*, 1983) was shown to be an active process. Adrenocorticotrophic hormone (ACTH) inhibits ascorbate uptake by cortical cells. The adrenal has two stores of ascorbic acid. Stress of administration of ACTH depletes about half the stores of ascorbate in the adrenals. Ascorbate and glucocorticoids released by the adrenal cortex enter the sinusoids supplying the chromaffin cells of the adrenal medulla. Medullary chromaffin cells also accumulate ascorbic acid by a saturable, energy-dependent process (Diliberto *et al.*, 1983). This uptake is inhibited by glucocorticoids at concentrations that occur in the adrenal portal system during stress. This inhibition has been suggested as a regulatory aspect of adrenal stress response to ensure augmented concentrations of ascorbic acid in the periphery (Levine & Pollard, 1983). The only ascorbate-dependent step of catecholamine biosynthesis is dopamine β-hydroxylase located in the catecholamine-containing chromaffin vesicle. Even though the rate of uptake of ascorbate by the chromaffin cells is high, the rate of transfer to the chromaffin vesicle is slow. The role of dopamine β-hydroxylase is dependent on the intravesicular concentration of ascorbate (Dhariwal *et al.*, 1989), suggesting that the intravesicular availability of ascorbate may be regulatory for the synthesis of norepinephrine. The predominantly nicotinic-receptor-mediated secretion of catecholamines from the adrenal medulla occurs by exocytosis; ascorbate is co-secreted, suggesting the possibility that ascorbate may be released from the chromaffin vesicle. However, the release of catecholamine and of ascorbate exhibit differential sensitivities to calcium channel blockers and secretagogues (Daniels *et al.*, 1983). It has been proposed that ascorbate within the cytosolic compartment is not protein-bound and is released by reversal of its uptake transporter (Knoth *et al.*, 1987). There is depletion of ascorbate from the adrenal medulla in the rat, following insulin-induced hypoglycemia, which is associated with neurogenic adrenomedullary secretion of catecholamines. Ascorbate from the chromaffin vesicle is co-secreted along with catecholamines and other vesicular components. However, the bulk of the ascorbate is released from multiple compartments of the chromaffin cell; this depletion of ascorbate is not neurogenic, but under hormonal control in the rat (Diliberto *et al.*, 1991).

Ascorbate modulates collagen production through its effect on prolyl hydroxylation. It has been reported that ascorbate may have an additional role. Murad *et al.* (1981) reported that, after prolonged exposure to ascorbate, collagen synthesis in cultured human fibroblasts increased eightfold with no significant change in the synthesis of non-collagen protein. This effect of ascorbate was unrelated to its cofactor function in collagen hydroxylations. Murad *et al.* (1981) suggested the possibility of a shift in the distribution of collagen-synthesizing ribosomes towards polysomes in the presence of ascorbate. In further work, Geesin *et al.* (1988) showed that, in cultured skin fibroblasts, ascorbic acid stimulated collagen production with no apparent change in the intracellular degradation of newly synthesized procollagen. They also found that the steady-state levels of type I (which accounts for approximately 85% of the collagenous protein produced by these cells) and type III procollagen mRNAs were increased in cells treated with ascorbic acid. A three- to fourfold increase in collagen synthesis was associated with a two- to threefold increase in the levels of mRNAs for both type I and type III procollagens. Geesin *et al.* (1988) suggest a translational control linked either to procollagen gene transcription or to mRNA degradation. Grinnell *et al.* (1989) cultured human foreskin fibroblasts for up to 6 weeks in a medium supplemented with ascorbic acid (50 μg ml^{-1}). The cells produced a new connective tissue matrix with *in-vivo*-like properties, which included complete processing of procollagen to collagen α-chains, covalent cross-linking of collagen and assembly of collagen fibrils into bundles. The high degree of molecular and topographical differentiation was a unique feature of long-term cultures. This was not seen in short-term cultures or in long-term cultures in the absence of ascorbate.

Thiamine binding proteins

A thiamine binding protein has been isolated from chicken egg white (Muniyappa & Adiga, 1979, 1980), egg yolk (Muniyappa & Adiga, 1981) and estrogenized chicken sera. The purified protein had a molecular mass of 38 kDa. The pattern of competition between thiamine derivatives and [^{14}C]thiamine for binding to the protein indicated that sequential addition of phosphate moieties to thiamine led to a progressive decrease in the ability of the vitamin derivatives to compete with [^{14}C]thiamine binding to the protein. It has been suggested that the pyrimidine moiety of thiamine is more important than the thiazole moiety with regard to the ligand–protein interaction. The egg yolk protein was indistinguishable from the egg white thiamine binding protein in physicochemical properties (Muniyappa & Adiga, 1981). The proteins are non-glycosylated. They both interact specifically with the riboflavin binding protein to form a complex without any effect on either of the ligand–protein bindings. A role for the egg yolk thiamine binding protein for the transport of thiamine to the developing embryo has been suggested. There is no information on the phosphorylation of the transported thiamine to the nucleus. This is intriguing in as much as the phosphorylated form of pyridoxine (vitamin B$_6$) is transported to the nucleus (Meisler & Thanassi, 1990). A thiamine diphosphate binding protein was isolated from rat liver. It was shown that the substrate for the synthesis of thiamine triphosphate was not free thiamine disphosphate but a protein-bound form (Nishino et al., 1983). Immunohistochemical distribution of the thiamine triphosphate suggests that it has a role in nerve conduction (Itokawa & Cooper, 1970; Fox & Duppel, 1975).

Pyridoxal phosphate binding proteins

Pyridoxal phosphate in plasma is bound to albumin with high affinity (Lumeng et al., 1974) and exists as a Schiff's base. Virtually none of the pyridoxal phosphate in plasma is dialyzable. This binding protects against the renal glomerular filtration of pyridoxal phosphate. Pyridoxal is less tightly bound to albumin than pyridoxal phosphate, and hence is susceptible to glomerular filtration. However, pyridoxal appearing in the glomerular filtrate is reabsorbed. The binding of pyridoxal phosphate to albumin also protects it against hydrolysis by alkaline phosphatase. Liver is the primary source of pyridoxal phosphate in plasma and possesses a unique transport mechanism which possibly releases pyridoxal phosphate as an albumin complex. The complex contains pyridoxal phosphate and albumin in a molar ratio of 2 : 1. The binding of pyridoxal to hemoglobin results in the accumulation of the vitamer in the erythrocyte. This makes it possible for pyridoxal to be a transport form of vitamin B_6. The role of protein binding in the regulation of the liver cellular pyridoxal phosphate has also been indicated (Li *et al.*, 1974). When hepatic cytosol is dialyzed exhaustively, only 50% of the pyridoxal phosphate is bound to proteins. This binding protects pyridoxal phosphate from hydrolysis by plasma-membrane-associated phosphatase activity in the liver. Thus, protein binding of pyridoxal phosphate within cells as well as in plasma is a significant modifier of pyridoxal phosphate by phosphatase. It has been proposed that the distribution of pyridoxal phosphate between protein-bound and free forms within the cell, together with the enzymatic hydrolysis of pyridoxal phosphate, maintains the concentration of this vitamer form (Bosron *et al.*, 1978). In liver cytosol, pyridoxal phosphate is found as five major protein-bound forms. Three of these proteins have been identified as phosphorylase, alanine aminotransferase and aspartate aminotransferase. This is emphasized by the finding that in vitamin B_6 deficiency the liver cytosolic concentrations of pyridoxal phosphate as well as the activities of these three enzymes decrease significantly. The other pyridoxal phosphate binding enzymes present in lesser concentrations would also account for a substantial portion of bound pyridoxal phosphate in liver.

Skeletal muscle is the major repository of vitamin B_6 in the body (Krebs & Fisher, 1964). About 60–90% of the total vitamin B_6 present in muscle is associated with glycogen phosphorylase, an enzyme which makes up about 5% of the total soluble protein of muscle. Pyridoxal phosphate is bound principally to glycogen phosphorylase. Phosphorylase requires pyridoxal phosphate for catalytic activity, but pyridoxal phosphate does not have a coenzyme function as the reduction of phosphorylase by sodium borohydride does not decrease enzyme activity. Pyridoxal phosphate binds to Lys_{679} in rabbit muscle glycogen phosphorylase with the 5′ phosphate adjacent to the substrate binding site (Sygusch *et al.*, 1977; Titani *et al.*, 1977). The function of muscle phosphorylase as a storage depot for pyridoxal phosphate has been demonstrated by the effect of vitamin B_6 deficiency on muscle phosphorylase activity (Black *et al.*, 1978; Takassi *et al.*, 1968).

The subcellular distribution of pyridoxal phosphate in rat liver was examined by Bosron *et al.* (1978). In later work the nuclear fraction of rat hepatoma-derived HTC cells has been shown to contain 8% of the total cellular pyridoxal phosphate (Meisler & Thanassi, 1990). In rats fed a diet adequate in vitamin B_6 the fraction of total pyridoxal phosphate found in the nuclei of liver cells was 21%; this increased to 39% in rats fed a vitamin B_6-deficient diet, indicating a conservation of the vitamin in the nuclear compartment (Bosron et al., 1978). Meisler & Thanassi (1990) have reported that pyridoxal phosphate is the only significant form of vitamin B_6 found in the nucleus and that there is one particularly prominent pyridoxal phosphate binding protein in the nucleoplasmic fraction of nuclear preparations obtained from HTC cells. This protein has an apparent molecular mass of 50–55 kDa on SDS–PAGE. The presence of pyridoxal phosphate in the nucleus could be of physiological significance. There is no information as to how the nucleus acquires pyridoxal phosphate or whether the phosphorylation of pyridoxal takes place in the nucleus, as it is generally accepted that phosphorylated vitamin B_6 vitamers do not cross the membranes of mammalian cells (Ink & Henderson, 1984).

The function of the protein-bound pyridoxal phosphate in cell nucleus is of much interest. Litwack and co-workers (Disorbo & Litwack, 1981; Schmidt & Litwack, 1982; Litwack, 1988) and Compton & Cidlowski (1986) have shown that pyridoxal phosphate affects steroid hormone activity by altering the interaction of steroid receptor complexes with DNA, chromatin and nuclei. Rat hepatoma cells grown in the presence of 5 mM pyridoxine have a significantly reduced glucocorticoid-dependent induction of tyrosine aminotransferase; and the enzyme activity is increased when the cells are grown in a medium deficient in pyridoxine (Disorbo & Litwack, 1981). The addition of pyridoxal phosphate to the medium containing mouse mammary gland explants resulted in a significant decrease of both dexamethasone binding to nuclear steroid receptor and dexamethasone-stimulated casein mRNA synthesis (Majumder et al., 1983). The possibility that pyridoxal phosphate might act as a modulator of protein–DNA interaction has been suggested (Meisler & Thanassi, 1990). The presence of vitamin-binding proteins such as those for retinoic acid, vitamin D, biotin and vitamin B_6 is significant. The nuclear binding proteins for biotin and pyridoxine (pyridoxal phosphate) may be involved in regulation of gene expression, as has been determined for retinoic acid and vitamin D.

Conclusion

Tocopherol binds to all serum lipoproteins. In view of the parallelism between the uptake of cholesterol and α-tocopherol, the LDL receptor-mediated uptake is indicated as one mode of cellular uptake.

However, internalization of the lipoproteins carrying α-tocopherol, such as HDL and LDL, is not necessary for cellular uptake. The characterization of the specific α-tocopherol binding protein from rat liver cytosol is significant, as it has been shown to have a role in transfer of tocopherol between intracellular membranes. It seems possible that this protein may be related to the tocopherol binding protein of erythrocyte membrane.

An ascorbate binding protein has not been isolated, although all evidence indicates its presence. The roles of ascorbic acid in prolyl hydroxylation, in collagen synthesis, in dopamine hydroxylation, in norepinephrine synthesis and in amidation of peptides are well understood. However, there are indications that ascorbate might be involved in the synthesis of the procollagens, possibly acting at the translation step as well as in the macromolecular assembly such as in the fabrication of connective tissue matrix. Cellular ascorbate has to interact with organelles involved in such assembly; the mechanism of action of ascorbate will be an area of future study.

Thiamine binding proteins from egg yolk have a specific function in the transport of thiamine to the developing oocyte. It is free thiamine that binds to this protein. In the nervous system there is a protein which specifically binds to thiamine diphosphate. Although its function as a substrate for thiamine triphosphate has been shown, the distribution of this protein would suggest that it might serve other functions, which have not yet been identified.

There are three types of protein binding of vitamin B_6. The binding of pyridoxal phosphate to albumin during circulation in blood protects it from renal glomerular filtration as well as hydrolysis by alkaline phosphatase. This later function is performed by pyridoxal phosphate binding to enzymes in the cell cytosol. The binding of pyridoxal phosphate to glycogen phosphorylase is unique. Pyridoxal phosphate is essential to maintain the active conformation of phosphorylase without its traditional coenzymatic function. Muscle phosphorylase contains more than half the vitamin B_6 in the body. There is no information as to how this protein evolved to be a vitamin B_6 storage depot. The identification of a nuclear pyridoxal phosphate binding protein might help in our understanding of the role of vitamin B_6 in modulating protein–DNA interactions.

References

Anderson, R.E., Kemp, J.W., Jee, W.S.S. & Woodbury, D.M. (1984). Ion-transporting ATPases and matrix mineralization in cultured osteoblast-like cells. *In vitro* **20**, 837–46.

Behrens, W.A. & Madere, R. (1982) Occurrence of a rat liver α-tocopherol binding protein *in vivo*. *Nutr. Rep. Int.* **25**, 1078–112.

Behrens, W.A. & Madere, R. (1983) Interrelationship and competition of α-and γ-tocopherol at the level of intestinal absorption, plasma transport and liver uptake. *Nutr. Res.* **3**, 891–7.

Behrens, W.A. & Madere, R. (1985) Transport of α- and γ-tocopherol in human plasma lipoproteins. *Nutr. Res.* **5**, 167–74.

Behrens, W.A. & Madere, R. (1986) Alpha- and gamma-tocopherol concentrations in human serum. *J. Am. Coll. Nutr.* **5**, 91–6.

Behrens, W.A. & Madere, R. (1987) Mechanism of absorption, transport and tissue uptake of RRR-α-tocopherol and d-γ-tocopherol in the white rat. *J. Nutr.* **117**, 1562–9.

Bellows, G.G., Aubin, J.E., Hersche, J.N.M. & Antosz, M.E. (1986) Mineralized bone nodules formed in vitro from enzymatically released rat calvaria cell populations. *Calcif. Tissue Int.* **38**: 143–54.

Bensley, E.H., Fowler, A.F., Creaghan, M.V., Moore, B.A. & McDonald, E.K. (1950) Plasma tocopherol in diabetes mellitus. *J. Nutr.* **40**, 323–7.

Bianchi, J. & Rose, R.C. (1986) Glucose-independent transport of dehydroascorbic acid in human erythrocytes. *Proc. Soc. Exp. Biol. Med.* **181**, 333–7.

Bieri, J.G. & Evarts, R.T. (1973) Tocopherols and fatty acids in American diets. *J. Am. Diet. Assoc.* **62**, 147–51.

Bieri, J.G. & Evarts, R.T. (1974) Gamma tocopherol: metabolism, biological activity and significance in human vitamin E nutrition. *Am. J. Clin. Nutr.* **27**, 980–6.

Bieri, J.G. & Prival, E.L. (1965) Serum vitamin E determined by thin-layer chromatography. *Proc. Soc. Exp. Biol. Med.* **120**, 554–7.

Bjornson, L.K., Kayden, H.J., Miller, E. & Moshell, A.N. (1976) The transport of α-tocopherol and β-carotene in human blood. *J. Lipid Res.* **17**, 343–52.

Black, A.L., Guirard, B.M. & Snell, E.E. (1978) The behavior of muscle phosphorylase as a reservoir for vitamin B_6 in the rat. *J. Nutr.* **108**, 670–7.

Bloj, B. & Zilversmit, D. (1977) Rat liver proteins capable of transferring phosphatidylethanolamine. *J. Biol. Chem.* **252**, 1613–19.

Bosron, W.F., Veitch, R.L., Lumberg, L. & Li. T. (1978) Subcellular localization and identification of pyridoxal 5'-phosphate-binding proteins in rat liver. *J. Biol. Chem.* **253**, 1488–92.

Bowers-Komro, K.M. & McCormick, D.B. (1990) Characterization of ascorbic acid uptake by isolated rat kidney cells. *J. Nutr.* **121**, 57–64.

Catignani, G.L. (1975) An α-tocopherol binding protein in rat liver cytoplasm. *Biochem. Biophys. Res. Commun.* **67**, 67–72.

Catignani, G.L. & Bieri, J.B. (1977) Rat liver α-tocopherol binding protein. *Biochim. Biophys. Acta* **497**, 349–57.

Chajek-Shaul, T., Friedman, G., Stein, O., Olivecrona, T. & Stein, Y. (1982) Binding of lipoprotein lipase to the cell surface is essential for the transmembrane transport of chylomicron cholesterylester. *Biochim. Biophys. Acta* **712**, 200–10.

Choi, J. & Rose, R.C. (1989) Transport and metabolism of ascorbic acid in human placenta. *Am. J. Physiol.* **257**, C110–13.

Compton, M.M. & Cidlowski, J.A. (1986) Vitamin B_6 and glucocorticoid action. *Endocrine Rev.* **7**, 140–8.

Daniels, A.J., Dean, G., Viveros, O.H. & Diliberto, E.J. (1983) Secretion of newly taken up ascorbic acid by adrenomedullary chromaffin cells originates from a compartment different from the catecholamine storage vesicle. *Molec. Pharmacol.* **23**, 437–44.

Darby, W.J., Ferguson, M.E., Furman, R.H., Lemley, J.M., Ballo, C.T. & Meneely, G.R. (1949) Plasma tocopherols in health and disease. *Ann. N.Y. Acad. Sci.* **52**, 328–33.

Davies, T., Kelleher, J., Smith, C.L. & Losowsky, M.S. (1971) The effect of orotic acid on the absorption, transport and tissue distribution of α-tocopherol in rat. *Int. J. Vit. Nutr. Res.* **41**, 360–7.

Dhariwal, K.R., Washko, P., Hartzell, W.O. & Levine, M. (1989) Ascorbic acid within chromaffin granules. *J. Biol. Chem.* **264**, 15404–9.

Diliberto, E.J., Daniels, A.J. & Viveros, O.H. (1991) Multicompartmental secretion of ascorbate and its dual role in dompamine β-hydroxylation. *Am. J. Clin. Nutr.* **54**, S1163–72.

Diliberto, E.J., Heckman, G.D. & Daniels, G.D. (1983) Characterization of ascorbic acid transport by adrenomedullary chromaffin cells. Evidence for Na⁺-dependent co-transport. *J. Biol. Chem.* **258**, 12886–94.

Disorbo, D.M. & Litwack, G. (1981) Changes in the intracellular levels of PL5′P affect the induction of TAT by glucocorticoids. *Biochem. Biophys. Res. Commun.* **99**, 1203–8.

Dixon, S.J. & Wilson, J.X. (1992) Transforming growth factor-β stimulates ascorbate transport activity in osteoblastic cells. *Endocrinology* **130**, 484–9.

Eipper, B.A., Park, L.P., Dickerson, I.M., Deutmann, H.T., Thiele, E.A., Rodriguez, H., Schofield, P.R. & Mains, R.E. (1987) Structure of the precursor to an enzyme mediating COOH-terminal amidation in peptide biosynthesis. *Molec. Endocrinol.* **1**, 777–90.

Fielding, C.J. (1978) Metabolism of cholesterol rich chylomicrons. Mechanism of binding and uptake of cholesteryl esters by the vascular bed of the perfused rat heart. *J. Clin. Invest.* **62**, 141–51.

Finn, F.M. & Johns, P.A. (1980) Ascorbic acid transport by isolated bovine adrenal cortical cells. *Endocrinology* **106**, 811–17.

Fox, J.M. & Duppel, W. (1975) The action of thiamine and its di- and triphosphate on the slow exponential decline of the ionic currents in the node of Ranvier. *Brain Res.* **89**, 287–302.

Friedman, S. & Kaufman, S. (1965) 3,4-dihydroxyphenylethylamine β-hydroxylase. *J. Biol. Chem.* **240**, 4763–73.

Gardiner, T.W., Armstrong-James, M., Caan, A.W. & Wightman, R.M. (1985) Modulation of neostriatal activity by iontophoresis of ascorbic acid. *Brain Res.* **344**, 181–5.

Geesin, J.C., Darr, D., Kaufman, R., Murad, S. & Pinnell, S.R. (1988) Ascorbic acid specifically increases Type 1 and Type III procollagen messenger RNA levels in human skin fibroblasts. *J. Invest. Dermatol.* **90**, 420–4.

Goldstein, J.I. & Brown, M.S. (1974) Binding and degradation of low density lipoprotein by cultured human fibroblasts. *J. Biol. Chem.* **249**, 5153–62.

Gontzea, I. & Nicolau, N. (1972) Relationship between serum tocopherol level and dyslipidemia. *Nutr. Rep. Int.* **5**, 225–31.

Grinnell, F., Fukamizu, H., Pawalek, P. & Nakagawa, S. (1989) Collagen processing, cross linking and fibril bundel assembly in matrix produced by fibroblasts in long-term cultures supplemented with ascorbic acid. *Exp. Cell. Res.* **181**, 483–91.

Guarnieri, C., Flamigni, F. & Caldarera, C.M. (1980) A possible role of rabbit heart cytosol tocopherol binding in the transfer of tocopherol into nuclei. *Biochem. J.* **190**, 469–71.

Gurusinghe, A., de Niese, M., Renaud, J. & Austin, L. (1988) The binding of lipoproteins to human muscle cells: binding and uptake of LDL, HDL, and α-tocopherol. *Muscle Nerve* **11**, 1231–9.

Hauswirth, J.W. & Nair, P.P. (1972) Some aspects of vitamin E in the expression of biological information. *Ann. N.Y. Acad. Sci.* **203**, 111–21.

Housmann, E. (1967) Cofactor requirement for the enzymatic hydroxylation of lysine in a polypeptide precursor of collagen. *Biochim. Biophys. Acta* **133**, 591–3.

Ink, S.L. & Henderson, L.M. (1984) Vitamin B_6 metabolism. *Ann. Rev. Nutr.* **4**, 455–70.

Itokawa, Y. & Cooper, J.R. (1970) Ion movements and thiamine in nervous tissue. *Biochim. Biophys. Acta* **196**, 274–84.

Kasparek, S. (1980) Chemistry of tocopherols and tocotrienols. In: *Vitamin E. A Comprehensive Treatise* (ed. L.J. Machlin), pp. 7–65. Marcel Dekker, New York.

Kessler, J.I., Stein, J., Dannacker, D. & Narcessian, P. (1970) Biosynthesis of low density lipoprotein by cell-free preparations of rat intestinal mucosa. *J. Biol. Chem.* **245**, 5281–8.

Kipp, D.E. & Schwarz, R.I. (1990) Effectiveness of isoascorbate versus ascorbate as an inducer of collagen synthesis in primary avian tendon cells. *J. Nutr.* **120**, 185–9.

Kitabchi, A.E. & Wimalasena, J. (1982a) Specific binding sites for D-α-tocopherol on human erythrocytes. *Biochim. Biophys. Acta* **684**, 200–6.

Kitabchi, A.E. & Wimalasena, J. (1982b). Demonstration of specific binding sites for ^{3}H-RRR-α-tocopherol on human erythrocytes. *Ann. N.Y. Acad. Sci.* **393**, 300–15.

Knoth, J., Viveros, O.H. & Diliberto, E.J. (1987) Evidence for the release of newly acquired ascorbate and alpha-aminoisobutyric acid from the cytosol of adrenomedullary chromaffin cells through specific transporter mechanisms. *J. Biol. Chem.* **262**, 14036–41.

Krebs, E.G. & Fisher, E.H. (1964) Phosphorylase and related enzymes of glycogen metabolism. *Vit. Horm.* **22**, 399–410.

Krishnamurthy, S. & Bieri, J.G. (1963) The absorption, storage, and metabolism of α-tocopherol-C^{14} in the rat and chicken. *J. Lipid Res.* **4**, 330–6.

Leonard, R.K., Auersperg, N. & Parkes, C.O. (1983) Ascorbic acid accumulation by cultured rat adrenocortical cells. *In vitro* **19**, 46–52.

Levine, M.A. & Pollard, H.B. (1983) Hydrocortisone inhibition of ascorbic acid transport by chromaffin cells. *FEBS Lett.* **158**, 134–8.

Li, T., Lumeng, L. & Veitch, R.L. (1974) Regulation of pyridoxal 5'-phosphate metabolism in liver. *Biochem. Biophys. Res. Commun.* **61**, 627–34.

Litwack, G. (1988) The glucocorticoid receptor at the protein level. *Cancer Res.* **48**, 2636–40.

Lumeng, L., Brasher, R.E. & Li. T. (1974) Pyridoxal 5'-phosphate in human plasma: source, protein-binding, and cellular transport. *J. Lab. Clin. Med.* **84**, 334–43.

Mains, R.E., Myers, A.C. & Eipper, B.A. (1985) Hormonal, drug, and dietary factors affecting peptidyl glycine alpha-amidating monooxygenase activity in various tissues of the adult male rat. *Endocrinology* **116**, 2505–15.

Majumder, P.K., Joshi, J.B. & Banerjee, M.R. (1983) Correlation between nuclear glucocorticoid receptor levels and casein gene expression in murine mammary glands *in vitro*. *J. Biol. Chem.* **258**, 6793–8.

Mann, G.V. & Newton, P. (1975) The membrane transport of ascorbic acid. *Ann. N.Y. Acad. Sci.* **258**, 243–52.

Meisler, N.T. & Thanassi, J.W. (1990) Pyridoxine-derived B$_6$ vitamers and pyridoxal 5′-phosphate-binding protein in cytosolic and nuclear fractions of HTC cells. *J. Biol. Chem.* **265**, 1193–8.

Morwi, H., Nakagawa, Y., Inoue, K. & Nojima, S. (1982) Enhancement of the transfer of α-tocopherol between liposomes and mitochondria by rat liver protein(s). *Eur. J. Biochem.* **117**, 537–42.

Muniyappa, K. & Adiga, P.R. (1979) Isolation and characterization of thiamine-binding protein from chicken egg white. *Biochem. J.* **177**, 887–94.

Muniyappa, K. & Adiga, P.R. (1980) Oestrogen-induced synthesis of thiamin-binding protein in immature chicks. *Biochem. J.* **186**, 201–10.

Muniyappa, K. & Adiga, P.R. (1981) Nature of thiamine-binding protein from chicken egg yolk. *Biochem. J.* **193**, 679–85.

Murad, S., Grove, D., Lindberg, K.A., Reynolds, G., Sivarajah, A. & Pinnell, S.R. (1981) Regulation of collagen synthesis by ascorbic acid. *Proc. Natl. Acad. Sci. USA* **78**, 2879–82.

Murphy, D.J. & Mavis, R.D. (1981) Membrane transfer of α-tocopherol-binding factors from the liver, lung, heart, and brain of the rat. *J. Biol. Chem.* **256**, 10464–8.

Nelsson-Ehle, P., Garginkel, A.S. & Schotz, M.C. (1980) Lipolytic enzymes and plasma lipoprotein metabolism. *Ann. Rev. Biochem.* **49**, 667–9.

Nishino, K., Itokawa, Y., Nishino, N., Piros, K. & Cooper, J.R. (1983) Enzyme system involved in the synthesis of thiamin triphosphate. I. Purification and characterization of protein-bound thiamin diphosphate:ATP phosphoryltransferase. *J. Biol. Chem.* **258**, 11871–8.

Njus, D. & Kelley, P.M. (1991) Vitamins C and E donate single hydrogen atoms *in vivo*. *FEBS Lett.* **284**, 147–51.

Ockner, R.K., Manning, J.A. & Kane, J.P. (1982) Fatty acid binding protein. *J. Biol. Chem.* **257**, 7872–8.

Padh, H. & Aleo, J.J. (1987) Characterization of ascorbic acid transport by 3T6 fibroblasts. *Biochim. Biophys. Acta* **901**, 283–90.

Padh, H. & Aleo, J.J. (1987) Activation of serum complement leads to inhibition of ascorbic acid transport. *Proc. Soc. Exp. Biol. Med.* **185**, 153–7.

Padh, H. & Aleo, J.J. (1989) Ascorbic acid transport by 3T6 fibroblasts. Regulation by and purification of human serum complement factor. *J. Biol. Chem.* **264**, 6065–9.

Patnaik, R.N. & Nair, P.P. (1975) Binding of D-α-tocopherol to rat nuclear compartment. *Experientia* **31**, 1023–4.

Patnaik, R.N. & Nair, P.P. (1977) Studies on the binding of D-α-tocopherol to rat liver nuclei. *Arch. Biochem. Biophys.* **178**, 333–41.

Peak, I.R., Windmueller, H.G. & Bieri, J.G. (1972) A comparison of the intestinal absorption, lymph and plasma transport and tissue uptake of α- and γ-tocopherols in the rat. *Biochim. Biophys. Acta* **260**, 679–88.

Peterkofsky, B. & Udenfriend, S. (1965) Enzymatic hydroxylation of proline in microsomal polypeptide leading to formation of collagen. *Proc. Natl. Acad. Sci. USA* **52**, 335–42.

Postel, S. (1956) Total free tocopherols in the serum of patients with thyroid disease. *J. Clin. Invest.* **35**, 1345–56.

Poukka, R.K.H. & Bieri, J.G. (1970) Blood α-tocopherol: erythrocyte and plasma relationships *in vitro* and *in vivo*. *Lipids* **5**, 757–61.

Prockop, D.J., Kivirikko, K.I., Tuderman, L. & Guzman, N.A. (1979) The biosynthesis of collagen and its disorders (first of two parts). *N. Engl. J. Med.* **301**, 13–23.

Rajaram, O.V., Fatterpaker, P. & Sreenivassan, A. (1974) Involvement of binding lipoproteins in the absorption and transport of α-tocopherol in the rat. *Biochem. J.* **140**, 509–16.

Rose, R.C. (1988) Transport of ascorbic acid and other water-soluble vitamins. *Biochim. Biophys. Acta* **947**, 335–66.

Rubinstein, H.M., Dietz, A.A. & Srinivasan, R. (1969) Relation of vitamin E and serum lipids. *Clin. Chim. Acta* **23**, 1–6.

Scharfmann, R., Leduque, P., Aratan-Spine, S., Dubois, P., Basmaciogullari, A. & Czernichow, P. (1988) Persistence of peptidylglycine α-amidating, monooxygenase activity and elevated thyrotropin-releasing hormone concentrations in fetal rat islets in culture. *Endocrinology* **123**, 1329–34.

Schmidt, T.J. & Litwack, G. (1982) Activation of the glucocorticoid-receptor complex. *Physiol. Rev.* **62**, 1131–92.

Silber, R., Winter, R. & Kayden, H.J. (1969) Tocopherol transport in the rat erythrocyte. *J. Clin. Invest.* **48**, 2089–95.

Simon, E.J., Gross, C.S. & Milhorat, A.T. (1956) The metabolism of vitamin E. I. The absorption and excretion of d-α-tocoperyl 5-methyl-C^{14}-succinate. *J. Biol. Chem.* **221**, 797–805.

Sygusch, J., Madsen, N.B., Kasvinsky, P.J. & Fletterick, R.J. (1977) Location of pyridoxal phosphate in glycogen phosphorylase a. *Proc. Natl. Acad. Sci. USA* **74**, 4757–61.

Takamsi, M., Fujioka, M., Wada, H. & Taguchi, T. (1968) Studies on pyridoxine deficiency in rat. *Proc. Soc. Exp. Biol. Med.* **129**, 110–17.

Taylor, S.L., Lamden, M.P. & Tappel, A.L. (1976) Sensitive fluorimetric method for tissue tocopherol analysis. *Lipids* **11**, 530–8.

Titani, K., Koide, A., Hermann, J., Ericsson, L.H., Kumar, S., Wade, R.K., Walsh, K.A., Neurath, H. & Fischer, E.H. (1977) Complete amino acid sequence of rabbit muscle glycogen phosphorylase. *Proc. Natl. Acad. Sci. USA* **74**, 4762—6.

Thorn, N.A., Nielsen, F.S. & Jeppesen, C.K. (1991) Characterization of ascorbic acid uptake by isolated ox neurohypophyseal nerve terminals and the influence of glucocorticoid and tri-iodothyronine on uptake. *Acta Physiol. Scand.* **141**, 97–106.

Thorn, N.A., Nielsen, F.S., Jeppesen, C.K., Farver, O. & Christensen, B.L. (1986) Uptake of ascorbic acid and dehydroascorbic acid by isolated nerve terminals and secretory granules from ox neurohypophyses. *Acta Physiol. Scand.* **128**, 629–38.

Toggenburger, G., Hausermann, B., Matsch, M., Genoni, B., Kessler, M., Neber, F., Hornig, D., O'Neill, B. & Semenza, G. (1981) Na^+-dependent, potential-sensitive L-ascorbate transport across brush border membrane vesicles from kidney cortex. *Biochim. Biophys. Acta* **646**, 433–43.

Traber, M.G. & Kayden, H.J. (1984) Vitamin E is delivered to cells via the high affinity receptor for low-density lipoprotein. *Am. J. Clin. Nutr.* **40**, 747–51.

Traber, M.G., Olivecrona, T. & Kayden, H.J. (1985) Bovine milk lipoprotein liapse transfers tocopherol to human fibroblasts during triglyceride hydrolysis in vitrol. *J. Clin. Invest.* **75**, 1729–34.

Washko, P., Rotrosen, D. & Levine, M. (1989) Ascorbic acid transport and accumulation in neutrophils. *J. Biol. Chem.* **264**, 18996–9002.

Wilson, J.X. (1989) Ascorbic acid uptake by a high-affinity sodium-dependent mechanism in cultured rat astrocytes. *J. Neurochem.* **53**, 1064–71.

Wilson, J.X. & Dixon, S.J. (1989) High-affinity sodium-dependent uptake of ascorbic acid by rat osteoblasts. *J. Membrane Biol.* **111**, 83–91.

Wilson, J.X., Jaworski, E.M., Kulaga, A. & Dixon, S.J. (1990) Substrate regulation of ascorbate transport activity in astrocytes. *Neurochem. Res.* **15**, 1037–43.

Wimalasena, J., Davis, M. & Kitabchi, A.E. (1982) Characterization and solubilization of the specific binding sites for d-α-tocopherol from human erythrocyte membranes. *Biochem. Pharmacol.* **31**, 3455–61.

Windmueller, H.G. & Levy, R.I. (1968) Production of β-lipoprotein by intestine in the rat. *J. Biol. Chem.* **243**, 4878–84.

Wirtz, K.W.A. (1974) Transfer of phospholipids between membranes. *J. Biol. Chem.* **344**, 95–117.

Yoshika, H., Nishino, K., Miyake, T., Ohshio, G., Kimura, T. & Hamashima, Y. (1987) Immunohistochemical localization of a new thiamine diphosphate-binding protein in the rat nervous system. *Neurosci. Lett.* **77**, 10–14.

Zhou, A., Nielsen, J.H., Farver, O. & Thorn, N.A. (1991) Transport of ascorbic acid and dehydroascorbic acid by pancreatic islet cells from neonatal rats. *Biochem J.* **274**, 739–44.

8

Biotin-binding Proteins

Krishnamurti Dakshinamurti and Jasbir Chauhan

Introduction

The discovery of biotin and the elucidation of its structure and role in metabolism involved diverse investigations spanning many decades. Isolation of crystalline biotin by Kogl & Tonnis (1936) and determination of its chemical structure and synthesis by Harris *et al.* (1943) were the highlights of the early history of research into this water-soluble vitamin. Biotin was shown to be *cis*-hexahydro-2-oxo-1H-thieno (3,4)imidazole-4-valeric acid (Figure 8.1a), with the (+) stereoisomer exhibiting significant biological activity. The role of biotin as the prosthetic group of the biotin-containing carboxylases was recognized in the

Figure 8.1. (*a*) Structure of biotin and biocytin. (*b*) Hydrolysis of biocytin or biotinyl peptides by biotinidase.

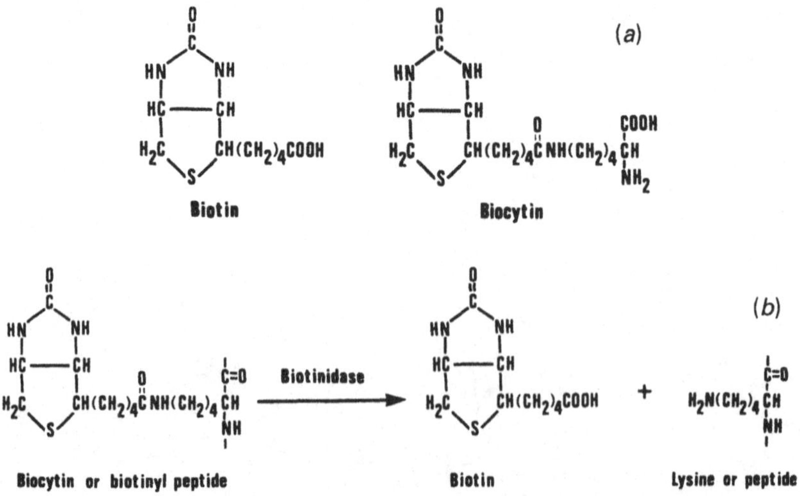

1960s. Biotin in most biological material was found to be protein-bound. Biocytin (biotinyl lysine) is released upon enzymatic digestion of biotin-containing proteins. It is cleaved by biotinidase into biotin and lysine (Figure 8.1*b*).

Biotin is covalently bound to a lysine residue of the carboxylases. The mechanism of action of biotin in the carboxylases is well understood (Lynen, 1967). This review focuses primarily on the non-covalent binding of biotin to various proteins. Such binding seems to be crucial to the uptake, transport and intracellular functions other than the prosthetic group function of this vitamin. Biotin-binding proteins of several species have been considered in relation to these functions.

Biotin-binding proteins have been classified into two groups: those proteins involved in biotin transport and those involved in recognizing biotin, resulting in a physiological function. In the category of transport proteins are the biotin-binding proteins of avian egg yolk as well as the animal biotinidase. The roles of biotin in altering the intracellular concentration of cGMP, the activity of RNA polymerase II, and its role in the induction or repression of specific proteins fall into the category of 'recognition' of an exogenous ligand and the 'transduction' of this message. These actions of biotin and the proteins involved in the 'recognition' could be considered to be analogous to the actions of some hormones and their specific receptors. Although a 'biotin receptor' has not yet been isolated, the identified non-prosthetic group functions of biotin would require such specific protein–biotin interactions.

Covalently bound biotin proteins

Biotin is best known for its classical role as the prosthetic group of biotin-containing enzymes. It generally serves as a carbon dioxide carrier and is covalently linked to the ϵ-amino group of a lysine residue in biotin enzymes. Biotin-containing enzymes can be divided into three classes: Class I biotin enzymes, carboxylases that carry out ATP-dependent carbon dioxide fixation into an acceptor such as pyruvate, acetyl CoA, propionyl CoA, 3-methylcrotonyl CoA, geranyl CoA, and urea; Class II biotin enzymes, decarboxylases that facilitate sodium transport in anaerobes coupled to the removal of CO_2 from β-keto acids and their thioesters, including oxaloacetate, methylmalonyl CoA, and glutaconyl CoA; and the Class III biotin enzyme transcarboxylase, which transfers a carboxyl group from a donor (oxaloacetate) to an acceptor (propionyl CoA). The mechanisms of biotin enzymes were reviewed in detail by Wood & Barden (1977) and by Knowles (1989). Only four biotin enzymes have been identified in higher organisms: acetyl-CoA carboxylase, propionyl CoA carboxylase, pyruvate carboxylase, and 3-methylcrotonyl-CoA carboxylase. Acetyl CoA carboxylase is a cytosolic enzyme whereas the other three are mitochondrial enzymes. The role of biotin enzymes in intermediary metabolism is shown in Figure 8.2.

Acetyl CoA carboxylase is recognized to be a regulatory enzyme of lipogenesis. Pyruvate carboxylase is a key regulatory enzyme of gluconeogenesis. It is also present in lipogenic tissues and participates in fatty acid synthesis by transporting acetyl groups via citrate and reducing groups via malate, from the mitochondria to the cytosol. In all tissues it has an anaplerotic role in the formation of oxaloacetate. Propionyl CoA carboxylase is a key enzyme in the catabolic pathway of odd chain fatty acids and some branched chain amino acids. It catalyzes the conversion of propionyl CoA to methylmalonyl CoA, which in turn enters the tricarboxylic acid cycle via succinyl CoA. β-Methylcrotonyl CoA carboxylase catalyzes the conversion of β-methylcrotonyl CoA to β-methylglutaconyl CoA, a key reaction in the degradative pathway of leucine. Of the four mammalian biotin enzymes, only acetyl CoA carboxylase and pyruvate carboxylase have regulatory features (Dakshinamurti & Chauhan, 1988). The amino acid sequences near the biotin of pyruvate carboxylase from sheep, chicken, and turkey livers (Rylatt *et al.*, 1977), of transcarboxylase from *Propionibacterium shermanii* (Maloy *et al.*, 1979), and of acetyl

Figure 8.2. Biotin carboxylase in cellular metabolism and accumulation of intermediary metabolites in individual carboxylase deficiencies.

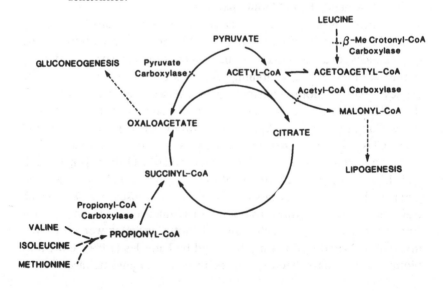

CoA carboxylase of *Escherichia coli* (Sutton *et al.*, 1977) show a great deal of homology in this region. In all cases, an Ala–Met–Bct–Met (Bct is the abbreviation for biocytin) sequence occurs, and in pyruvate carboxylase and transcarboxylases the identity extends to Ala–Met–Bct–Met–Glu–Thr. It has been suggested that this conservation provides evidence that these biotin carboxylases and transcarboxylase may have evolved from a common ancestor (Wood & Barden, 1977). Furthermore, the sequence may be involved in activating the biotin and/or orienting it so that it is in an effective carboxyl carrier position between the substrate sites. The sequence may also be important in designating the specific lysine of the protein that is to be biotinated post-translationally by the holocarboxylase synthetase (Wood & Kumar, 1985).

Non-covalently bound biotin-binding proteins

Apart from the carboxylases in which biotin is attached co-valently to the apocarboxylase, there are a group of proteins which bind to biotin non-covalently. Both avidin, the biotin-binding protein of hen egg white, and strepavidin, the extracellular biotin-binding protein of *Streptomyces avidinii*, have exceedingly high affinities for biotin, with a K_d of 10^{-15}M, the strongest non-covalent binding known between a protein and a low-molecular-mass ligand. There are other biotin-binding proteins such as the egg yolk biotin-binding proteins, biotinidase, biotin holocarboxylase synthetase, antibody to biotin, and nuclear biotin-binding protein, with progressively lower affinities for biotin. Biotin-binding proteins vary considerably along the evolutionary scale. In *E. coli*, biotin holocarboxylase synthetase, a biotin-binding protein, has been shown to regulate the synthesis of the enzymes of the biotin biosynthesis pathway (Eisenberg, 1973, 1985; Eisenberg *et al.*, 1982).

In the laying hen there are three biotin-binding proteins. In addition to avidin, which is present in the albumen of the egg, there are two biotin-binding proteins in egg yolk. Other vitamins such as riboflavin have only one species of binding protein which is present in both the egg albumen and the yolk. The presence of distinct biotin-binding proteins in the yolk suggests a specialized role for biotin in the regulation of the development of the embryo (Mandella *et al.*, 1978; White, 1985; White & Whitehead, 1987).

In mammals only two proteins have so far been shown to bind non-covalently to biotin. They are biotin holocarboxylase synthetase and biotinidase. Attention has recently been focused on the clinical significance of inherited defects in these two proteins in relation to the subtypes of the multiple carboxylase deficiency syndrome (Sweetman & Nyhan,

1986). This disease manifests itself in a neonatal or infantile form due to the deficiency of holocarboxylase synthetase, and in a late-onset or juvenile form due to the deficiency of biotinidase.

Avidins

Avidin is a minor constituent of egg white. Feeding hen egg white to rats produces a nutritional deficiency identified as biotin deficiency. Avidin forms a very stable non-covalent complex with biotin; the complex is resistant to proteolysis in the gastrointestinal tract (Gyorgy *et al.*, 1941). The interaction between biotin and avidin is the strongest non-covalent binding known, with a dissociation constant of 10^{-15} M. The reason for the uniqueness of the biotin binding property of avidin is not known. Avidin has been isolated from hen egg white (Eakin *et al.*, 1940; Melamed & Green, 1963) and the oviduct of several avian species (Hertz & Sebrell, 1942). It is a basic glycoprotein with a molecular mass of 67 kDa and an isoelectric pH of 10.5 (Green, 1975). It is a tetramer of four identical subunits, each with 128 amino acid residues and one biotin-binding site (DeLang & Huang, 1971). The carbohydrate moiety accounts for about 10% of the molecular mass and consists of four to five mannose and three N-acetylglucosamine residues per subunit (DeLang, 1970). The composition and structural arrangement of the carbohydrate moiety of avidin has been shown to be heterogenous (Burch & White, 1982).

Strepavidin is a non-glycosylated neutral protein secreted by *Streptomyces avidinii* (Chailet & Wolf, 1964). Its binding to biotin is remarkably similar to that of avidin, with a dissociation constant of 4×10^{-14} M (Green, 1975). Both avidin and strepavidin are of similar molecular mass and tetrameric with each subunit containing one biotin-binding site. However, they lack immunological cross-reactivity and evolutionary relatedness. The absence of similar proteins elsewhere in microorganisms might suggest a chance transfection rather than an ancient lineage (Green, 1990). The amino acid composition and primary sequences of these two proteins are very different. The gene for strepavidin has been cloned and sequenced (Argarana *et al.*, 1986). The molecular mass of strepavidin has been demonstrated to be much higher than was earlier reported. The smaller form is the major constituent of commercial preparations. This seems to be the result of processing by extracellular proteases, which selectively cleave native strepavidin at both N and C terminals to give a subunit of 125–127 amino acids (Bayer *et al.*, 1986). There is about 33% sequence identity between this form and avidin, seen as a series of short interrupted stretches (Argarana *et al.*, 1986).

The nature of the biotin-binding sites of avidin and strepavidin has been extensively studied. A homotypic non-glycosylated avidin tetramer was isolated from commercial avidin preparations (Hiller *et al.*, 1987). This non-glycosylated avidin tetramer is as efficient as the glycosylated form in binding biotin, indicating that the oligosaccharide moiety is not essential for binding activity. The oligosaccharide side chain may have an effect on the secondary, tertiary or quaternary structures of avidin. It may also be involved in the uptake of avidin by cell-surface mannose receptors or by subcellular organelles such as lysosomes.

The role of tryptophan in biotin binding was deduced from biotin-induced difference spectra and supported by the sensitivity of avidin to oxidation by N-bromosuccinimide. Gitlin *et al.* (1988) found that modification of an average of one tryptophan residue per avidin subunit resulted in a complete loss of biotin-binding activity. This could be accomplished by the modification of either Trp_{70} or Trp_{110}. Lysine has also been implicated as part of the binding site. Upon reaction of avidin with 1-fluoro-2,4-dinitrobenzene, one dinitrophenol group is introduced per subunit with complete loss of biotin-binding activity (Gitlin *et al.*, 1988). The involvement of both lysine and tryptophan residues of avidin in biotin binding is evident from the homology between avidin and strepavidin. The Trp–Lys sequences (70–80 and 120–121) of strepavidin appear in positions analogous to those (70–71 and 110–111) in avidin. In further work, Gitlin *et al.* (1990) have shown that the single tyrosine (Tyr_{33}) in avidin and the analogous residue (Tyr_{43}) in the longest single homologous stretch in strepavidin contribute to the biotin-binding site of this protein. This role of tyrosine is supported by an X-ray study of strepavidin, which indicates a critical interaction of Tyr_{43} with the ureido ring of biotin (Weber *et al.*, 1989).

Avidin is a secretory protein produced by the oviduct and concentrated in the egg white. It is induced by progesterone. Avidin is found in a variety of avian, reptilian and amphibian species. A bactericidal role has been ascribed for avidin in the egg. In view of the presence in the egg of other proteins with antimicrobial action, the possibility of other biological functions for avidin needs to be explored. Various reports suggest that, in addition to being an oviductal secretory protein, avidin is also synthesized in areas of tissue damage and inflammation caused by diverse agents such as toxic doses of actinomycin D, septic infection or thermal injury in the chicken (Elo, 1980). This progesterone-independent synthesis of avidin is inhibited by anti-inflammatory steroids (Norback *et al.*, 1982). Macrophages isolated from the chicken yolk sac have been shown to synthesize and secrete avidin (Korpela, 1984*a*).

In chickens infected with leukemia virus OK10, there was an increased concentration of avidin in tumor tissue. In chicken embryo fibroblast cultures, avidin was present in both Rous sarcoma virus-transformed and non-transformed damaged cells, but transformation increased the synthesis of avidin quite significantly. The production of avidin may represent a response to cell damage; transformation might activate this process (Korpela, 1984b). A selective antimicrobial role as the sole biological function of avidin is not convincing, as some microbes synthesize biotin themselves. To date there is no information on the biological role of avidin in injured or transformed cells. An avidin-like protein has not been detected in mammalian species.

The avidin–biotin system has been used in a wide range of biological studies including isolation, localization, cytochemistry, immuno-chemistry and diagnostics as well as for probing genes (Wilcheck & Bayer, 1990). In view of some non-specific adsorption of avidin, the bacterial protein strepavidin has been used instead. Strepavidin is non-glycosylated and neutral; however, it has been shown to bind at low concentrations and with high affinity to cell surfaces. This binding ability was traced to the presence of a separate RYD (Arg–Tyr–Asp) sequence in the protein molecule (Alon et al., 1990) and can be displaced by fibronectin as well as by RGD (Arg–Gly–Asp) or RYD containing peptides. Strepavidin itself can displace fibronectin from cell surfaces. The RYD sequence of strepavidin mimics the RGD sequence which is a universal recognition domain for adhesion receptors, the receptors for fibronectin, fibrinogen and other adhesive proteins.

Non-avidin biotin-binding proteins

In addition to avidin and strepavidin, the two examples of exceptionally strong affinity, there are other biotin binding proteins with lower affinities for biotin. These include the egg yolk biotin-binding proteins, biotin-holocarboxylase synthetase, biotinidase, the nuclear biotin-binding protein and the monoclonal antibody to biotin.

Biotin-holocarboxylase synthetase

Biotin carboxylases are synthesized in the form of the apopro-teins which undergo post-translational covalent modification by the addition of biotin, the prosthetic group, to the ϵ-amino group of lysine of the apoprotein. This covalent attachment of biotin to a specific lysine in the apoenzyme is catalyzed by biotin-holocarboxylase synthetase in a two step reaction (equations 8.1 and 8.2).

$$ATP + biotin\ (B) + holoenzyme\ synthetase\ (HS) \rightarrow$$

$$B-AMP-HS + PP_i. \quad (8.1)$$

$$B-AMP-HS + apoenzyme \rightarrow holoenzyme + AMP + HS. \quad (8.2)$$

Biotin-holocarboxylase synthetase has been partially purified from *Propionibacterium shermanii* (Lane *et al.*, 1969) and rabbit liver (McAllister & Coon, 1966). The ATP, divalent cation and biotin requirements of the reaction were very similar in all species. Biotinyl adenylate completely replaced the biotin requirement. The synthetase which biotinates the transcarboxylase (biotin-[methylmalonyl-CoA-carboxyltransferase] synthetase, EC 6.3.4.9) has been purified from *Propionibacterium shermanii* (Otsuka *et al.*, 1988). The synthetase catalyzes the biotination of apocarboxylase from heterologous sources. The biotin-holocarboxylase synthetases from higher organisms have not yet been purified to homogeneity. It is generally assumed that there is only one synthetase with no substrate specificity. Thus, the synthetase of one species could biotinate the apocarboxylase from other species.

Many biotin mutants of *Escherichia coli* have been isolated to help establish the biotin biosynthetic pathway and its regulation. The biotin operon is the most extensively studied among the vitamins. The reaction sequence for the biosynthesis of biotin by *E. coli* and the genes involved in this synthesis have been reported by Eisenberg (1985). The biotin (*bio*) locus in *E. coli* was shown to be located in the same region as *gal* and *att*, and the order was determined to be *gal, att, bio*. The biotin locus is composed of five consecutive genes in the order A,B,F,C and D (representing the genes coding for various enzymes of biotin biosynthesis) and three unlinked genes, *bioH*, *bioR* (a regulatory gene coding for the biotin repressor protein) and *bioP* (a permeability gene). Transcription of this cluster of genes is from two 'face to face' partly overlapping promoters. One transcript includes the *bioB F C D* genes, whereas the second transcript includes the *bioA* gene and an open reading frame (ORF) of unknown function (Otsuka *et al.*, 1988). Transcription from both promoters was blocked *in vivo* upon addition of high levels of biotin to the medium of *E. coli* cultures. Starvation of *E. coli* biotin auxotrophs resulted in increased transcription. This system resembles the classical regulation of the tryptophan (trp) operon of *E. coli* by tryptophan. In the trp operon, the most extensively studied system (Yanofsky, 1981) the trp-tRNA acts as a negative modulator at the attenuator site, terminating transcription of the trp operon under conditions of tryptophan excess or as a substrate for protein synthesis when the levels are low. In the biotin operon the repressor is the biotin–protein ligase and the corepressor is biotinyl AMP. Maximal rates of biotin operon transcription are associated with conditions of biotin limitation. The repression of the biotin operon was studied by using a coupled transcription–translation system. The addition of a repressor preparation by itself inhibited by a few per cent the synthesis of diaminopelargonic acid aminotransferase, which is coded by the *bioA* gene. The addition of biotin (120 nM) along with the repressor resulted in complete repression (Prakash & Eisenberg, 1979).

In extensive studies of the genetic and biochemical analysis of the *bir* mutant of *E. coli*, Barker & Campbell (1982*a,b*) showed that this mutant required a high concentration of biotin for growth. Complementation analysis of *bir* and *bioR* phenotypes showed complete overlap of the two genes. Biochemical analysis showed that the *bir* gene coded for the enzyme biotin-holocarboxylase synthetase, which activated biotin to biotinyl AMP (B-AMP), which in turn was used to biotinate a lysine residue of the acceptor protein (apo biotin carboxyl carrier protein, BCCP), which is a component of acetyl coenzyme A carboxylase, the major or only biotin-containing enzyme of *E. coli*. The studies of Barker & Campbell (1982*a,b*) showed that a partially purified preparation of this enzyme could protect the operator site from Taq I restriction enzyme but only in the presence of B-AMP. This indicated that the holoenzyme synthetase also possessed repressor activity in the presence of the corepressor B-AMP. Definitive evidence was presented for the bifunctional nature of a homogenous preparation of the biotin repressor protein (Eisenberg *et al.*, 1982), which is a monomeric protein of molecular mass 37–44 kDa with an isoelectric point of 7.2. The protein binds biotin with a K_d value of 1.3×10^{-7} M. This value was decreased to 1.1×10^{-9} M with biotinyl 5'-adenylate. All three activities – biotin binding, biotin activation and *in vitro* repression – were coincidental during the purification of this protein. The repressor protein (R holocarboxylase synthetase, HCS) carries out the following series of reactions:

$$ATP + Biotin\ (B) + R(HCS) \rightarrow B.AMP.R(HCS) + PPi. \quad (8.3)$$

$$B.AMP.R(HCS) + apoenzyme \rightarrow$$

$$holoenzyme + AMP + R(HCS). \quad (8.4)$$

$$B.AMP.R(HCS) + Operator(O) \rightarrow B.AMP.R(HCS).O. \quad (8.5)$$

The common link in reactions 8.4 and 8.5 is the formation of the B.AMP.R(HCS) complex in reaction 8.3. The complex either participates in holoenzyme synthesis or inhibits initiation of transcription by binding to the operator site. The concentration of this complex is dependent on the intracellular concentration of biotin and the apoenzyme. Increased concentrations of *birA* protein (R/HCS) resulted in increased repression at a given biotin concentration. On the contrary, an increased supply of the apoenzyme caused derepression at a given biotin concentration (Cronan, 1989). Thus the protein-bound biotin (or more appropriately B-AMP) has a dual function, that of a ligase or repressor.

Biotin-binding proteins involved in biotin transport

Egg yolk biotin-binding proteins

The proteins of egg yolk differ from those of egg white with regard to their time and place of synthesis. The entire egg white is secreted by the magnum of the oviduct in 4–6 h after ovulation. The yolk proteins, in contrast, are serum proteins synthesized by the liver. Most of the yolk proteins accumulate in the ovarian follicle in the one week period before ovulation of the follicle. There are high-affinity binding proteins for riboflavin, thiamine, cholecalciferol, retinol, vitamin B_{12} and iron. Vitamin and minerals are transported protein-bound to the oocyte. The transport proteins such as the riboflavin- and thiamine-binding proteins and transferrin, present in egg yolk and egg white respectively, are products of their respective single genes. The exception seems to be the biotin-binding proteins.

White and co-workers (Mandella *et al.*, 1978) isolated a biotin-binding protein (BBP I) from hen egg yolk which was shown to be distinct from avidin. The purified protein has a molecular mass of 74 kDa and is a tetramer of identical subunits (Murthy & Adiga, 1984). It is a glycoprotein with a PI of 4.6. BBP-I binds to biotin with a K_d of 1×10^{-12} M.

BBP I differs from avidin in the temperature-dependent biotin exchange. Avidin binds to biotin very tightly and does not exchange, whereas the yolk biotin-binding protein exchanges biotin at the body temperature of the bird. Thus biotin could be made available to the developing embryo from its reserve in the yolk. Only about 15% of the biotin of the whole egg is associated with avidin, under normal conditions of biotin nutrition of the bird. The rest is associated with the yolk BBP. Avidin is not saturated with biotin, except when the biotin intake of the laying hen is very high. The synthesis of avidin by the hen is not responsive to the nutritional status of the bird with respect to biotin. White and co-workers (White & Whitehead, 1987; Mesler *et al.*, 1978) showed that the plasma of laying hens and egg yolk contained another biotin-binding protein (BBP II) which was stable to 45 °C whereas BBP I was stable at 65 °C. Both are tetrameric isoproteins. They are normally saturated with biotin, and together account for most of the biotin of hen plasma and yolk. Yolk BBPs have a nutritional role, that of transporting biotin to the developing embryo. BBP II is preferentially deposited in the yolk and is specifically involved in the transport of biotin to the oocyte (White & Whitehead, 1987). Both BBP I and BBP II cross-react with anti BBP II serum. The maximal production of BBP I is attained at a lower level of dietary biotin (about 50 μg kg^{-1}) than for BBP II (about 250 μg kg^{-1}). At this higher level of dietary biotin the production of BBP II is several times higher than that of BBP I. The observation that the amounts of these proteins are limited by dietary biotin within the normal dietary range would suggest that biotin is required for the synthesis, stability or secretion of these proteins. Plasma concentration of BBP I is maintained at a constant level except when dietary biotin is severely restricted. Even under conditions of dietary deficiency BBP I is saturated with biotin.

The synthesis of both BBP I and BBP II is stimulated during egg laying, indicating that the gene(s) for these two proteins is (are) induced by estrogen. The difference in their response to dietary biotin content as well as in their distribution between hen and chick plasma would suggest that they are either products of different genes of differentially modified products of the same gene. Both estrogen and the specific ligand for these proteins, biotin, are involved in the regulation of their synthesis. The mechanism of action of biotin in this is not known; it could be at the transcriptional or translational level.

A biotin-binding protein analogous to BBP II has been characterized in the serum of the pregnant or estrogenized female rat (Seshagiri & Adiga, 1987). This protein is not present in the sera of normal male rats. The molecular mass of this protein is 66 kDa and the isoelectric pH is 4.1. It exhibits immunological cross-reactivity with the purified egg yolk biotin-binding protein, BBP II. The similarity between these proteins extends to their acidic glycoprotein nature, estrogen dependence for synthesis and secretion, and similar tryptic peptides. The physiological significance of the biotin-binding protein during rodent embryonic development was demonstrated by selective immunoneutralization of the maternal vitamin carrier. This led to a drastic decrease in the transport of biotin to the embryo, resulting in early embryonic mortality without affecting maternal biotin nutrition.

Biotinidase

Acid hydrolysis of biotin proteins releases free biotin, whereas proteolytic hydrolysis yields biotin peptides. The smallest among these is biocytin, ε-N-biotinyl-L-lysine. Thoma & Peterson (1954) described an enzyme from pig liver which liberated biotin from a peptide digest of liver. They proposed the name biotinidase (biotin-amide amidohydro-lase, EC 3.5.1.12) for this enzyme, which also hydrolyzed the synthetic substrate N-(d-biotinyl)-p-aminobenzoic acid. At about the same time Wright *et al.* (1954) described an enzyme in blood plasma that hydrolyzed biocytin. After this initial description and partial purification from hog kidney (Knappe *et al.*, 1963), there was very little interest in this enzyme until Wolf *et al.* (1983a,b) showed that biotinidase deficiency is the primary defect in patients with the late-onset variant of the multiple carboxylase deficiency syndrome. This enzyme has been partially puri-fied from bacterial sources as well as from hog liver (Pispa, 1965). Rat liver biotinidase has also been purified to homogeneity and to a specific activity comparable to that of the human serum enzyme (K. Dakshina-murti *et al.*, manuscript submitted). The rat liver enzyme is associated with membranes although not an integral part of the membrane struc-ture. The enzymes from these two sources are similar in many of their properties such as optimal pH, K_m for substrate, acidic nature and inhibition by chaotropic agents as well as by sulfhydryl agents. The molecular mass of the rat liver enzyme is 61 kDa whereas that of the human serum enzyme is 68 kDa. The major difference between these two proteins is in the heat stability of the rat liver enzyme. It appears that the liver is the site of synthesis of the serum enzyme. Craft *et al.* (1985) purified human plasma biotinidase to a specific activity of 361 U mg^{-1} protein; we have purified human serum biotinidase to homogeneity (Chauhan & Dakshinamurti, 1986). The enzyme is a glycoprotein mono-mer with a molecular mass of 68 kDa. It contains sialic acid residues; however, these are not required for enzyme activity. With biocytin, the natural substrate, the maximal activity of the enzyme is in the pH range 4.5 – 6.0, although with the most commonly used synthetic substrate, N-(d-biotinyl)-p-aminobenzoic acid, the optimum pH is in the range 6.0 – 7.5. Biotinidase is not a general proteolytic enzyme. It has specific structural requirements in the substrate for hydrolysis. Biotinidase inhibits biotinidase activity in the millimolar range (Chauhan &

Dakshinamurti, 1986). The physiological significance of this inhibition is very little, as biotin is present in the nanomolar concentration range in human plasma (Baker, 1985).

A biotinidase clone of 1.25 kb was isolated from a human liver gt_{11} library (Chauhan *et al.*, 1988). The poly(A^+) RNA corresponding to the clone is 1.8 kb. This partial clone was further sequenced by the dideoxy-chain termination method. Although the predicted amino acid sequence from the biotinidase clone does not share extensive homology with avidin and strepavidin, there is conservation of sequence around tryptophan residues which has been shown to be critical for biotin-binding.

It is possible that for biotinidase and egg yolk biotin-binding proteins tryptophan residues in the active center are required for hydrophobic binding of biotin to these enzymes. The sequence for strepavidin has been published and shows a considerable homology to avidin around the tryptophan residues (Argarama *et al.*, 1986). Although the sequence for *E. coli* biotin-holocarboxylase synthetase is known, it is not known whether this protein requires the tryptophan residues for biotin binding. All the other biotin-binding proteins – biotinidase, avidin, strepavidin, and egg-yolk biotin-binding proteins – require tryptophan residues for biotin binding. We have done tentative sequence comparison between biotinidase, avidin, and strepavidin. It was suggested that the lysine near the tryptophan might have a role in biotin binding. Two of three lysines (near tryptophan) are conserved in strepavidin (avidin 9, 71 and 111; strepavidin 80 and 121). In biotinidase two lysines are conserved (positions 123 and 159) and we have found a sequence KWNV (position 123), which, in reverse order, is highly conserved in biotinidase and avidin. Another striking feature which we have observed is the occurrence of asparagine residues near the tryptophan in these proteins (biotinidase 135, 237, 264; avidin 12, 69; strepavidin 23, 81, 118). It is possible that there may be homology in the amino acid sequence around the tryptophan residues in all the known biotin-binding proteins.

Role of biotinidase as a biotin-carrier protein

Based on investigations with HeLa cells and human fibroblasts, we suggested that biotin transport in mammalian cells may involve pinocytosis through the functioning of specific circulating proteins (Dakshinamurti & Chalifour, 1981; Chalifour & Dakshinamurti, 1983). Although earlier studies have indicated that human albumin and α- and β-globulin can bind biotin (Frank *et al.*, 1970), we found this binding to be non-specific. It was reported that a glycoprotein in human serum can bind biotin (Vallotton *et al.*, 1965; Gehrig & Leuthardt, 1976). We have isolated this protein, using an agarose–biotin column, but it does not bind biotin specifically.

Human serum was fractionated on a Sephadex G-150 column and analyzed for biotinidase and biotin-binding activity (Chauhan & Dakshinamurti, 1988). Both activities coincided (Figure 8.3), suggesting that biotinidase is the only protein which binds biotin in human serum. These findings were corroborated by binding studies on fractions of human serum analyzed on DEAE–Sephacel, hydroxyapatite and octyl Sepharose columns. On all column separations biotin-binding activity coincided with biotinidase activity. Biotin-binding studies with human serum support the notion that biotinidase might be a biotin-carrier protein in human serum.

We determined the binding of biotin to pure human serum biotinidase. Figure 8.4a shows the concentration-dependent binding of [^{3}H]d-biotin to biotinidase and the Scatchard transformation of the data (1949). The Scatchard plot shows two linear parts indicating that two classes of sites with high and low affinities, respectively, for biotin are present (0.5 nM,

Figure 8.3. Human serum (15 ml) was applied onto a Sephadex G-150 column (2.5 cm × 100 cm) and equilibrated with 0.05 M phosphate buffer, pH 6.0, containing 1 mM β-mercaptoethanol, 1 mM EDTA, 0.15 M NaCl, and 0.02% sodium azide. The same buffer was used for elution at the 20 ml h^{-1} flow rate, and 3 ml fractions were collected. Biotin-binding activity (squares) and biotinidase activity (triangles) were determined. (From Chauhan & Dakshinamurti (1988). Used with permission of the publisher.)

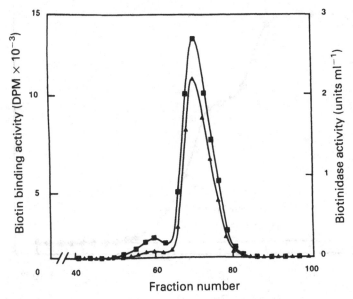

Figure 8.4. Concentration-dependent binding of [³H]-d-biotin to purified serum biotinidase. (*a*) Samples of purified biotinidase were incubated with increasing concentrations of [³H]-d-biotin (0.1 nM – 10 μM). A parallel set contained the amounts of [³H]-d-biotin plus a 1000-fold excess of cold d-biotin for the determination of non-specific binding. Each point is the mean of duplicate determinations. Linear regression of the points gave K_d values of 0.5 and 50 nM. (*b*) Competitive inhibition of biotin binding to human serum biotinidase. Binding of human serum biotinidase with 10^{-5} M N-bromosuccinimide (circles) or p-chloromercurobenzoate (dashed line). (From Chauhan and Dakshinamurti (1988). Used with permission of the publisher.)

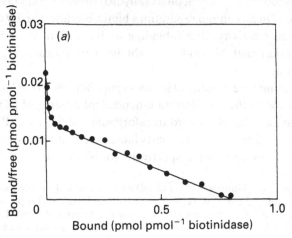

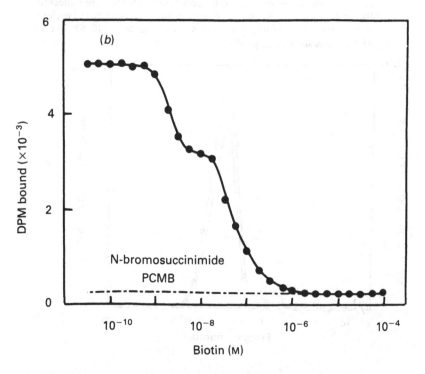

50 nm). Biotin binding was also investigated using displacement curve analysis. As shown in Figure 8.4*b* the displacement curve also shows both high- and low-affinity binding sites. The high-capacity, low-affinity binding corresponds to that of other vitamin-binding proteins. For example, thiamine-binding protein binds thiamine with a K_d for thiamine of 5.5 × 10^{-7} M (Nishimura *et al.*, 1984). Desthiobiotin and iminobiotin displace the bound biotin in the micromolar range. The biotin-binding activity of biotinidase can be completely inhibited with N-bromosuccinimide and p-chloromercuribenzoate (Figure 8.4*b*), suggesting that tryptophan and cysteine residues are required for biotin binding. It is likely that the cysteine residues in biotinidase are required for the formation of an enzyme–acyl complex for high-affinity, low-capacity binding. The formation of an enzyme–acyl complex was shown for biotinidase.

Baumgartner *et al.* (1985) suggested that the decreased concentration of plasma biotin in biotinidase-deficient patients may be due to decreased concentrations of biotinidase or to a specific biotin-binding protein in plasma. Other indirect evidence for the existence of a biotin-binding protein comes from the study of epileptic patients on long-term therapy with anticonvulsants (Krause *et al.*, 1985). Significantly low levels of serum biotin were observed in these patients, suggesting that these anticonvulsants probably compete with biotin for the biotin-binding protein in circulation. The anticonvulsants used in their study share with biotin the common structural feature of a cyclic carbamide group. It is this group of biotin that is involved in protein binding (Green, 1975). We have shown that all of these anticonvulsants compete with biotin for the biotin-binding activity of biotinidase (Chauhan & Dakshinamurti, 1988). Our results indicate that biotin in human serum is essentially non-covalently bound to protein(s) and that biotinidase is the only protein which exchanges with [^{3}H]d-biotin.

Role of biotinidase in intestinal absorption of biotin

There are conflicting reports on the intestinal absorption of biotin, with suggestions that the mode of uptake of biotin differs among species. Earlier investigations using the everted sac technique (Turner & Hughes, 1962; Spencer & Brody, 1964) reported that the absorption of biotin in the rat occurred by passive diffusion. Spencer & Brody (1964) observed that the hamster intestine transported biotin against a concentration gradient. In later work *Berger et al.* (1972) presented evidence for a sodium-dependent saturable process for the uptake of biotin by the proximal part of the hamster intestine. However, the K_m given for biotin transport (1.0 mm) was in the non-physiological range.

The uptake of biotin by human cell lines was reported by us (Dakshina-murti & Chalifour, 1981; Chalifour & Dakshinamurti, 1982a,b, 1983) to be saturable. Based on the binding of avidin and avidin–biotin complex to rat liver plasma membrane (Chalifour & Dakshinamurti, 1983) we suggested that avidin in these systems was mimicking a natural biotin-binding carrier involved in biotin transport (Dakshinamurti *et al.*, 1985). This was corroborated by the findings of Cohen & Thomas (1982) in their study using fully differentiated 3T3-Li cells. Bowers-Komro & McCor-mick (1985) have indicated that biotin was transported in hepatocytes by a sodium-dependent process but was not associated with a definitive saturable process. Gore *et al.* (1986), who used isolated rat mucosal cells in their study, concluded that at physiological concentrations the uptake of biotin by these cells was a passive phenomenon. However, both Bowman *et al.* (1986) and Said & Redha (1987) have concluded that, at concentrations less than 5 μM, absorption of biotin by rat jejunal seg-ments proceeded largely by a saturable process. Although a carrier has been proposed for biotin uptake, no attempt has been made to identify this carrier.

The uptake of biotin and biocytin in rat jejunal segments is shown in Figure 8.5. The biphasic transport of biotin and biocytin in the rat small

Figure 8.5. Uptake of (*a*) biotin and (*b*) biocytin in rat jejunal segments. The uptake of biotin or biocytin in 2.0 cm jejunal segments was determined as a function of the concentration of biotin. The data shown are the means ± SE of three experiments. (From Dakshinamurti *et al.* (1987). Used with permission of the publisher.)

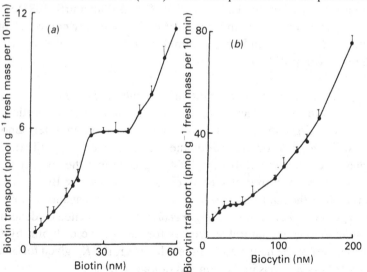

intestine observed in this study suggests that when the biotin concentration in the gut is below 50 nM, the saturable uptake mechanism would operate so that enough biotin is made available to the animal. If such a system indeed operated in other mammals, including man, it would have a tremendous advantage in view of the fluctuating amounts of biotin ingested in the diet. This is born out by the rarity of primary biotin deficiency in humans. The late-onset type of multiple carboxylase deficiency was shown to be due to the deficiency of biotinidase (Wolf *et al.*, 1983*b*). It is suggested in certain cases of late-onset type multiple carboxylase deficiency that these patients lack the system for absorbing biotin in the nanomolar range (Thoene *et al.*, 1983). Furthermore, it has been shown that these patients respond only to pharmacological doses of biotin (Roth *et al.*, 1982) indicating that only the saturable portion of biotin transport system is defective in these patients.

In order to identify the nature of the biotin-binding carrier protein in rat intestine, we fractionated solubilized brush border membrane and cytosol by sucrose density gradient centrifugation. Figure 8.6 shows the

Figure 8.6. (*a*) Cytosol and (*b*) detergent-treated, dialyzed preparation of brush border proteins were separated by sucrose density gradient centrifugation. Cytosol and solubilized brush border proteins (0.3 ml) were layered in separate tubes on top of a 5–20% gradient of sucrose (12.5 ml) containing 0.1 M potassium phosphate buffer, pH 6.0, 1 mM 2-mercaptoethanol, and 1 mM EDTA. Centrifugation was at 3°C for 22 h at 35 000 rpm in a Beckman SW-40 Ti rotor. At the end of the run, fractions were collected from the bottom of the gradient and analyzed for biotinidase (circles) and biotin-binding activity (squares). See text for further details. (From Dakshinamurti *et al.* (1987). Used with permission of the publisher.)

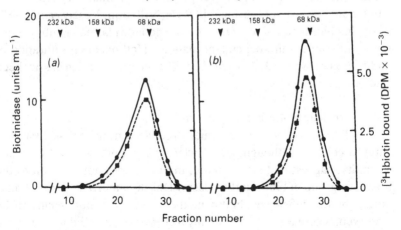

biotinidase activity and biotin-binding profiles of the separated proteins from the two preparations. In both fractions biotinidase activity migrated slightly ahead of albumin (68 kDa). The biotin-binding activity in cytosol and brush border preparations migrated to a position identical to the biotinidase peak, and there was good correlation at each point between the two activities. Furthermore, when we used N-bromosuccinimide and p-chloromercuribenzoate, which are inhibitors of biotinidase activity (Craft *et al.*, 1985; Chauhan & Dakshinamurti, 1986, 1988), both biotinidase and biotin-binding activities were inhibited, suggesting that tryptophan and cysteine residues play a role in biotin-binding activity. These results suggest that biotinidase is the only protein in brush border membrane which binds biotin. Based on this preliminary *in vitro* study it is suggested that biotinidase *in vivo* may have a role in the transport of biotin.

Avidin–biotin receptors from hepatic plasma membrane

Rat liver plasma membrane and rat intestinal mucosa plasma membrane were examined for their ability to bind to avidin–[^{3}H]biotin complex (Chalifour & Dakshinamurti, 1983). There was no specific binding of the avidin–biotin complex to rat intestinal mucosa plasma membrane. Avidin–[^{3}H]biotin complex bound specifically to liver plasma membrane with a K_d of approximately 35 nM and B_{max} of 136 pmol mg^{-1} membrane protein. Binding was not directed by the oligosaccharide chain of avidin, as indicated by the lack of inhibition of binding by simple and complex carbohydrate structures. Avidin does not seem to bind to liver plasma membrane at or near the biotin-binding site. As avidin is not a normal circulating protein in the rodent it should be mimicking a natural ligand in its specific binding to liver plasma membrane.

In binding experiments using d-[8,9-^{3}H(N)]biotin, Vesely *et al.* (1987) found that biotin bound to its receptor on hepatic plasma membrane in a time- and temperature-dependent manner. Half-maximal saturation of binding was between 10^{-9} and 10^{-10} M. The receptor protein has not been isolated.

Nuclear biotin-binding protein

Biotin is present in cell nuclei prepared from various tissues or cultured cells even though such nuclei do not possess any CO_2-fixing capability, suggesting that the biotin in nuclei does not function as the prosthetic group of biotin carboxylases (Dakshinamurti & Mistry, 1963). During biotin deficiency, biotin in the nuclear fraction seems to be conserved, whereas it is preferentially lost from other cellular organelles.

Biotin in nuclei has been shown to be non-covalently bound to a protein. A biotin-binding protein from rat liver nucleus has been isolated (Dakshinamurti *et al.*, 1985). This protein binds reversibly to biotin *in vitro*, with a maximal binding of 3.54 pmol per microgram of protein and with a dissociation constant for biotin of 2.2×10^{-7} M. Polyacrylamide gel electrophoresis in the presence of sodium dodecyl sulfate indicates an apparent subunit molecular mass of 60 kDa for this protein.

This report of a nuclear biotin-binding protein resembles that of Meisler & Thanassi (1990) on the identification of pyridoxine-derived pyridoxal phosphate (PLP) in the nucleus and the demonstration of a nuclear pyridoxal phosphate-binding protein in rat hepatoma-derived HTC cells. The fraction of total PLP found in the nuclei of liver cells was 21% in rats fed a diet adequate in vitamin B_6 and 39% in rats fed a diet deficient in vitamin B_6, indicating a conservation of the active vitamin in the nucleus, analogous to our observation regarding the distribution of biotin in the cell nucleus. The possibility that the presence of PLP in the nucleus is of physiological significance has been suggested Meisler & Thanassi (1990). It is known that PLP modulates steroid hormone activity by altering the interaction of steroid receptor complex with DNA, chromatin and nuclei (Schmidt & Litwack, 1982; Litwack, 1988). Cells grown in the presence of 5 mM pyridoxine have a decreased glycocorticoid-dependent induction of enzymes such as tyrosine aminotransferase (Disorbo & Litwack, 1981). It has been suggested that, in addition, the pyridine ring of PLP might mimic the base portion of a nucleotide, with the phosphate group being common to both PLP and a nucleotide. It is possible that PLP might act as a small molecular modulator of protein–DNA interactions. The observations on the nuclear protein biotin- and pyridoxal phosphate-binding proteins are akin to those on the nuclear retinoic acid-binding protein (Takahashi & Beritman, 1989).

Monoclonal antibody to biotin

Monoclonal antibodies to biotin have been prepared using biotin linked to keyhole limpet hemocyanin (KLH) as the antigen (Dakshinamurti *et al.*, 1986; Dakshinamurti & Rector, 1990). Spleen cells obtained from mice immunized with biotin–KLH were fused with the myeloma cell line NS-1. The resulting hybridomas were screened for the production of antibodies to biotin using an enzyme-linked immunosorbent assay. Clones producing antibodies to biotin were isolated by limiting dilution methods. Four cell lines, each derived from a different fusion, were chosen for the production of monoclonal antibodies. One clone, no. 33, produced antibody of high affinity that bound both free and haptenic biotin antigens as well as biocytin.

Non-prosthetic group functions of biotin

Apart from its role in carboxylases, biotin has been implicated in other areas of metabolism where its role is not explained on the basis of its function as the prosthetic group of known biotin-containing enzymes. Biotin has been shown to have an important role in the regulation of the biotin operon in *E. coli*. Biotinyl AMP-holocarboxylase synthetase functions as the repressor for the biotin operon (Eisenberg *et al.*, 1982; Cronan, 1989). In the avian system, biotin has been shown to regulate the synthesis of both the egg yolk BBP-I and BBP-II by transcriptional control that overrides hormonal control. In mammals the concentrations of various proteins and enzymes are regulated by the biotin status of the animal (Dakshinamurti & Litwack, 1970; Boeckx & Dakshinamurti, 1970). It is likely that transcriptional control, as for some hormones, is involved in these as well. Biotin seems to enhance the activities of guanylate cyclase and RNA polymerase II. These effects seem to be direct and on the activities of the enzymes rather than on the concentration of these enzymes.

Requirement for biotin by cells in culture

A requirement for biotin by cells in culture would be expected in view of the obligatory involvement of biotin in the metabolism of carbohydrates and lipids and in the further utilization of deaminated residues of certain amino acids. However, various earlier reports have claimed that cells in culture do not require biotin (Eagle, 1955; Swim & Parke, 1958; Holmes, 1959; Dupree *et al.*, 1962). Keranen (1972) reported that HeLa cells grown in biotin-deficient medium contained more biotin than those in a biotin-supplemented medium, perhaps owing to the ability of these transformed cells to synthesize biotin. Using biotin-depleted fetal bovine serum (FBS; Dakshinamurti & Chalifour, 1981) and Eagle's minimum essential medium, we demonstrated a requirement for biotin by HeLa cells, human fibroblasts, and Rous sarcoma virus-transformed baby hamster kidney cells based on the viability, biotin content, and activities of biotin-dependent and -independent enzymes (Chalifour & Dakshinamurti, 1982*a*,*b*). There was a drastic reduction in the viability of HeLa cells starting with the fourth passage, and following the sixth passage in this medium no further cell growth was observed (Figure 8.7).

In a further study (Bhullar & Dakshinamurti, 1985) we have shown that there was a significant decrease in the incorporation of leucine into protein of the homogenate or cytosol of biotin-deficient HeLa cells compared with cells grown in a biotin-supplemented medium. When biotin was added to the biotin-deficient medium there was a twofold increase in the incorporation of leucine into proteins. Based on experiments using puromycin and cordycepin, we concluded that the appearance of new RNA in the cytoplasm is an event brought about when biotin-deficient cells are supplemented with exogenous biotin. Normal cells in G_1 arrest due to serine starvation start incorporating [^{3}H]thymidine into DNA as soon as serine is restored to the medium. Biotin-deficient HeLa cells under similar conditions do not incorporate [^{3}H]thymidine into DNA even when serine is restored to the medium. However, within 4 h of supplementation of biotin to the biotin-deficient medium, the incorporation of [^{3}H]thymidine into DNA reaches a maximum (Figure 8.8). Cells cultured in a medium deficient in growth factor or nutrients do not multiply, owing to arrest of growth in the G_1 phase of the cell cycle (Allen

Figure 8.7. Viability of HeLa cells cultured in a biotin-depleted medium. HeLa cells were subcultured in Eagle's minimum essential medium containing 5% FBS (filled circles) or biotin-deficient FBS (open circles). The cells were carried through successive passages; viability was determined using the trypan blue exclusion test. Results are the average of three experiments. Bars indicate one standard deviation. (From Dakshinamurti & Chalifour (1981). Used with permission of the publisher.)

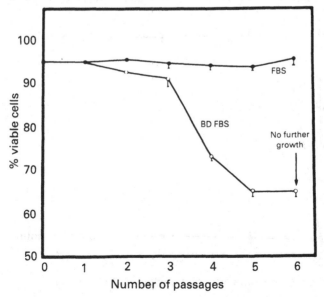

& Moskowitz, 1973). Net synthesis of macromolecules during the G_1 phase is necessary for entry into the S phase. Progression through the cell cycle is dependent on the synthesis of specific proteins, and it is possible that the inhibition of the synthesis of such proteins would block the cell cycle. Significantly, the incorporation of thymidine into DNA reaches a maximum within 4 h of the addition of biotin to the deficient cells. By this time there is stimulation of protein synthesis. We speculate that these two phenomena are related and that the growth-promoting effect of biotin might be achieved through the stimulation of the synthesis of certain proteins.

Of the biotin-dependent carboxylases of significance in animals, acetyl CoA carboxylase is required for lipogenesis and hence membrane genesis. Pyruvate carboxylase is essential for the generation of oxaloacetate, for the maintenance of the tricarboxylic acid cycle (Nakano *et al.*, 1982) and gluconeogenesis. Propionyl CoA carboxylase and methylcrotonyl CoA carboxylase are required for the further metabolism of certain amino acid residues. Thus, under culture conditions, biotin should subserve only the lipogenic and dicarboxylic acid needs of the cell.

Figure 8.8. Incorporation of [^{3}H]thymidine in DNA of biotin-supplemented and -deficient cells. Biotin-deficient cells were exposed to 2 ng ml^{-1} of biotin. The control (deficient) cells were left in the deficient medium. At various time intervals biotin-supplemented and control biotin-deficient cells were removed, and the incorporation of [^{3}H]thymidine into DNA was determined. (From Bhullar & Dakshinamurti (1985). Used with permission of the publisher.)

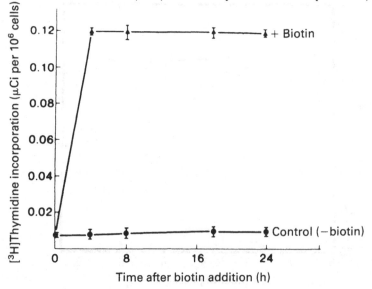

However, serum lipids are taken up by cells (Wisniesky *et al.*, 1973) and saturated fatty acids are without growth-promoting activity for cells such as simian virus-transformed 3T3 cells under conditions of biotin deficiency (Messmer & Young, 1977).

Cheng & Moskowitz (1982) have shown that G_1-arrested Rous sarcoma virus-transformed (RST) baby hamster kidney cells can be maintained in culture continuously in basal medium supplemented with biotin only. Cells multiplied continuously in basal medium supplemented with delipidized serum and serum lipid extract but not in basal medium supplemented with serum lipid extract alone. These results suggest that cell growth depended on some biotin-mediated activity whose level had decreased when cells were grown in a biotin-deficient medium. It has been claimed that RST cells growing in a medium containing biotin produced a non-dialyzable factor that stimulated cell multiplication (Moskowitz & Cheng, 1985). We have shown (Bhullar & Dakshinamurti, 1985) that the addition of biotin to the culture medium of HeLa cells results in enhanced protein synthesis, DNA synthesis, and cell growth. It is possible that the growth-promoting effect of biotin might be achieved through stimulation of the synthesis of certain proteins, highlighting the non-prosthetic group function of biotin.

A requirement for high-density lipoprotein (HDL) has been shown for the growth of Madin–Darby canine kidney cells exposed to Dulbecco's modified Eagle's medium (DMEM) supplemented with transferrin when cells were exposed to a mixture (1:1) of DMEM and F-12 medium (Cohen & Gospodarowicz, 1985; Gospodarowicz & Cohen, 1988). The components of the F-12 medium responsible for support of growth in the absence of HDL are biotin, which is absent from DMEM, and choline, which is present in insufficient concentration in DMEM. It is of particular significance in this context that the HDL fraction of plasma has a considerable amount of biotin associated with it (authors' unpublished observations). This biotin is not dialyzable and is probably non-specifically attached to the protein. Regardless of the nature of the association, this would then explain the HDL requirement, which can be replaced by biotin, as a specific requirement for biotin itself. Collins *et al.* (1987) reported that biotin is essential for the expression of the asialoglycoprotein receptor in HepG2 cells.

Role of biotin in cell differentiation

It is known that various transformed cell lines produce transforming growth factors that enable them to grow in a serum-free medium supplemented with transferrin and insulin (Kaplan *et al.*, 1982). The 3T3-

L1 subline derived from the 3T3 mouse fibroblast cell line has the capacity to differentiate in the resting state into a cell type having the characteristics of adipocytes (Green & Kehinde, 1974). When they reach confluence and start to differentiate, they greatly increase their rate of triglyceride synthesis and accumulate the product. The increase in lipogenic rate parallels a coordinate rise in the activities of the key enzymes of the fatty acid biosynthetic pathway (Mackall & Lane, 1977; Mackall *et al.*, 1976). The increases in many enzyme activities have been shown to result from specific translatable mRNA (Spiegelman & Farmer, 1982; Wise *et al.*, 1984). This increase correlated with a marked rise in nuclear run-off transcription rates for these mRNAs during differentiation (Bernlohr *et al.*, 1985). The process of differentiation can be accelerated in a number of ways, such as by increasing the amount of serum in the culture medium or by adding certain hormones, such as insulin, and other chemical agents, such as biotin, 1-methyl-3-isobutylxanthine, or dexamethasone (Rosen *et al.*, 1979). When serum in the cultures was extensively dialyzed, the morphological and enzyme changes characteristic of adipocyte differentiation were induced without deposition of fat (Kuri-Harcuch *et al.*, 1978). It was presumed that dialysis removed all biotin from the serum and that the role of biotin was linked to the increase in acetyl CoA carboxylase. However, it is not clear from their work why it would take 24–48 h for the lipid-synthesizing enzymes to increase following the addition of biotin to the biotin-deficient cell culture medium; if this were only dependent on the synthesis of holoacetyl CoA carboxylase from the apoenzyme, it would have been completed in 1 h. The delay suggests that perhaps some factor required for the induction of the whole set of lipogenic enzymes (which, with the exception of acetyl CoA carboxylase, are not biotin enzymes) might be formed under the influence of biotin. In addition, the 3T3-L1 subline had a greater capacity than did the very slowly differentiating M_2 subline to remove labeled palmitate from the culture medium and to incorporate it into triglyceride (Kasturi & Wakil, 1983). Esterification was the principal route of fatty acid utilization by both pre-adipose cells and cells committed to differentiation. The frequency of adipocyte conversion is a heritable condition. Genetic and epigenetic factors regulate the level of the lipogenic multienzyme system. To express differentiation, the attainment of confluency is accompanied by the presence of an adipogenic factor in the culture medium (Kuri-Harcuch & Green, 1978; Kuri-Harcuch & Marsch-Morino, 1983). This is present in most animal sera but not in the serum of the domestic cat. It is significant that cat serum contains less than one tenth of the biotin content of bovine serum.

After interaction with the adipogenic factor, cells undergo DNA replication and cell division before they express the differentiated state. Addition of cytosine arabinoside to the medium prevented cells from undergoing conversion. The molecular events initiated by the interaction of the cells with the adipogenic factor remain to be identified. The temporal relationship between DNA synthesis and differentiation after susceptible cells are stimulated with adipogenic serum, and the inhibition of differentiation by cytosine arabinoside, supports the interpretation that DNA synthesis is required by 3T3-L1 cells for differentiation. As shown by us (Dakshinamurti & Chalifour, 1981) biotin is not completely removed from fetal calf serum by extensive and repeated dialysis or by treatment with avidin–Sepharose. It is possible that, at the low concentration present in extensively dialyzed serum, biotin is still available for the reactions that have to do with differentiation. Higher concentrations of biotin would, however, be required to biotinate acetyl CoA apocarboxylase to form the holoenzyme.

Biotin and testicular function

In early work, Shaw & Phillips (1942) reported that the testes of biotin-deficient rats appeared very small in relation to the size of the animal and that the seminiferous tubules were very small and showed signs of degeneration. Delost & Terroine (1956) performed a more systematic study and found that the testes of biotin-deficient rats were smaller and of lower mass than those of normal animals. There was evidence of delayed spermatogenesis and a decreased number of spermatozoa. Delost & Terroine (1956) reported that these disturbances in the testes were due to biotin deficiency alone, since animals on a restricted but normal diet did not show any of these abnormalities (Terroine, 1960). In our study (Paulose et al., 1989), testicular function in biotin-deficient rats responded to luteinizing hormone or biotin administration by increasing testosterone production. The morphological abnormalities of the biotin-deficient rat testes were reserved by biotin treatment alone, whereas continuous testosterone supply through implantation throughout the biotin-deficient period was without any effect. This paralleled the effects of biotin and testosterone on protein synthesis in the testes. The stimulation of testicular protein synthesis by testosterone required normal biotin status.

Biotin and embryological development

Congenital malformations have been reported in chick embryos of domestic fowl maintained on a biotin-deficient diet (Cravens et al.,

1944). Kennedy and Palmer (1945) found that female rats given a biotin-deficient diet maintained a normal estrus cycle and progressed through gestation normally until the 11–13 days of intrauterine life. From then on, and particularly around days 20 and 21 of gestation, there was a high rate of fetal resorption, which coincided with the period of most rapid fetal growth. During this period the transfer of biotin from the mother to the fetus was very high. Rose *et al.* (1956) found that in the postnatal period lactation was arrested by the fifteenth day post-partum.

The role of biotin in embryonic growth and development has been re-evaluated by Watanabe (1990). Maternal biotin deficiency was strongly teratogenic in mice even when dams did not exhibit any typical clinical signs of biotin deficiency (Watanabe, 1983). In a further study (Watanabe & Endo, 1984) it was found that at mid-gestation biotin-deficient embryos weighed less than normal ones and had external malformations such as micrognathia and micromelia. There was a marked reduction in the size of the palatal process on day 15.5 of gestation. This effect may be due to altered proliferation of mesenchyme. The incidence of embryos with delayed formation of the palatal process (89.5%) was in close agreement with that of fetuses with cleft palates. The same may be said for the proliferation of mesenchyme in the limb bud. The primary effect of biotin deficiency on craniofacial and limb development is caused by an abnormality in cell proliferation in the mesenchyme.

In a recent study (T. Watanabe *et al.*, in preparation) of the development of the palatal process in culture, it was found that after 72 h of organ culture (from 12.5 d to 15.5 d of embryo age) more than 90% of the explants from normal mouse embryos (biotin-sufficient) were at stage 6 of development. The corresponding figure for explants from biotin-deficient embryos cultured in a biotin deficient medium was 6.5%. If the deficient explants were cultured in a medium containing biotin (10^{-8} M) the percentage at stage 6 of development rose to over 30. Administration of biotin (20 μg) to biotin-deficient dams 24 h prior to removal of the embryos resulted in 33% of the explants at state 6 of development when cultured in a biotin-deficient medium and more than 50% at stage 6 if cultured in medium containing biotin (10^{-7} M).

In view of the role of the biotin carboxylases in the major metabolic pathways, it was possible that the teratogenicity of biotin deficiency could be due to the accumulation of intermediary or secondary metabolites such as propionic acid, lactic acid, 3-hydroxypropionic acid, 2-methyl citrate, propionyl glycine, β-hydroxyisovaleric acid or β-methylcrotonyl glycine as a result of individual carboxylase deficiencies. However, there was no detrimental effect of any of the organic acid intermediates or secondary metabolites on palatal closure of the explants when these compounds were added to the organ culture medium at a concentration of 10^{-4} M. Also, the addition of avidin to the organ culture medium did not significantly decrease the percentage of explants at stage 6 of development. These results highlight the continuous requirement for biotin during the proliferation of the mesenchyme. Embryological malformations could not be ascribed to a deficient activity of the biotin carboxylases; this finding emphasizes the role of biotin in the synthesis of proteins, including growth factors, during organogenesis.

Biotin and defense mechanisms

Administration of biotin to normal rats increases immunological reactivity. An increase in hemolysin production in response to inoculation with sheep erythrocytes has been shown (Petrelli & Marsili, 1969). Biotin exerts a protective action against bacterial infection in splenectomized rats. Biotin administration to normal rats produces an increase in activity of the reticuloendothelial system, as assessed by various criteria (Petrelli & Marsili, 1971). Biotin increases the *in vivo* and *in vitro* phagocytosis of *Staphylococcus epidermidis* by rat and human neutrophils. Incubation with biotin increases significantly the number of lymphocytes carrying B and T markers from biotin-deficient guinea pigs (Petrelli *et al.*, 1981).

Effects of biotin on guanylate cyclase and RNA polymerase II enzyme activities

Vesely (1981, 1982) and Vesely *et al.* (1984) reported that biotin and its analogs at micromolar concentrations enhanced guanylate cyclase activity in various rat tissues and suggested a role for biotin in the activation of this enzyme. Spence & Koudelka (1984) studied the relationship between the biotin-mediated increase in glucokinase activity and cellular level of cGMP as well as the activity of guanylate cyclase. Increases in intracellular cGMP as well as in guanylate cyclase were reported. It should be noted that these effects were elicited at a micromolar concentration of biotin, which is about two orders of magnitude higher than the normal cellular concentration of biotin.

The effects of biotin on cell growth are seen at physiological concentrations of this vitamin (Singh & Dakshinamurti, 1988). We have examined the mechanism of action of biotin in this. The addition of physiological concentrations of biotin to the growth medium of biotin-deficient HeLa cells and fibroblasts resulted in increases in guanylate cyclase activity, as well as in the intracellular concentration of cGMP.

The major role of cGMP in cells is connected with cell growth, DNA and RNA synthesis, and possibly in malignant transformation of cells. Using Novikoff hepatoma, Zeilberg & Goldberg (1977) showed that as the synchronized cells entered mitosis cellular cGMP levels rose and cAMP levels decreased. Similar changes in cyclic nucleotide concentrations were observed after growth induction by fibroblast growth factor (Rudland *et al.*, 1974). Agents that increase lymphocyte cGMP levels stimulate transformation of these cells (Schumm *et al.*, 1974). An increase in intracellular cGMP has been shown during rat liver regeneration, which is regulated by insulin and glucagon (Earp, 1980). In all the above cases the growth factors produced an increase in the activity of guanylate cyclase *in vivo*. None of these factors stimulated the activity of guanylate cyclase *in vitro*. Their mode of action would seem to be distinct from the *in vitro* stimulatory effect of agents like azide, nitroprusside and nitrosamine (Braughler *et al.*, 1979) as well as biotin and its analogs when used at pharmacological concentrations.

RNA polymerase II synthesizes mRNA precursors. The activity of this enzyme in nuclear lysate from biotin-deficient HeLa cells and fibroblasts was significantly lower than in similar preparations from biotin-supplemented cells. Addition of biotin to the culture medium at a concentration as low as 10^{-8} M resulted in a significant increase in enzyme activity, which reached a plateau by 4 h following addition of biotin to the medium.

Increased RNA polymerase II activity is found in rapidly growing mammalian cells. cGMP stimulation of nuclear DNA-dependent RNA polymerase has been shown in lymphocytes (Johnson & Haddon, 1975), rat mammary gland (Anderson *et al.*, 1975) and fetal calf liver cells cultured in the presence of sheep erythropoietin (Canas & Congote, 1984; White & George, 1981). The addition of physiological concentrations of biotin to the growth medium of biotin-deficient cells results in a stimulation of guanylate cyclase activity and an increase in intracellular cGMP as well as in RNA polymerase II activity. We contend that the stimulation of the synthesis of certain proteins (growth factors) by biotin might be mediated by its effect on cGMP.

Effect of biotin on the regulation of specific proteins

Role of biotin in the induction of asialoglycoprotein receptor
The asialoglycoprotein receptor is characteristic of fully differentiated hepatocytes. The human hepatoblastoma line HepG2 expresses

maximum receptor activity only in confluent cultures (Theilman *et al.*, 1983). Collins *et al.* (1988) have recently shown that HepG2 cells grown to confluency in minimal essential medium (MEM) made 10% with respect to fetal bovine serum (FBS) demonstrated an asialoglycoprotein receptor with ligand-binding characteristics and molecular mass comparable to the receptor purified from human liver. However, dialysis or ultrafiltration, with removal of the low-molecular-mass fraction of FBS, dramatically reduced expression of asialoglycoprotein receptor, under conditions where protein synthesis and total cellular protein content were comparable to those of control cells. The addition of biotin or biocytin to medium supplemented with dialyzed FBS during logarithmic growth supported normal expression of asialoglycoprotein receptor by confluent HepG2 cells (Collins *et al.*, 1988). Biotin was most effective at or about a final concentration of 10^{-8} M. Isolation of mRNA from HepG2 revealed no difference in asialoglycoprotein receptor transcripts when cells were grown in MEM–20% FBS or MEM–10% dialyzed FBS. Thus, a biotin-dependent post-transcriptional event permits the ultimate expression of asialoglycoprotein receptor by HepG2 cells.

Role of biotin in the induction of pyruvate kinase and phosphofructokinase

The effect of biotin on two glycolytic enzymes, pyruvate kinase and phosphofructokinase, as well as on the bifunctional enzyme phosphohexose isomerase, in normal, diabetic, biotin-treated and insulin-treated diabetic rats is shown in Table 8.1 (Dakshinamurti *et al.*, 1970). In the diabetic rat, liver pyruvate kinase was about a third of the normal level. Biotin treatment increased pyruvate kinase activity by 130% over the diabetic level. Administration of insulin or insulin plus biotin resulted in a further stimulation to 175% over the diabetic level. Phosphofructokinase activity of the diabetic rat liver was about 65% of the normal level. This enzyme activity was restored to normal levels by biotin, insulin, or both given together. The activities of liver phosphohexose isomerase in normal, diabetic, and biotin-treated diabetic rats showed no difference between these groups.

Role of biotin in the induction of biotin-binding proteins of egg yolk

The biotin content of hens' eggs was shown to be directly related to the amount of biotin in the diet (Brewer & Edwards, 1972). After two

Table 8.1. *Effect of biotin on hepatic glycolytic enzyme activities in the diabetic rat*

Treatment	Pyruvate kinase		Phosphofructokinase		Phosphohexose isomerase	
	Specific activity[a] ($10^6 \times$ units mg^{-1} protein)	Percentage stimulation	Specific activity ($10^6 \times$ units mg^{-1} protein)	Percentage stimulation	Specific activity ($10^2 \times$ units mg^{-1} protein)	Percentage stimulation
Normal group						
1 Nil	137 ± 21	—	190 ± 17	—	182 ± 17	—
Diabetic group						
2 Nil	49 ± 8^b	—	126 ± 30^b	—	188 ± 6^b	—
3 biotin (200 µg)	64 ± 7^c	131	182 ± 27^c	145	188 ± 13	nil
4 Insulin (4 units)	86 ± 20^c	175	192 ± 53^c	152	—	—
5 Insulin (4 units) + biotin (200 µg)	84 ± 17^c	175	189 ± 39^c	143	—	—

[a]Specific activity is given as mean ± s.d.; [b]$p < 0.001$ with respect to group 1; [c]$p < 0.001$ with respect to 2.
Source: Dakshinamurti *et al.* (1970). Used with the permission of the publisher.

weeks on a biotin-deficient diet, a hen lays eggs whose hatchability is reduced to zero and whose biotin content is 10% of that of a normal hen's egg (Couch *et al.*, 1949). A single hen's egg contains about 10 μg of biotin. Of this about 1 μg is in the albumen and bound to avidin, where it occupies 15% or less of the binding sites; the remainder is in egg yolk, bound to the two biotin-binding proteins BBPI and BBPII. It was further demonstrated that the concentrations of BBPI and BBPII in egg yolk were directly related to dietary biotin concentrations. At low dietary concentrations of biotin, BBPI is the major transporter, at higher dietary concentrations, BBPII predominates. It has been shown that avidin (Korpela, 1984*b*) and BBPI (White & Whitehead, 1987) can be induced by various sex hormones. The induction of both of the egg yolk biotin binding proteins by biotin (White & Whitehead, 1987) would suggest that biotin regulates the biotin-binding proteins of egg yolk by transcriptional control that overrides the hormonal control.

Role of biotin in the induction of glucokinase activity

The only protein which had been shown to be specifically induced by biotin is glucokinase. Biotin is not a part of this protein. Earlier work from this laboratory (Dakshinamurti & Cheah-Tan, 1968*a,b*) demonstrated that liver glucokinase activity was altered in response to the biotin status of rats. Biotin also plays a role in the precocious development of glucokinase in young rats (Dakshinamurti & Hong, 1969). In all these experiments increase in enzyme activity was associated with protein synthesis. The synthesis of glucokinase is under developmental, nutritional and hormonal control (Weinhouse, 1976; Meglasson & Matschinsky, 1984). Glucokinase activity in liver decreases when rats are starved or fed a carbohydrate-free diet. On refeeding a normal diet or glucose there is a remarkable recovery, reaching essentially normal glucokinase levels within a few hours (Sharma *et al.*, 1963; Ureta *et al.*, 1970). Spence & Koudelka (1984), using primary cultures of rat hepatocytes, have shown that glucokinase activity increased by 50% during the first 4 h following insulin addition to the medium, remained constant for an additional 2–3 h and was followed by a second increase in activity. In our earlier work as well as in that of Spence & Koudelka (1984) the effect on glucokinase synthesis was observed at pharmacological concentrations of biotin. Using physiological concentrations of biotin we have shown that biotin enhances glucokinase activity in biotin-deficient rats. Spence & Koudelka (1984) indicated that biotin increased the amount of translatable mRNA coding for glucokinase.

Role of biotin in the induction of glucokinase mRNA in starved rats

To study the *in vivo* regulation of glucokinase mRNA by biotin we utilized starved rats. We have previously shown (Dakshinamurti & Cheah-Tan, 1968*a*) that in starved rats there was a decrease in liver glucokinase activity and that biotin administration resulted in a three- to fourfold increase in enzyme activity. The time course of glucokinase mRNA induction was followed after biotin administration (Chauhan & Dakshinamurti, 1991). There was a 0.6-fold decrease in albumin mRNA in the starved rat; after biotin administration it returned to normal (Figure 8.9). In addition, in the starved rat there was a 6.2-fold increase in hepatic phosphoenolpyruvate carboxykinase (PEPCK) mRNA in comparison with normal-fed rats. At the same time, the concentration of

Figure 8.9. Effect of biotin on glucokinase, albumin and phosphoenolpyruvate carboxykinase mRNAs in liver. A, RNA blot probed with rat albumin cDNA; B, probed with phosphoenolpyruvate carboxykinase cDNA; C, probed with glucokinase cDNA. 5 μg poly(A)$^+$ RNA was loaded per lane for blots A and B, and 10 μg per lane for blot C. Lane 1, normal control; lane 2, fasted for 24 h, no biotin treatment; lane 3, fasted, 1 h after biotin; lane 4, fasted, 2 h after biotin; lane 5, fasted, 4 h after biotin; lane 6, fasted, 6 h after biotin; lane 7, fasted, 8 h after biotin; lane 8, fasted, 12 h after biotin. (From Chauhan & Dakshinamurti (1991). Used with permission of the publisher.)

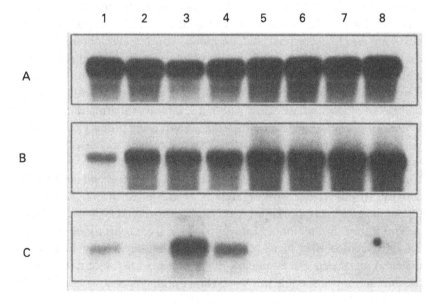

Table 8.2. *Time course of biotin effect on glucokinase activity and mRNA amount in liver of starved rats*

The areas under the absorbance peaks corresponding to the mRNA bands were calculated and expressed in arbitrary units. Data are given as mean of four separate determinations.

Time after biotin administration to starved rats (h)	Glucokinase activity (milliunits mg^{-1} protein)	Relative amount of glucokinase mRNA (arbitrary units)
Normal fed control	14.1 ± 3.2	0.99 ± 0.01
0 (starved 24 h)	4.2 ± 1.5	0.20 ± 0.12
1	6.8 ± 1.9	3.92 ± 1.10
2	16.6 ± 4.6	1.22 ± 0.53
4	14.9 ± 3.8	not detectable
6	14.1 ± 4.2	not detectable
8	12.8 ± 3.1	not detectable
12	10.9 ± 2.4	0.05 ± 0.01

Source: Chauhan & Dakshinamurti (1991). Used with permission of the publisher.

glucokinase mRNA of biotin-injected starved rats increased 3.9-fold in comparison to normal-fed rats, 1 h after injection of biotin. Following this increase there was rapid decay of glucokinase mRNA which was hardly detectable 4 h after biotin administration. The induction of glucokinase mRNA at 1 h following biotin administration to starved rats was 19-fold in comparison with starved rats not receiving the biotin injection. The induction of glucokinase mRNA by biotin is therefore relatively rapid and marked.

To correlate induction of glucokinase mRNA with glucokinase activity, we measured glucokinase activity in the cytosolic fraction of liver homogenates (Table 8.2). Within 2 h after biotin injection the glucokinase mRNA concentration of the biotin-injected starved rats had fallen to about 30% of the concentration at 1 h, and by 4 h glucokinase mRNA was not detectable. From this we concluded that the half life of the glucokinase mRNA was of the order of 30 min under the experimental conditions. The glucokinase enzyme activity reached its maximum by 2 h following biotin administration and stayed high for the next 6–8 h. Iynedjian *et al.* (1988) have shown that, following insulin treatment of rats, the amount (and activity) of glucokinase increased in a time-dependent fashion, after an initial lag of 4 h, to reach 65% of the non-diabetic control amount in 24 h. In contrast, glucokinase mRNA was shown to increase until 8 h and subsequently the concentration of the mRNA decayed rapidly so that little message was left after 16 h and virtually none after 24 h. Although the actions of insulin and biotin are parallel, the effect of biotin on glucokinase induction seems to be more rapid than that of insulin.

The rapid and marked increase of glucokinase mRNA after biotin administration suggests a biotin effect at the transcriptional level. To examine this further, we carried out 'run on' transcriptional experiments with isolated liver nuclei to estimate the relative rates of transcription of the glucokinase gene at various times following biotin administration to the starved rat. As shown in Figure 8.10A, the transcription of the actin gene, included as internal control, was uninfluenced by biotin administration. Biotin administration increased glucokinase gene transcription by about 6.7-fold and decreased that of phosphoenolpyruvate carboxykinase by about 2.6-fold (Table 8.3). The increased transcription of the glucokinase gene within 45 min of biotin addition is not simply due to an increase in overall transcriptional efficiency, because the transcription of the β-actin gene is unaffected by biotin administration. The difference

Figure 8.10. Biotin-induced transcription of glucokinase gene. In A, nuclei were isolated at specified times after biotin administration. The radiolabeled transcripts were hybridized to filter-bound plasmids containing cDNA for phosphoenolpyruvate carboxykinase (PEPCK), glucokinase and actin, respectively. Specific transcripts from the three genes were detected by autoradiography of the filters. The vector plasmids Bluescript and pBR322 were also bound to the filters as control for background hybridization. B. Effect of biotin on glucokinase mRNA. RNA was isolated from the same rat livers as used for nuclear run-on assay in A. (From Chauhan and Dakshinamurti (1991). Used with permission of the publisher.)

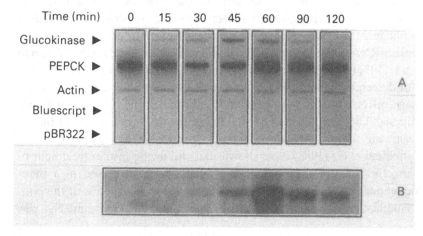

Table 8.3 *Effect of biotin on gene transcription in liver as shown by run-on transcription assays with isolated nuclei*

Starved rats were treated with biotin for the time shown. The areas under absorbance peaks corresponding to individual spots were calculated and expressed in arbitrary units. Appropriate exposures of autoradiograms were used to allow valid comparison between different genes. Data are the means of three separate assays using liver nuclei from different animals. Fold effects were calculated by using a value of 1 to the transcriptional rate in non-treated starved rats (0 min).

	Rate of gene transcription					
	Glucokinase		PEPCK		β-actin	
Time after biotin (min)	arbitrary units	-fold	arbitrary units	-fold	arbitrary units	-fold
0	1.56 ± 0.41	1.0	101.6 ± 21.2	1.0	8.0 ± 1.5	1.0
15	1.83 ± 0.53	1.2	93.2 ± 9.3	0.9	7.2 ± 1.6	0.9
30	3.71 ± 0.78	2.4	39.3 ± 10.9	0.4	7.4 ± 1.2	0.9
45	10.51 ± 3.96	6.7	76.7 ± 16.7	0.7	7.1 ± 1.9	0.9
60	8.06 ± 2.55	5.2	107.6 ± 20.9	1.1	7.9 ± 1.2	1.0
90	2.10 ± 0.85	1.3	125.2 ± 22.4	1.2	7.8 ± 0.9	1.0
120	0.51 ± 0.22	0.3	132.5 ± 21.4	1.3	6.5 ± 1.8	0.8

Source: Chauhan & Dakshinamurti (1991). Used with permission of the publisher.

between induction of mRNA and rate of transcription may be due to the stabilization of glucokinase mRNA by biotin. As shown in Figure 8.10B, glucokinase mRNA was increased as early as 30 min after biotin administration and the maximum effect was observed at 1 h. The glucokinase transcriptional rate was at a maximum at 45 min, after which there was a decrease in transcriptional rate, and it was back to the pre-stimulated level by 2 h. When longer time intervals were studied no significant change in the rate of transcription was observed compared with the pre-stimulated level.

It has recently been shown that the glucokinase gene uses different promoters, separated by 12 kb, in liver and pancreatic β cells (Magnuson & Shelton, 1989; Iynedjian *et al.*, 1989). The use of alternative splice sites and tissue-specific promoters for the glucokinase gene offers a mechanism for the known differences in the regulation of glucokinase. If such a mechanism is operative for biotin it is likely that in liver the proximal promoter region is involved. Insulin and glucagon are the most important regulators of glucokinase gene expression. Hoppner & Seitz (1989) have

Table 8.4. *Time course of the effect of insulin and biotin on liver and kidney PEPCK mRNA amounts in diabetic rats*

The height of the absorbance peaks corresponding to the mRNA bands were calculated and expressed in arbitrary units. Values represent the means of three or four separate experiments.

	Treatment							
	Insulin				Biotin			
	Liver		Kidney		Liver		Kidney	
Time after treatment (h)	PEPCK mRNA	%	PEPCK mRNA	%	PEPCK mRNA	%	PEPCK mRNA	%
0	82 ± 7	100	66 ± 12	100	72 ± 14	100	49 ± 7	100
1	22 ± 9	27	57 ± 6	86	61 ± 22	85	49 ± 6	100
2	18 ± 12	22	58 ± 3	83	26 ± 6	36	52 ± 10	106
3	9 ± 2	11	55 ± 11	83	11 ± 2	15	59 ± 1	120
5	29 ± 20	35	53 ± 10	80	51 ± 4	71	50 ± 8	102
7	72 ± 11	88	55 ± 11	83	not determined		not determined	

shown that thyroid hormones are permissive for the induction of glucokinase. There are significant differences between the actions of thyroid hormones and biotin in this. Differences in glucokinase mRNA levels due to different thyroid states were observed only in the re-fed and not in starved rats. Biotin administration produces a significant increase in glucokinase mRNA in the starved rat. The intracellular signaling mechanism used by biotin to control gene activity is not yet understood.

Role of biotin in the repression of phosphoenolpyruvate carboxykinase mRNA in diabetic rats

Phosphoenolpyruvate carboxykinase (PEPCK) is present in a high specific activity in liver, kidney, cortex and white adipose tissue (Hanson & Garber, 1972). In liver and kidney it participates in gluconeogenesis and is generally considered to be rate-limiting (Ui et al., 1973). Both enzyme activity and enzyme synthesis are subject to dietary and multihormonal controls (Tilghman et al., 1976; Granner & Pilkis, 1990; Cimbala et al., 1982; Watford & Mapes, 1990). In both fasted and diabetic rats, hepatic PEPCK activities are markedly increased. Refeeding a high-carbohydrate diet to fasted rats decreased PEPCK activity markedly. The decrease in PEPCK synthesis in animals fed a high-carbohydrate diet is directly due to a decrease in PEPCK mRNA which,

Table 8.5 *Biotin-induced suppression of hepatic PEPCK gene transcription*

Time after biotin injection (min)	Rate of gene transcription			
	PEPCK	%	PEPCK	%
0	70 ± 14	100	12 ± 9	100
15	44 ± 7	63	13 ± 8	105
30	32 ± 6	45	14 ± 6	113
45	51 ± 11	72	16 ± 4	130
60	63 ± 11	90	15 ± 3	125

in turn, is due to the repression by insulin of transcription of the PEPCK gene (Granner *et al.*, 1983; Sasaki *et al.*, 1984; Iynedjian & Hanson, 1977; Lamers *et al.*, 1982).

It is well known that insulin has reciprocal effects on the enzymes of glycolysis and gluconeogenesis (Granner & Pilkis, 1990). As we have shown that the administration of biotin caused a rapid and marked induction of glucokinase mRNA in starved rats, we investigated the effect of biotin on the regulation of PEPCK in the liver and kidney of diabetic rats (Dakshinamurti & Li, 1993).

The time course of repression of hepatic PEPCK mRNA by biotin is shown in Figure 8.11. The relative abundance of PEPCK mRNA in the diabetic rat is about 10-fold higher in comparison with that of the non-diabetic control (Table 8.4). After administration of biotin, hepatic PEPCK mRNA started to decrease at 1 h and dropped to 15% of the non-biotin-injected control level, at 3 h after injection. The level of hepatic mRNA recovered significantly by 5 h after biotin injection. On the contrary, no change was found in the concentration of kidney PEPCK (Table 8.4). In parallel studies on insulin administration to diabetic rats, the concentration of hepatic PEPCK mRNA decreased in the 2–5 h time period and the inhibition was nearly fully reversed by 7 h after insulin injection. Again, no change was observed in kidney PEPCK mRNA levels following insulin treatment of the diabetic rat. The concentration of insulin in serum was not affected by the administration of biotin to diabetic rats.

The run-on transcription experiments were used to estimate the relative rates of liver PEPCK gene transcription at various time intervals following biotin administration to diabetic rats. The results are shown in Table 8.5. Hybridization with [α-^{32}P]UTP-labeled RNA transcripts to vector PBR322 DNA was negligible. The transcription of the actin gene, included as internal control, was not influenced by biotin administration. Biotin suppressed the transcription rates of the hepatic PEPCK gene by 55% at 30 min; the rate of transcription then gradually increased back to the original level.

Both insulin and biotin share the same tissue specificity of regulating hepatic but not renal PEPCK. PEPCKs from liver and kidney are immunochemically similar (Longshaw & Pogson, 1972; Iynedjian *et al.*, 1975), but each enzyme has a unique pattern of hormonal regulation. The selective effect of biotin on hepatic rather than renal PEPCK provides additional information on the diverse mechanisms of hepatic and renal PEPCK regulation.

Figure 8.11. Differential effects of biotin on PEPCK mRNAs in liver and kidney. A, 28S rRNA of rat liver. B, PEPCK mRNA of rat liver. C, 28S rRNA of rat kidney. D, PEPCK mRNA of rat kidney. Lane 1, normal rat; lane 2, diabetic control; lane 3, diabetic rat 1 h after biotin treatment; lane 4, diabetic rat 2 h after biotin treatment; lane 5, diabetic rat 3 h after biotin treatment; lane 5, diabetic rat 5 h after biotin treatment.

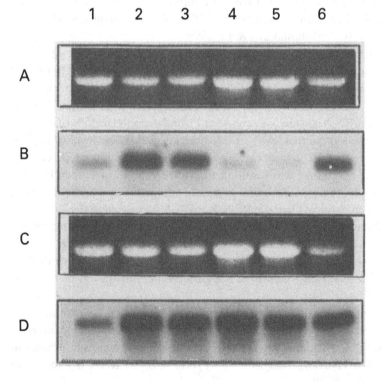

It is to be noted that the inhibition of transcription by biotin is dominant over other stimulatory effects. The elevated plasma glucagon characteristic of the fasting or diabetic condition induces PEPCK synthesis through enhanced transcription of the PEPCK gene. There are many similarities between biotin and insulin in their action on the enzymes of glucose metabolism. Both induce the mRNA that encodes glucokinase, a key glycolytic enzyme, and repress the mRNA that encodes PEPCK, a key gluconeogenic enzyme.

The effect of biotin on hepatic PEPCK, similar to its effect on hepatic glucokinase (Chauhan & Dakshinamurti, 1991) was seen at a concentration of biotin higher than the physiological concentration of this vitamin. Hence the question as to the physiological significance of the inhibition of the PEPCK gene cannot be answered now. This is the first demonstration that a water-soluble vitamin, biotin, exerts both positive and negative transcriptional control over key enzymes of glucose metabolism. The molecular mechanism responsible for the decrease of hepatic PEPCK in the diabetic rat has been investigated. The rapidity of the effect of biotin would indicate a direct effect, such as is produced by insulin on the rate of transcription of the PEPCK gene. The intracellular signaling mechanism used by biotin to control gene activity is not understood yet.

Conclusion

In biotin proteins the vitamin is covalently bound to a lysine residue of the apoprotein. Apart from the enzymes in which biotin is the prosthetic group, there are a number of proteins which bind non-covalently to biotin. These proteins cover the entire range from the tightly bound avidins to other proteins which can exchange their bound biotin at body temperature. The major functions of these proteins are in the transport of biotin between organs or to the developing embryo. The identity of the *E. coli* repressor protein for the biotin operon with the biotin-holocarboxylase synthetase indicates the non-prosthetic group role of this vitamin. In higher organisms, this aspect of biotin is emphasized by its role in cell multiplication and differentiation, as well as in its role in regulating the synthesis of specific proteins which include the regulatory enzymes of glucose metabolism, the asialoglycoprotein receptor, and the egg yolk biotin-binding proteins. Glucokinase and phosphoenolpyruvate carboxykinase are regulated by biotin at the transcriptional level. In this respect the action of biotin is comparable to that of some hormones. It is possible that biotin-responsive elements are present on glucokinase and PEPCK genes. Although such a receptor for biotin has not yet been identified, the presence of a nuclear biotin-binding protein might suggest that either this or a similar protein may be directly involved in regulating these specific genes. Other possibilities include the generation of an intracellular messenger such as cGMP by biotin, which in turn could be the regulator. However, such a possibility is weakened by the specificity, magnitude and time course of the effect of biotin in regulating these genes. Whereas the covalently bound biotin enzymes might be classified as metabolic initiators, the biotin-responsive elements would be initiators of gene action. The identification of such proteins will be an active area of future investigation.

The authors' work quoted in this chapter was supported by grants from the Medical Research Council of Canada.

References

Allen, R.W. & Moskowitz, M. (1973) Regulation of the rate of protein synthesis in BHK21 cells by exogenous serine. *Exp. Cell Res.* **116**, 139–52.

Alon, R., Bayer, E.A. & Wilcheck, M. (1990) Streptavidin contains an RYD sequence which mimics the RGD receptor domain of fibronectin. *Biochem. Biophys. Res. Commun.* **170**, 1236–41.

Anderson, K.M., Mendelson, I.S. & Gusik, G. (1975) Solubilized DNA-dependent nuclear RNA polymerases from the mammary glands of late-pregnant rats. *Biochim. Biophys. Acta* **343**, 56–66.

Argarana, C.E., Kuntz, I.D., Birken, S., Axel, R. & Cantor, C.R. (1986) Molecular cloning and nucleotide sequence of the strepavidin gene. *Nucleic Acids Res.* **14**, 1871–81.

Baker, H. (1985) Assessment of biotin status. *Ann. N.Y. Acad. Sci.* **447**, 129–32.

Barker, D.F. & Campbell, A.M. (1982*a*) The *birA* gene of *Escherichia coli* encodes a biotin holocarboxylase synthetase. *J. Molec. Biol.* **146**, 451–67.

Barker, D.F. & Campbell, A.M. (1982*b*) Genetic and biochemical characterization of the *birA* gene and its product: Evidence for a direct role of biotin holoenzyme synthetase in repression of the biotin operon in *Escherichia coli. J. Molec. Biol.* **146**, 469–92.

Baumgartner, E.R., Suormala, T., Wick, H., Bausch, J. & Bonjour, J.P. (1985) Biotinidase deficiency associated with renal loss of biocytin and biotin. *Ann. N.Y. Acad. Sci.* **447**, 272–87.

Bayer, E.A., Ben-Hur, H., Gitlin, G. & Wilchek, M. (1986) An improved method for the single-step purification of streptavidin. *J. Biochem. Biophys. Methods* **13**, 103–12.

Berger, E., Long, E. & Semenza, G. (1972) The sodium activation of biotin absorption in hamster small intestine. *Biochim. Biophys. Acta* **255**, 873–87.

Bernlohr, D.A., Bolanowski, M.A., Kelly, T.J. Jr & Lane, M.D. (1985) Evidence for an increase in transcription of specific mRNA during differentiation of 3T3-L1 preadipocytes. *J. Biol. Chem.* **260**, 5563–7.

Bhullar, R.P. & Dakshinamurti, K. (1985) The effects of biotin on cellular functions in HeLa cells. *J. Cell. Physiol.* **122**, 425–30.

Boeckx, R.L. & Dakshinamurti, K. (1970) Biotin mediated protein biosynthesis. *Biochem. J.* **140**, 549–56.

Bowers-Komro, D.M. & McCormick, D.B. (1985) Biotin uptake by isolated rat liver hepatocytes. *Ann. N.Y. Acad. Sci.* **447**, 350–8.

Bowman, B.B., Sellhub, J. & Rosenberg, D.E. (1986) Intestinal absorption of biotin in the rat. *J. Nutr.* **116**, 1266–71.

Braughler, J.M., Mittal, C.K. & Murad, F. (1979) Purification of soluble guanylate cyclase from rat liver. *Proc. Natl. Acad. Sci. USA* **76**, 219–22.

Brewer, L.E. & Edwards, H.M.J. (1972) Studies on the biotin requirement of broiler breeders. *Poult. Sci.* **51**, 519–624.

Burch, R.C. & White, H.B. (1982) Compositional and structural heterogeneity of avidin glycopeptides. *Biochemistry* **21**, 5334–41.

Canas, P.E. & Congote, L.F. (1984) Effect of cGMP on RNA polymerase II activities in fetal calf liver nuclei. *Exp. Biol.* **43**, 5–11.

Chailet, L. & Wolf. F.J. (1964) The properties of strepavidin, a biotin-binding protein produced by streptomycetes. *Arch. Biochem. Biophys.* **106**, 1–5.

Chalifour, L.E. & Dakshinamurti, K. (1982*a*). The requirement of human fibroblasts in culture. *Biochem. Biophys. Res. Commun.* **104**, 1047–53.

Chalifour, L.E. & Dakshinamurti, K. (1982*b*). The characterization of the uptake of avidin-biotin complex by HeLa cells. *Biochim. Biophys. Acta* **721**, 64–9.

Chalifour, L.E. & Dakshinamurti, K. (1983) The partial characterization of the binding of avidin-biotin complex to rat plasma membrane. *Biochem. J.* **210**, 121–8.

Chauhan, J. & Dakshinamurti, K. (1986) Purification and characterization of human serum biotinidase. *J. Biol. Chem.* **261**, 4268–75.

Chauhan, J. & Dakshinamurti, K. (1988) Role of human serum biotinidase as biotin-binding protein. *Biochem. J.* **256**, 265–70.

Chauhan, J. & Dakshinamurti, K. (1991) Transcriptional regulation of the glucokinase gene by biotin in starved rats. *J. Biol. Chem.* **266**, 10035–8.

Chauhan, J., Dakshinamurti, K., Dodd, J.G. & Matusik, R.J. (1988) Cloning and sequence of cDNA of human liver biotinidase. *Can. Fed. Biol. Soc.* **31**, 184 (abstr).

Cheng, D.K.S. & Moskowitz, M. (1982) Growth stimulation of Rous sarcoma virus-transformed BHK cells by biotin and serum lipids. *J. Cell. Physiol.* **113**, 487–93.

Cimbala, M.A., Lamers, W.H., Nelson, K., Monahan, J.E., Warren, H.Y. & Hanson, R.W. (1982) Rapid changes in the concentration of phosphoenolpyruvate carboxykinase mRNA in rat liver kidney. Effects of insulin and cyclic AMP. *J. Biol. Chem.* **257**, 7629–36.

Cohen, D.C. & Gospodarowicz, D. (1985) Biotin and choline replace the growth requirement of Madin-Darby canine kidney cells for high density lipoprotein. *J. Cell. Physiol.* **124**, 96–106.

Cohen, N.D. & Thomas, M. (1982) Biotin transport into fully differentiated 3T3-L1 cells. *Biochem. Biophys. Res. Commun.* **108**, 1508–16.

Collins, J.C., Morell, A.G. & Stockert, R.J. (1987) Biotin is required for expression of the asialoglycoprotein receptor (ASGP-R) in HepG2. *J. Cell Biol.* **105**, 330 (abstr).

Collins, J.C., Paietta, E., Green, R., Morell, A.G. & Stockert, R.J. (1988) Biotin-dependent expression of asialoglycoprotein receptor in HepG2. *J. Biol. Chem.* **263**, 11280–3.

Couch, J.R., Craven, W.W., Elvehjem, C.A. & Halpin, J.G. (1949) Studies on the function of biotin in the domestic fowl. *Arch. Biochem.* **21**, 77–86.

Craft, D.V., Goss, N.H., Chandramouli, N. & Wood, H.G. (1985) Purification of biotinidase from human plasma and its activity on biotinyl peptides. *Biochemistry* **24**, 2471–6.

Cravens, W.W., McGibbon, W.H. & Sebesta, E.E. (1944) Effect of biotin deficiency on embryonic development in the domestic fowl. *Anat. Rec.* **90**, 55–64.

Cronan, J.E. (1989) The *E. coli bio operon* operon: transcriptional repression by an essential protein modification enzyme. *Cell* **58**, 427–9.

Dakshinamurti, K., Bhullar, R.P., Scoot, A., Rector, E.S., Delespess, G. & Sehon, A.H. (1986) Production and characterization of a monoclonal antibody to biotin. *Biochem. J.* **237**, 477–82.

Dakshinamurti, K. & Chalifour, L.E. (1981) The biotin requirement of HeLa cells. *J. Cell Physiol.* **107**, 427–38.

Dakshinamurti, K., Chalifour, L.E. & Bhullar, R.P. (1985) Requirement for biotin and the function of biotin in cells in culture. *Ann. N.Y. Acad. Sci.* **447**, 38–55.

Dakshinamurti, K. & Chauhan, J. (1988) Regulation of biotin enzymes. *Ann. Rev. Nutr.* **8**, 211–33.

Dakshinamurti, K. & Cheah-Tan, C. (1968a) Liver glucokinase of the biotin deficient rat. *Can. J. Biochem.* **46**, 75–80.

Dakshinamurti, K. & Cheah-Tan, C. (1968b) Biotin-mediated synthesis of hepatic glucokinase in the rat. *Arch. Biochem. Biophys,* **127**, 17–21.

Dakshinamurti, K. & Hong, H.C. (1969) Regulation of key hepatic glycolytic enzymes. *Enzymol. Biol. Clin.* **11**, 423–8.

Dakshinamurti, K. & Li, W. (1993) Transcriptional regulation of liver phosphoenolpyruvate carboxykinase by biotin in diabetic rats. *Mol. Cell. Biochem.*, in press.

Dakshinamurti, K. & Litwack, S. (1970) Biotin and protein synthesis in rat liver. *J. Biol. Chem.* **245**, 5600–5.

Dakshinamurti, K. & Mistry, S.P. (1963) Tissue and intracellular distribution of biotin-$C^{14}OOH$ in rats and chick. *J. Biol. Chem.* **238**, 294–6.

Dakshinamurti, K. & Rector, E.S. (1990) Monoclonal antibodies to biotin. In: *Methods in Enzymology*, vol. 184 (ed. M. Wilchek & E.A. Bayer), pp. 111–19. Academic Press, San Diego, California.

Dakshinamurti, K., Tarrago-Litvak, L. & Hong, H.C. (1970) Biotin and glucose metabolism. *Can. J. Biochem.* **48**, 493–500.

DeLang, R.J. (1970) Egg white avidin I. Amino acid composition; sequence of the amino- and carboxyl-terminal cyanogen bromide peptides. *J. Biol. Chem.* **245**, 907–16.

DeLang, R.J. & Huang, T.S. (1971) Egg white avidin III. Sequence of the 78-residue middle cyanogen bromide peptide. Complete amino acid sequence of protein subunit. *J. Biol. Chem.* **246**, 698–709.

Delost, P. & Terroine, T. (1956) Les troubles endocriniens dans la carence en biotin. *Arch. Sci. Physiol.* **10**, 17–51.

Disorbo, D.M. & Litwack, G. (1981) Changes in the intracellular levels of pyridoxal 5'-phosphate affect the induction of tyrosine aminotransferase by glucocorticoids. *Biochem. Biophys. Res. Commun.* **99**, 1203–8.

Dupree, L.T., Sanford, K.K., Westfall, B.B. & Covalensky, A.B. (1962) Influence of serum protein on determination of nutritional requirement of cells in culture. *Exp. Cell. Res.* **28**, 381–405.

Eagle, H. (1955) The minimum requirement of the L and HeLa cells in tissue culture, the production of specific vitamin deficiencies and their cure. *J. Exp. Med.* **102**, 595–600.

Eakin, R.E., Snell, E.E. & Williams, R.J. (1940) A constituent of raw egg white capable of inactivating biotin *in vitro. J. Biol. Chem.* **136**, 801–2.

Earp, H.S. (1980) The role of insulin, glucagon and cAMP in the regulation of hepatocyte guanylate cyclase activity. *J. Biol. Chem.* **255**, 8979–82.

Eisenberg, M.A. (1973) Biotin: biogenesis, transport, and their regulation. *Adv. Enzymol.* **38**, 317–72.

Eisenberg, M.A. (1985) Regulation of biotin operon in *E. coli. Ann. N.Y. Acad. Sci.* **447**, 335–49.

Eisenberg, M.A., Prakash, O. & Hsiang, S.C. (1982) Purification and properties of the biotin repressor. A bifunctional protein. *J. Biol. Chem.* **257**, 15167–73.

Elo, H.A. (1980) Occurrence of avidin-like biotin-binding capacity in various vertebrate tissues and its induction by tissue injury. *Comp. Biochem. Physiol.* **67B**, 221–4.

Frank, O., Luisada-Opper, A.V., Feingold, S. & Baker, H. (1970) Vitamin binding by humans and some animal proteins. *Nutr. Rep. Int.* **1**, 161–8.

Gehrig, D. & Leuthardt, F. (1976) A biotin-binding glycoprotein from human plasma: Isolation and characterization. *Proc. Int. Cong. Biochem.* **10**, 208.

Gitlin, G., Bayer, E.A. & Wilchek, M. (1988) Studies on the biotin-binding site of avidin. Tryptophan residues involved in the active site. *Biochem. J.* **250**, 291–4.

Gitlin, G., Bayer, E.A. & Wilchek, M. (1990) Studies on the biotin-binding site of avidin and strepavidin. Tyrosine residues are involved in the binding site. *Biochem. J.* **269**, 527–30.

Gore, J., Hoinard, C. & Maingault, P. (1986) Biotin uptake by isolated rat intestinal cells. *Biochim. Biophys. Acta* **856**, 357–61.

Gospodarowicz, D. & Cohen, D.C. (1988) MDCK cell growth in defined medium: Roles of high density lipoproteins, biotin, pyruvate and an autocrine growth factor. *FASEB J.* **2**, A726 (abstr).

Granner, D., Andreone, T., Sasaki, K. & Beale, E. (1983) Inhibition of transcription of the phosphoenolpyruvate carboxykinase gene by insulin. *Nature* **305**, 549–51.

Granner, D. & Pilkis, S. (1990) The genes of hepatic glucose metabolism. *J. Biol. Chem.* **265**, 10173–6.

Green, N.M. (1975) Avidin. *Adv. Protein Chem.* **29**, 85–133.

Green, N.M. (1990) Avidin and streptavidin. *Methods Enzymol.* **184**, 51–67.

Green, H. & Kehinde, O. (1974) Subline of mouse 3T3 cells that accumulate lipid. *Cell (Cambridge, Mass.)* **1**, 113–16.

Gyorgy, P., Rose, C.S., Eakin, R.E., Snell, E.E. & Williams, R.J. (1941) Egg white injury as result of nonabsorption of inactivation of biotin. *Science* **93**, 477–8.

Hanson, R.W. & Garber, A.J. (1972) Phospheonolpyruvate carboxykinase. I. Its role in gluconeogenesis. *Am. J. Clin. Nutr.* **25**, 1010–21.

Harris, S.A., Wolf, D.E., Mozingo, R. & Folkes, K. (1943) Synthetic biotin. *Science* **97**, 447–8.

Hertz, R. & Sebrell, W.H. (1942) Occurrence of avidin in oviduct and secretion of genital tract of several species. *Science* **92**, 257.

Hiller, Y., Gershoni, J.M., Bayer, E.A. & Wiolchek, M. (1987) Biotin binding to avidin. Oligosaccharide side chain not required for ligand association. *Biochem. J.* **248**, 167–71.

Holmes, R. (1959) Long-term cultivation of human cells (Chang) in chemically defined medium and effect of added peptone fraction. *J. Biophys. Biochem. Cytol.* **6**, 535–6.

Hoppner, W. & Seitz, H. (1989) Effect of thyroid hormones on glucokinase gene transcription in rat liver. *J. Biol. Chem.* **264**, 20643–7.

Iynedjian, P.B., Ballard, F.J. & Hanson, R.W. (1975) The regulation of phosphoenolpyruvate carboxykinase (GTP) synthesis in rat kidney cortex. The role of acid-base balance and glucocorticoids. *J. Biol. Chem.* **250**, 5596–603.

Iynedjian, P.B., Gjinovci, A. & Renold, A.E. (1988) Stimulation by insulin of glucokinase gene transcription in liver of diabetic rats. *J. Biol. Chem.* **263**, 740–4.

Iynedjian, P. & Hanson, R.W. (1977) Increase in level of functional messenger RNA coding for phosphoenolpyruvate carboxykinase (GTP) during induction by cyclic adenosine 3′:5′-monophosphate. *J. Biol. Chem.* **252**, 655–62.

Iynedjian, P.B., Pilot, P., Nouspikel, T., Milburn, J.L., Quaade, C., Hughes, S., Ulca, C. & Newgard, C.B. (1989) Differential expression and regulation of the glucokinase gene in liver and islets of Langerhans. *Proc. Natl. Acad. Sci. USA* **86**, 7838–42.

Johnson, L.D. & Haddon, J.M. (1975) cGMP and lymphocyte proliferation: Effects on DNA-dependent RNA polymerase I and II activities. *Biochem. Biophys. Res. Commun.* **65**, 1498–505.

Kaplan, P.L., Anderson, M. & Ozanne, B. (1982) Transforming growth factor(s) production enables cells to grow in the absence of serum: An autocrine system. *Proc. Natl. Acad. Sci. USA* **79**, 485.

Kasturi, R. & Wakil, S.J. (1983) Increased synthesis and accumulation of phospholipids during differentiation of 3T3-L1 cells into adipocytes. *J. Biol. Chem.* **258**, 3559–4832.

Kennedy, C. & Palmer, L.S. (1945) Biotin deficiency in relation to reproduction and lactation. *Arch. Biochem.* **7**, 9–13.

Keranen, A.J.A. (1972) The biotin synthesis of HeLa cells *in vitro*. *Cancer Res.* **32**, 119–24.

Knappe, J., Brummer, W. & Biderbick, K. (1963) Reinigung und Eigenschaften der Biotinidase aus Schweinenieren und *Lactobacillus casei*. *Biochem. Z.* **338**, 599–613.

Knowles, J.R. (1989) The mechanism of biotin-dependent enzymes. *Ann. Rev. Biochem.* **58**, 195–221.

Kogl, F. & Tonnis, B. (1936) Uber des Bios-Problem. Darstellung von krystallisiertem Biotin aus Schweinenieren und Lactobacillus casei. *Hoppe-Seyler's Z. Physiol. Chem.* **242**, 43–73.

Korpela, J. (1984*a*) Chicken macrophages synthesize and secrete avidin in culture. *Eur. J. Cell. Biol.* **33**, 105–11.

Korpela, J. (1984*b*) Avidin, a high affinity biotin-binding protein, as a tool and subject to biological research. *Medical Biol.* **62**, 5–26.

Kuri-Harcuch, W. & Green, H. (1978) Adipose conversion of 3T3 cells depends on a serum factor. *Proc. Natl. Acad. Sci. USA* **75**, 6107–9.

Kuri-Harcuch, W. & Marsch-Morino, M. (1983) DNA synthesis and cell division related to adipose differentiation of 3T3 cells. *J. Cell Physiol.* **114**, 39–44.

Kuri-Harcuch, W., Wise, L.S. & Green, H. (1978) Interruption of adipose conversion of 3T3 cells by biotin deficiency: Differentiation without triglyceride accumulation. *Cell (Cambridge, Mass.)* **14**, 53–9.

Krause, K.H., Bonjour, J.P., Berlit, P. & Kochen, W. (1985) Biotin status of epileptics. *Ann. N.Y. Acad. Sci.* **447**, 297–313.

Lamers, W., Hanson, R.W. & Meisner, H. (1982) cAMP stimulates transcription of the gene for cytosolic phosphoenolpyruvate carboxykinase in rat liver nuclei. *Proc. Natl. Acad. Sci. USA* **79**, 5137–41.

Lane, M.D., Young, D.L. & Lynen, F. (1969) The enzymatic synthesis of holotranscarboxylase from apotranscarboxylase and (+)-biotin. I. Purification of the apoenzyme and synthetase; characteristics of the reaction. *J. Biol. Chem.* **239**, 2858–64.

Litwack, G. (1988) The glucocorticoid receptor at the protein level. *Cancer Res.* **48**, 2636–40.

Longshaw, I.D. & Pogson, C.I. (1972) The effect of steroids and ammonium chloride acidosis on phosphoenolpyruvate carboxykinase in rat kidney cortex. I. Differentiation of the inductive process and characterization of enzyme activities. *J. Clin. Invest.* **51**, 2277–83.

Lynen, F. (1967) The role of biotin-dependent carboxylation in biosynthetic reactions. *Biochem. J.* **102**, 381–400.

Magnuson, M.A. & Shelton, K.D. (1989) An alternate promotor in the glucokinase gene is active in pancreatic β cell. *J. Biol. Chem.* **264**, 15936–42.

Maloy, W.L., Bowien, B.U., Zwolinski, G.K., Kumar, G.K., Wood, H.G., Ericsson, L.H. & Walsh, K.A. (1979) Amino acid sequence of the biotinyl subunit from transcarboxylase. *J. Biol. Chem.* **254**, 11615–22.

Mandella, R.D., Mesler, H.W. & White, H.B. III. (1978) Relationship between biotin-binding proteins from chicken plasma and egg yolk. *Biochem. J.* **175**, 629–33.

McAllister, H.C. & Coon, M.J. (1966) Further studies on the properties of liver propionyl coenzyme A holocarboxylase synthetase and specificity of holocarboxylase formation. *J. Biol. Chem.* **241**, 2855–61.

Mackall, J.C. & Lane, M.D. (1977) Role of pyruvate carboxylase in fatty acid synthesis: Alterations during preadipocyte differentiation. *Biochem. Biophys. Res. Commun.* **79**, 720–5.

Mackall, J.C., Student, A.K., Polakis, S.E. & Lane, M.D. (1976) Induction of lipogenesis during differentiation in a "preadipocyte" cell line. *J. Biol. Chem.* **251**, 6462–4.

Meglasson, M.D. & Matschinsky, F.M. (1984) New perspective on pancreatic islet glucokinase. *Am. J. Physiol.* **246**, E1-13.

Meisler, N.T. & Thanassi, J.W. (1990) Pyridoxine-derived B_6 vitamers and pyridoxal 5'-phosphate-binding protein in cytosolic and nuclear fractions of HTC cells. *J. Biol. Chem.* **265**, 1193–8.

Melamed, M.D. & Green, N.M. (1963) Avidin II. Purification and composition. *Biochem. J.* **89**, 591–9.

Mesler, H.W., Camper, S.A. & White, H.B., III. (1978) Biotin-binding protein from egg yolk. A protein distinct from egg white avidin. *J. Biol. Chem.* **253**, 6979–82.

Messmer, T.O. & Young, D.V. (1977) The effects of biotin and fatty acids on SV 3T3 cells growth in the presence of normal calf serum. *J. Cell Physiol.* **90**, 265–270.

Moskowitz, M. & Cheng, D.K.S. (1985) Stimulation of growth factor production in cultured cells by biotin. *Ann. N.Y. Acad. Sci.* **447**, 212–21.

Murthy, C.V.R. & Adiga, P.R. (1984) Purification of biotin-binding protein from chicken egg yolk and comparison with avidin. *Biochim. Biophys. Acta* **786**, 222–30.

Nakano, E.T., Ciampi, N.A. & Young, D.V. (1982) The identification of a serum viability factor for SV3T3 cells as biotin and its possible relationship to the maintenance of krebs cycle activity. *Arch. Biochem. Biophys.* **215**, 556–63.

Nishimura, H., Yoshoka, K. & Iwashima, A. (1984) A method for determining binding kinetics applied to thiamine-binding protein. *Analyt. Biochem.* **139**, 373–6.

Norback, I., Joensuu, T. & Tuhimaa, P. (1982) Effects of glucocorticoids and disodium chromoglycate on avidin production in chick tissues. *J. Endocrinol.* **92**, 283–91.

Otsuka, A.J., Buoncristiani, M.R., Howard, P.J., Flamm, J., Johnson, C., Yamamoto, R., Uchida, K., Cook, C., Ruppert, J. & Matsuzaki, J. (1988) The *Escherichia coli* biotin biosynthetic enzyme sequences predicted from the nucleotide sequence of the bio operon. *J. Biol. Chem.* **263**, 19577–85.

Paulose, C.S., Thliveris, J., Viswanathan, M. & Dakshinamurti, K. (1989) Testicular function in biotin-deficient adult rats. *Horm. Metabol. Res.* **21**, 661–5.

Petrelli, F. & Marsili, G. (1969) Ricerche concernenti il ruolo della biotina nella produzione di anticorpi. *Acta Vitaminol. Enzymol.* **23**, 86–7.

Petrelli, F. & Marsili, G. (1971) Studies on the relationship between the function of biotin and activity of the RES in the rat. *J. Reticuloendothel. Soc.* **9**, 86–95.

Petrelli, F., Moretti, P. & Campanati, G. (1981) Studies on the relationships between biotin and the behaviour of B and T lymphocytes in the guinea-pig. *Experientia* **37**, 1204–6.

Pispa, J. (1965) Animal biotinidase. *Ann. Med. Exp. Biol. Fenn.* **43**(suppl. 5), 5–39.

Prakash, O. & Eisenberg, M.A. (1979) Biotinyl 5'-adenylate: Corepressor role in the regulation of the biotin genes of *Escherichia coli* K-12. *Proc. Natl. Acad. Sci. USA* **76**, 5592–5.

Rose, M.R., Commural, R. & Mazella, O. (1956) Heriditary deficiencies. *Arch. Sci. Physiol.* **10**, 381–421.

Rosen, O.M., Smith, C.J., Hirsch, A., Lai, E. & Rubin, C.S. (1979) Recent studies of the 3T3-L1 adipocyte-like line. *Recent Prog. Horm. Res.* **35**, 477–99.

Roth, K.S., Yang, W., Allan, L., Saunders, M., Gravel, R.A. & Dakshinamurti, K. (1982) Prenatal administration of biotin in biotin responsive multiple carboxylase deficiency. *Pediatr. Res.* **16**, 126–9.

Rudland, P.S., Gospodarowicz, D. & Seifert, W. (1974) Activation of guanylate cyclase and intracellular cyclic GMP by fibroblast growth factor. *Nature* **250**, 771–3.

Rylatt, D.B., Keech, D.B. & Wallace, J.C. (1977) Pyruvate carboxylase: isolation of the biotin-containing tryptic peptide and the determination of its primary sequence. *Arch. Biochem. Biophys.* **183**, 113–22.

Said, S.M. & Redha, R. (1987) A carrier-mediated system for transport of biotin in rat intestine *in vitro*. *Am. J. Physiol.* **252**, G52–5.

Susaki, K., Cripe, T.P., Koch, S.R., Andreone, T.L., Peterson, D.D., Beale, E.G. & Granner, D.K. (1984) Multihormonal regulation of phosphoenolpyruvate carboxykinase gene transcription. The dominant role of insulin. *J. Biol. Chem.* **259**, 15242–51.

Scatchard, G. (1949) The attractions of proteins for small molecules and ions. *Ann. N.Y. Acad. Sci.* **51**, 660–72.

Schmidt, T.J. & Litwack, G. (1982) Activation of glucocorticoid-receptor complex. *Physiol. Rev.* **62**, 1131–92.

Schumm, D.E., Morris, H.P. & Webb, T.E. (1974) Early biochemical changes in phytohemagglutinin-stimulated peripheral blood lymphocytes from normal and tumor bearing rats. *Europ. J. Cancer,* **10**, 107–13.

Seshagiri, P.B. & Adiga, P.R. (1987) Isolation and characterization of a biotin-binding protein from the pregnant-rat serum and comparison with that from the chicken egg yolk. *Biochim. Biophys. Acta.* **916**, 474–81.

Sharma, C., Manjeshwar, R. & Weinhouse, S. (1963) Effects of diet and insulin on glucose-adenosine triphosphate phosphotransferase of rat liver. *J. Biol. Chem.* **238**, 3840–5.

Shaw, J.H. & Phillips, P.H. (1942) Pathological studies of acute biotin deficiency in the rat. *Proc. Soc. Exp. Biol. Med.* **51**, 406–7.

Singh, I. & Dakshinamurti, K. (1988) Stimulation of guanylate cyclase and RNA polymerase II activities in HeLa cells and fibroblasts by biotin. *Molec. Cell. Biochem.* **79**, 47–55.

Spence, J.T. & Koudelka, A.P. (1984) Effects of biotin upon the intracellular level of cGMP and the activity of glucokinase in cultured rat hepatocytes. *J. Biol. Chem.* **259**, 6393–6.

Spencer, R.P. & Brody, K.R. (1964) Biotin transport by small intestine of rat, hamster, and other species. *Am. J. Physiol.* **206**, 653–7.

Spiegelman, B.M. & Farmer, S.R. (1982) Decrease in tubulin and actin gene expression prior to morphological differentiation of 3T3 adipocytes. *Cell (Cambridge, Mass.)* **29**, 53–60.

Sutton, M.R., Fall, R.R., Nervi, A.M., Alberts, A.M., Vagelos, P.R. & Bradshaw, R.A. (1977) Amino acid sequences of *Escherichia coli* carboxyl carrier protein. *J. Biol. Chem.* **252**, 3934–40.

Sweetman, L. & Nyhan, W.L. (1986) Inheritable biotin-treatable disorders and their associated phenomena. *Ann. Rev. Nutr.* **6**, 317–343.

Swim, H.E. & Parke, R.F. (1958) Vitamin requirements of uterine fibroblasts, strain U12-79; their replacement by related compounds. *Arch. Biochem. Biophys.* **78**, 46–53.

Takahashi, N. & Beritman, T.R. (1989) Retinoic acid acylation (retinoylation) of a nuclear protein in the human acute myeloid leukemia cell line HL60. *J. Biol. Chem.* **264**, 5159–63.

Terroine, T. (1960) Physiology and biochemistry of biotin. *Vitam. Horm. (N.Y.)* **18**, 1–42.

Thoene, J.G., Lemons, R. & Baker, H. (1983) Imparied intestinal absorption of biotin in juvenile multiple carboxylase deficiency. *N. Engl. J. Med.* **308**, 639–42.

Thomas, R.W. & Peterson, W.H. (1954) The enzymatic degradation of soluble bound biotin. *J. Biol. Chem.* **210**, 569–79.

Theilman, L., Teicher, L., Schildkraut, C.S. & Stockert, R.J. (1983) Growth-dependent expression of a cell surface glycoprotein. *Biochim. Biophys. Acta* **762**, 475–7.

Tilghman, S.M., Hanson, R.W. & Ballard, F.J. (1976) Hormonal regulation of phosphoenolpyruvate carboxykinase (GTP) in mammalian tissues. In: *Gluconeogenesis: Its Regulation in Mammalian Species* (ed. R.W. Hanson & M.S. Mehlman, pp. 47–91. Wiley-Interscience, New York.

Turner, J.B. & Hughes, D.E. (1962) The absorption of some B-group vitamers by surviving rat intestine preparations. *Q. J. Exp. Physiol. Cogn. Med. Sci.* **47**, 107–123.

Ui, M., Claus, T.H., Exton, J.H. & Park, C.R. (1973) Studies on the mechanism of action of glucagon on gluconeogenesis. *J. Biol. Chem.* **248**, 5344–9.

Ureta, T., Radojkovic, J. & Niemeyer, H. (1970) Inhibition by catecholamines of the induction of rat liver glucokinase. *J. Biol. Chem.* **245**, 4819–24.

Vallotton, M., Hess-Sander, U. & Leuthardt, F. (1965) Fixation spontanée de la biotine à une proteine dans le serum humain. *Helv. Chim. Acta* **48**, 126–33.

Vesely, D.L. (1981) Human and growth hormones enhance guanylate cyclase activity. *Am. J. Physiol.* **240**, E79–82.

Vesely, D.L. (1982) Biotin enhances guanylate cyclase activity. *Science* **216**, 1329–30.

Vesely, D.L., Kemp, S.F. & Elders, M.J. (1987) Isolation of a biotin receptor from hepatic plasma membranes. *Biochem. Biophys. Res. Commun.* **143**, 913–16.

Vesely, D.L., Wormser, H.C. & Abramson, H.N. (1984) Biotin analogs activate guanylate cyclase. *Molec. Cell. Biochem.* **60**, 109–14.

Watanabe, T. (1983) Teratogenic effect of biotin deficiency in mice. *J. Nutr.* **113**, 574–81.

Watanabe, T. (1990) Micronutrients and congenital anomalies. *Cong. Anom.* **30**, 79–92.

Watanabe, T. & Endo, A. (1984) Teratogenic effects of avidin-induced biotin deficiency in mice. *Teratology* **30**, 91–4.

Watford, M. & Mapes, R.E. (1990) Hormonal and acid-base regulation of phosphoenolpyruvate carboxykinase mRNA levels in rat kidney. *Arch. Biochem. Biophys.* **282**, 399–403.

Weber, P.C., Ohlendorf, D.H., Wendoloski, J.J. & Salemme, F.R. (1989) Structural origins of high-affinity biotin binding to streptavidin. *Science* **243**, 85–8.

Weinhouse, S. (1976) Regulation of glucokinase in liver. *Curr. Top. Cell Regul.* **11**, 1–50.

White, H.B. III (1985) Biotin-binding proteins and biotin transport to oocytes. *Ann. N.Y. Acad. Sci.* **447**, 202–11.

White, L.D. & George, W.J. (1981) Increased concentrations of cGMP in fetal liver cells stimulated by erythroprotein. *Proc. Soc. Exp. Biol. Med.* **166**, 186–93.

White, H.B. III & Whitehead, C.C. (1987) Role of avidin and other biotin-binding proteins in the deposition and distribution of biotin in chicken eggs: discovery of a new biotin-binding protein. *Biochem. J.* **241**, 677–84.

Wilchek, M. & Bayer, E.A. (eds) (1990) Avidin-biotin technology. *Methods in Enzymology,* vol. 184. Academic Press, New York. 746 pp.

Wise, L.S., Sul, H.S. & Rubin, C.S. (1984) Coordinate regulation of the biosynthesis of ATP-citrate lyase malic enzyme during adipocyte differentiation. Studies on 3T3-L1 cells. *J. Biol. Chem.* **259**, 4827–32.

Wisniesky, B.J., Williams, R.E. & Fox, C.F. (1973) Manipulations of fatty acid composition in animal cells grown in culture. *Proc. Natl. Acad. Sci. USA* **70**, 3669–73.

Wolf, B., Grier, R.E., Allen, R.J., Goodman, S.I. & Kien, C.L. (1983*a*) Biotinidase deficiency: The enzymatic defect in late-onset multiple carboxylate deficiency. *Clin. Chem. Acta* **131**, 273–81.

Wolf, B., Grier, R.E., Allen, R.J., Goodman, S.I., Kien, C.L., Parker, W.D., Howell, D.M. & Hurst, D.L. (1983*b*) Phenotype variation in biotinidase deficiency. *J. Pediatr.* **103**, 233–7.

Wood, H.G. & Barden, R.E. (1977) Biotin enzymes. *Ann. Rev. Biochem.* **46**, 385–413.

Wood, H.G. & Harman, F.R. (1980) Comparison of the biotination of apotranscarboxylase and its aposubunit. Is assembly essential for biotination? *J. Biol. Chem.* **255**, 7379–409.

Wood, H.G. & Kumar, G.K. (1985) Transcarboxylase: Its quaternary structure and the role of the biotinyl subunit in the assembly of the enzyme and in catalysis. *Ann. N.Y. Acad. Sci.* **447**, 1–21.

Wright, L.D., Driscoll, C.A. & Boger, W.P. (1954) Biocytinase, an enzyme concerned with hydrolytic cleavage of biocytin. *Proc. Soc. Exp. Biol. Med.* **86**, 335–7.

Yanofsky, C. (1981) Attenuation in the control of expression of bacterial operons. *Nature* **289**, 751–8.

Zeilberg, C.E. & Goldberg, N.D. (1977) Cell-cycle related changes of 3', 5'-cyclic GMP levels in Novikoff hempatoma cells. *Proc. Natl. Acad. Sci. USA* **74**, 1052–6.

ACTH adrenocorticotrophic hormone
APL acute promyelocytic leukemia
ASGP-R asialoglycoprotein receptor
B biotin
B-AMP biotinyl AMP
BBP biotin-binding protein
BCCP biotin carboxyl carrier protein
Bct biocytin
BSA bovine serum albumin
CAT chloramphenicol acetyltransferase
cbl cobalamin
CG chorionic gonadotropin
CGL chronic granulocytic leukemia
CRABP cellular retinoic acid-binding protein
CRALBP cellular retinal-binding protein
CRBP cellular retinol-binding protein
CSF cerebrospinal fluid
$1,25 (OH_2)D_3$ 1,25-dihydroxyvitamin D_3
DMEM Dulbecco's modified Eagle's medium
DNP dinotrophenyl
EC embryonic carcinoma
ELISA enzyme-linked immunosorbent assay
FAD flavin adenine dinucleotide
FBP folate binding protein
FDS fetal distress syndrome
FMN flavin mononucleotide
FPGS folate polyglutamate synthetase
GPI glycosylphosphatidylinositol

HCS holocarboxylase synthetase
HDL high-density lipoprotein
HRE hormone response element
HS holoenzyme synthetase
IF intrinsic factor
IgG immunoglobulin G
IRBP interphotoreceptor retinol-binding protein
IUGR intrauterine growth retardation
KLH keyhole limpet hemocyanin
LDL low-density lipoprotein
MAb monoclonal antibody
MEN minimal essential medium
MTX methotrexate
NAF nuclear accessory factor
NMR nuclear magnetic resonance
OC osteocalcin
OP osteopontin
ORF open reading frame
PAM peptidylglycyl-α-amino monooxygenase
PAS periodic acid–Schiff
PEG polyethylene glycol
PEPCK phosphoenolpyruvate carboxykinase
PLP pyridoxal phosphate
PPAR peroxisome proliferator activated receptor
R repressor protein
RA retinoic acid
RAR retinoic acid receptor
RARE retinoic acid response element
RBC red blood cell/s
RBP retinol-binding protein or riboflavin-binding protein
RCP riboflavin carrier protein
RfBP riboflavin-binding protein
RIA radioimmunoassay
RST Rous sarcoma virus-transformed [cells]
RXR retinoid X receptor
RXRE retinoid X response element
SDS sodium dodecyl sulfate
SDS–PAGE sodium dodecyl sulfate – polyacrylamide gel electrophoresis
T3R thyroid hormone receptor
TC transcobalamin
TGFB transforming growth factor

TRE thyroid hormone response element
TREp palindromic thyroid hormone response element
TTR transthyretin
UT untranslated
VD_3R vitamin D_3 receptor
VDR 1,25-dihydroxyvitamin D_3 receptor
VDRE vitamin D_3-responsive element
VLDL very low-density lipoprotein
ZPA zone of polarizing activity

INDEX